Handbook of
MICROWAVE TECHNOLOGY FOR FOOD APPLICATIONS

FOOD SCIENCE AND TECHNOLOGY

A Series of Monographs, Textbooks, and Reference Books

EDITORIAL BOARD

Senior Editors
Owen R. Fennema University of Wisconsin–Madison
Marcus Karel Rutgers University (emeritus)
Gary W. Sanderson Universal Foods Corporation (retired)
Pieter Walstra Wageningen Agricultural University
John R. Whitaker University of California–Davis

Additives **P. Michael Davidson** University of Tennessee–Knoxville
Dairy science **James L. Steele** University of Wisconsin–Madison
Flavor chemistry and sensory analysis **John Thorngate** University of Idaho–Moscow
Food engineering **Daryl B. Lund** Cornell University
Health and disease **Seppo Salminen** University of Turku, Finland
Nutrition and nutraceuticals **Mark Dreher** Mead Johnson Nutritionals
Processing and preservation **Gustavo V. Barbosa-Cánovas** Washington State University–Pullman
Safety and toxicology **Sanford Miller** University of Texas–Austin

1. Flavor Research: Principles and Techniques, *R. Teranishi, I. Hornstein, P. Issenberg, and E. L. Wick*
2. Principles of Enzymology for the Food Sciences, *John R. Whitaker*
3. Low-Temperature Preservation of Foods and Living Matter, *Owen R. Fennema, William D. Powrie, and Elmer H. Marth*
4. Principles of Food Science
    Part I: Food Chemistry, *edited by Owen R. Fennema*
    Part II: Physical Methods of Food Preservation, *Marcus Karel, Owen R. Fennema, and Daryl B. Lund*
5. Food Emulsions, *edited by Stig E. Friberg*
6. Nutritional and Safety Aspects of Food Processing, *edited by Steven R. Tannenbaum*
7. Flavor Research: Recent Advances, *edited by R. Teranishi, Robert A. Flath, and Hiroshi Sugisawa*
8. Computer-Aided Techniques in Food Technology, *edited by Israel Saguy*
9. Handbook of Tropical Foods, *edited by Harvey T. Chan*
10. Antimicrobials in Foods, *edited by Alfred Larry Branen and P. Michael Davidson*

41. Handbook of Cereal Science and Technology, *Klaus J. Lorenz and Karel Kulp*
42. Food Processing Operations and Scale-Up, *Kenneth J. Valentas, Leon Levine, and J. Peter Clark*
43. Fish Quality Control by Computer Vision, *edited by L. F. Pau and R. Olafsson*
44. Volatile Compounds in Foods and Beverages, *edited by Henk Maarse*
45. Instrumental Methods for Quality Assurance in Foods, *edited by Daniel Y. C. Fung and Richard F. Matthews*
46. *Listeria*, Listeriosis, and Food Safety, *Elliot T. Ryser and Elmer H. Marth*
47. Acesulfame-K, *edited by D. G. Mayer and F. H. Kemper*
48. Alternative Sweeteners: Second Edition, Revised and Expanded, *edited by Lyn O'Brien Nabors and Robert C. Gelardi*
49. Food Extrusion Science and Technology, *edited by Jozef L. Kokini, Chi-Tang Ho, and Mukund V. Karwe*
50. Surimi Technology, *edited by Tyre C. Lanier and Chong M. Lee*
51. Handbook of Food Engineering, *edited by Dennis R. Heldman and Daryl B. Lund*
52. Food Analysis by HPLC, *edited by Leo M. L. Nollet*
53. Fatty Acids in Foods and Their Health Implications, *edited by Ching Kuang Chow*
54. *Clostridium botulinum*: Ecology and Control in Foods, *edited by Andreas H. W. Hauschild and Karen L. Dodds*
55. Cereals in Breadmaking: A Molecular Colloidal Approach, *Ann-Charlotte Eliasson and Kåre Larsson*
56. Low-Calorie Foods Handbook, *edited by Aaron M. Altschul*
57. Antimicrobials in Foods: Second Edition, Revised and Expanded, *edited by P. Michael Davidson and Alfred Larry Branen*
58. Lactic Acid Bacteria, *edited by Seppo Salminen and Atte von Wright*
59. Rice Science and Technology, *edited by Wayne E. Marshall and James I. Wadsworth*
60. Food Biosensor Analysis, *edited by Gabriele Wagner and George G. Guilbault*
61. Principles of Enzymology for the Food Sciences: Second Edition, *John R. Whitaker*
62. Carbohydrate Polyesters as Fat Substitutes, *edited by Casimir C. Akoh and Barry G. Swanson*
63. Engineering Properties of Foods: Second Edition, Revised and Expanded, *edited by M. A. Rao and S. S. H. Rizvi*
64. Handbook of Brewing, *edited by William A. Hardwick*
65. Analyzing Food for Nutrition Labeling and Hazardous Contaminants, *edited by Ike J. Jeon and William G. Ikins*
66. Ingredient Interactions: Effects on Food Quality, *edited by Anilkumar G. Gaonkar*
67. Food Polysaccharides and Their Applications, *edited by Alistair M. Stephen*
68. Safety of Irradiated Foods: Second Edition, Revised and Expanded, *J. F. Diehl*
69. Nutrition Labeling Handbook, *edited by Ralph Shapiro*

11. Food Constituents and Food Residues: Their Chromatographic Determination, *edited by James F. Lawrence*
12. Aspartame: Physiology and Biochemistry, *edited by Lewis D. Stegink and L. J. Filer, Jr.*
13. Handbook of Vitamins: Nutritional, Biochemical, and Clinical Aspects, *edited by Lawrence J. Machlin*
14. Starch Conversion Technology, *edited by G. M. A. van Beynum and J. A. Roels*
15. Food Chemistry: Second Edition, Revised and Expanded, *edited by Owen R. Fennema*
16. Sensory Evaluation of Food: Statistical Methods and Procedures, *Michael O'Mahony*
17. Alternative Sweeteners, *edited by Lyn O'Brien Nabors and Robert C. Gelardi*
18. Citrus Fruits and Their Products: Analysis and Technology, *S. V. Ting and Russell L. Rouseff*
19. Engineering Properties of Foods, *edited by M. A. Rao and S. S. H. Rizvi*
20. Umami: A Basic Taste, *edited by Yojiro Kawamura and Morley R. Kare*
21. Food Biotechnology, *edited by Dietrich Knorr*
22. Food Texture: Instrumental and Sensory Measurement, *edited by Howard R. Moskowitz*
23. Seafoods and Fish Oils in Human Health and Disease, *John E. Kinsella*
24. Postharvest Physiology of Vegetables, *edited by J. Weichmann*
25. Handbook of Dietary Fiber: An Applied Approach, *Mark L. Dreher*
26. Food Toxicology, Parts A and B, *Jose M. Concon*
27. Modern Carbohydrate Chemistry, *Roger W. Binkley*
28. Trace Minerals in Foods, *edited by Kenneth T. Smith*
29. Protein Quality and the Effects of Processing, *edited by R. Dixon Phillips and John W. Finley*
30. Adulteration of Fruit Juice Beverages, *edited by Steven Nagy, John A. Attaway, and Martha E. Rhodes*
31. Foodborne Bacterial Pathogens, *edited by Michael P. Doyle*
32. Legumes: Chemistry, Technology, and Human Nutrition, *edited by Ruth H. Matthews*
33. Industrialization of Indigenous Fermented Foods, *edited by Keith H. Steinkraus*
34. International Food Regulation Handbook: Policy • Science • Law, *edited by Roger D. Middlekauff and Philippe Shubik*
35. Food Additives, *edited by A. Larry Branen, P. Michael Davidson, and Seppo Salminen*
36. Safety of Irradiated Foods, *J. F. Diehl*
37. Omega-3 Fatty Acids in Health and Disease, *edited by Robert S. Lees and Marcus Karel*
38. Food Emulsions: Second Edition, Revised and Expanded, *edited by Kåre Larsson and Stig E. Friberg*
39. Seafood: Effects of Technology on Nutrition, *George M. Pigott and Barbee W. Tucker*
40. Handbook of Vitamins: Second Edition, Revised and Expanded, *edited by Lawrence J. Machlin*

Handbook of
MICROWAVE TECHNOLOGY FOR FOOD APPLICATIONS

edited by

ASHIM K. DATTA
Cornell University
Ithaca, New York

RAMASWAMY C. ANANTHESWARAN
The Pennsylvania State University
University Park, Pennsylvania

MARCEL DEKKER, INC.   NEW YORK · BASEL

ISBN: 0-8247-0490-8

This book is printed on acid-free paper.

Headquarters
Marcel Dekker, Inc.
270 Madison Avenue, New York, NY 10016
tel: 212-696-9000; fax: 212-685-4540

Eastern Hemisphere Distribution
Marcel Dekker AG
Hutgasse 4, Postfach 812, CH-4001 Basel, Switzerland
tel: 41-61-261-8482; fax: 41-61-261-8896

World Wide Web
http://www.dekker.com

The publisher offers discounts on this book when ordered in bulk quantities. For more information, write to Special Sales/Professional Marketing at the headquarters address above.

Copyright © 2001 by Marcel Dekker, Inc. All Rights Reserved.

Neither this book nor any part may be reproduced or transmitted in any form or by any means, electronic or mechanical, including photocopying, microfilming, and recording, or by any information storage and retrieval system, without permission in writing from the publisher.

Current printing (last digit):
10 9 8 7 6 5 4 3 2 1

PRINTED IN THE UNITED STATES OF AMERICA

To Anasua, Ankurita, and Amita
and
Lata, Rohit, and Nikhil

Preface

Microwave heating of food has existed since 1949, and almost all households in the United States own a domestic microwave oven. *The Handbook of Microwave Technology for Food Applications* provides food product and process developers, food science and engineering researchers and educators and food regulatory agencies with a comprehensive source of information for microwave heating and processing for the first time. Microwave processing involves complex interactions between wide-ranging disciplines, such as electromagnetics, dielectric properties, heat transfer, moisture transfer, solid mechanics, fluid flow, food chemistry, food microbiology, and packaging. This book links these disciplines and is a comprehensive treatise on microwave processing that demystifies the microwave heating of food.

In recent years, there has been a lack of significant growth in the development of microwave food products and processes. Unpredictable results from ineffective use have prevented consumers from enjoying the increased benefits of the technology. In addition, product and process development is often inefficient, frustrating, and expensive. Lack of sufficient and unified knowledge of this complex and radically different heating process has been the primary contributor to its unpredictability. The Handbook guides product and process developers toward achieving higher quality and reliability. It supplies information on utilizing microwaves for rapid, selective, and more uniform heating to provide safe, wholesome, high-quality, and convenient foods in today's busy world.

ORGANIZATION OF THE BOOK

Fundamentals of microwaves are covered in Chapter 1, followed by details of cavity (oven) heating in Chapter 2. Dielectric properties that determine the microwave absorption in food are introduced in Chapter 3. Depending on the electromagnetics and the dielectric properties of food, heat is generated inside the food—this leads to thermal diffusion, moisture transfer, and flow (in liquid systems). Temperature and moisture changes due to microwave absorption are discussed in Chapter 4. Heating leads to biochemical changes; changes in flavor are discussed in Chapter 5, while Chapter 6 covers microbial and enzymatic changes. Chapter 7 reviews hardware components of a microwave system. Chapter 8 discusses instrumentation and measurement techniques. Description of microwave processes in the industry and practical guidelines for effective home use of microwave oven are discussed in Chapters 9 and 10, respectively. Chapters 11 and 12 discuss microwaveable product and process development, covering guiding principles for heating of complex food systems and packaging. The last chapter reviews safety in microwave food processing.

FUTURE DEVELOPMENTS

Future developments in microwave heating of food are likely to be in the areas of equipment, processes, and product development. Combination ovens such as microwaves combined with infrared are already available. Dramatic redesign of ovens, enabling, for example, phase-control heating and variable-frequency heating, and fringing field applicators, may hold promise for the future. Although most consumers in the United States have access to a microwave oven, significant growth can be expected in global markets in the coming years. Process developments, such as microwave sterilization, are being investigated with renewed vigor. With better understanding of the dielectric properties and flavor systems and how they change in microwave heating, product development should see significant changes. Finally, with the space station becoming a reality, we will certainly see activities related to microwave food heating in space, since microwaves are an ideal method of heating in that environment.

ACKNOWLEDGMENTS

We deeply acknowledge the contribution of the individual authors, without whose participation it would be impossible to have a comprehensive treatise on such an interdisciplinary subject matter. We also greatly appreciate the comments made by several reviewers. External reviewers who reviewed one or more chapters include William Atwell (Pillsbury Technology Center, Minneapolis, Minnesota), John Bows (Unilever Research, United Kingdom), Allen R. Edison (University of Nebraska, Lincoln), Hua-Feng Huang (formerly with DuPont, Wilmington, Delaware), Mike Kent (K & S Associates, Biggar, Scotland), Andrzej Kraszewski, Kurt C. Lawrence, and Samir Trabelsi (USDA, ARS, Athens, Georgia), Mustapha Merabet (Nestlé Research Center, Switzerland), John Osepchuk (Full Spectrum Consulting, Concord, Massachusetts), Richard Edgar (formerly with Raytheon), G. S. V. Ragahavan and Hosahalli Ramaswamy (McGill University, Canada), Deepay Mukerjee (RF Technologies Corporation, Lewiston, Maine), John Roberts (Cornell University, Geneva, New York), Georges Roussy (Laboratories de Spectroscopie et des Techniques Microondes, France), and Craig Saltiel (Synergetic Technologies, Delmar, New York). The administrative assistance of Sue Fredenburg at Cornell University is much appreciated.

Ashim K. Datta
Ramaswamy C. Anantheswaran

Contents

Preface   *v*
Contributors   *xv*

PART I. FUNDAMENTAL PHYSICAL ASPECTS OF MICROWAVE ABSORPTION AND HEATING

1. **Electromagnetics: Fundamental Aspects and Numerical Modeling**   **1**
   David Dibben

   Introduction   1
   Fundamental Electromagnetics   2
   Analytical Modeling of Microwave Applicators   18
   Numerical Modeling of Cavities   18
   Application of Modeling to Microwave Heating   23
   Conclusion   27
   List of Symbols   27
   References   28

2. **Electromagnetics of Microwave Heating: Magnitude and Uniformity of Energy Absorption in an Oven**   **33**
   Hua Zhang and Ashim K. Datta

   Major Electromagnetic Issues in Microwave Heating of Foods   33

Electromagnetic Fields Inside a Domestic Microwave Oven	34
Magnitude and Uniformity of Energy Absorption: Food Factors	36
Magnitude and Uniformity of Energy Absorption: Oven Factors	54
Future Concepts and Developments in Oven Design	59
Appendix	62
References	63

3. Dielectric Properties of Food Materials and Electric Field Interactions 69
Stuart O. Nelson and Ashim K. Datta

Introduction	69
Definition of Terms and Basic Principles	70
Variation of Dielectric Properties	72
Measurement Principles and Techniques	78
Dielectric Behavior of Food Materials	81
Dielectric Properties: Data Complications	106
Acknowledgments	107
List of Symbols	107
References	107

4. Fundamentals of Heat and Moisture Transport for Microwaveable Food Product and Process Development 115
Ashim K. Datta

Introduction	115
Nature of Microwave Heating	116
Lambert's Law as a Simplified Description of Microwave Power Absorption	119
Heat Transport in Microwave Heating: General Description	121
Heat Transport in Microwave Heating of Solids	125
Heat Transport in Microwave Heating of Liquids	129
Effect of Changes in Temperature and Frequency on Heat Transport	132
Some Unit Operations Involving Primarily Heat Transport	133
Moisture Transport in Microwave Heating of Solid Food	141
Coupling of the Temperature and Moisture Variation of Dielectric Properties During Processing	153
Quality Improvement	158
Computer-Aided Engineering of Heat and Mass Transfer Processes	163
Acknowledgments	165
List of Symbols	165
References	166

Contents

PART II. CHEMICAL AND BIOLOGICAL CHANGES DUE TO HEATING

5. Generation and Release of Food Aromas Under Microwave Heating — 173
Varoujan A. Yaylayan and Deborah D. Roberts

- Introduction — 173
- Generation of Maillard Aromas Under Microwave Heating — 175
- Aroma Release During Microwave Heating of Food Products — 181
- Development of New Products/Processes to Optimize Aroma Formation and Minimize Aroma Release During Microwave Processing of Food Products — 184
- References — 186

6. Bacterial Destruction and Enzyme Inactivation During Microwave Heating — 191
Ramaswamy C. Anantheswaran and Hosahalli S. Ramaswamy

- Introduction — 191
- Kinetics of Destruction During Microwave Heating — 192
- Thermal, Nonthermal, and Microwave Enhanced Effects Due to Microwave Heating — 195
- Factors Affecting Microbial Destruction During Microwave Heating — 196
- Impact of Microwave Heating on Injury of Bacteria — 204
- Microwave Inactivation of Enzymes — 206
- Conclusion — 209
- References — 210

PART III. PROCESSING SYSTEMS AND INSTRUMENTATION

7. Consumer, Commercial, and Industrial Microwave Ovens and Heating Systems — 215
Richard H. Edgar and John M. Osepchuk

- Historical Introduction — 215
- Power Sources for Microwave Heating — 226
- Microwave Applicators and Cavities — 233
- Review of Available Oven Systems and Properties — 252
- Power and Efficiency Considerations — 266
- Uniformity Considerations — 269
- Controls and Sensors — 271
- Trends and Outlook — 272
- References — 275

Contents

8. Measurement and Instrumentation — 279
Ashim K. Datta, Henry Berek, Douglas A. Little, and Hosahalli S. Ramaswamy

Introduction	279
Measurement of Electric Field	280
Point Measurement of Temperature	282
Measurement of Surface Heating Patterns	288
Measurement of Internal Temperature Profiles Using Magnetic Resonance Imaging	291
Measurement of Cook or Sterilization Values (Time-Temperature History Effects)	292
Set-Point Temperature Control in a Microwave Oven	293
Measurement of Moisture Loss and Moisture Profiles During Microwave Heating	295
References	296

PART IV. PROCESSES AT INDUSTRY AND HOME

9. Microwave Processes for the Food Industry — 299
Robert F. Schiffmann

Introduction	299
Meat and Poultry Processing	303
Tempering	309
Baking	312
Drying	320
Pasteurization and Sterilization	325
Future of Microwave Processing in the Food Industry	330
Conclusion	334
References	335

10. Basic Principles for Using a Home Microwave Oven — 339
Carolyn Dodson

Introduction	339
Cooking Patterns, Power Levels, and Temperature Correlation	340
What Affects Microwave Cooking Time?	341
Effects of Containers, Covers, and Shielding	344
Various Processes at Home	346
Cooking Various Food Types	349
Converting Directions from Conventional Heating	350
References	352

Contents

PART V. PRODUCT AND PROCESS DEVELOPMENT

11. Ingredient Interactions and Product Development for Microwave Heating — 355
Triveni P. Shukla and Ramaswamy C. Anantheswaran

Introduction — 355
Microwave Energy — 356
Microwave Oven — 356
Interaction of Food Components with Microwaves — 357
Food Product Design for Microwave Heating — 377
Product Performance Testing — 380
Advanced Technologies for Microwaveable Food Product Development — 382
References — 389

12. Packaging Techniques for Microwaveable Foods — 397
Timothy H. Bohrer and Richard K. Brown

Introduction — 397
Passive Packages — 399
Active Packages — 407
Environmental Considerations — 441
Some Remaining Challenges — 443
Summary — 444
Appendix: Patents — 444
References — 467

PART VI. SAFETY

13. Safety in Microwave Processing — 471
Gregory J. Fleischman

Introduction — 471
Uniformity of Thermal Treatment in Conventional and Microwave Heating — 472
Industrial and Commercial Production of Microbiologically Safe Foods — 476
Chemical Migration — 487
Operational Safety Considerations — 489
Summary — 494
References — 494

Index — 499

Contributors

Ramaswamy C. Anantheswaran   Department of Food Science, The Pennsylvania State University, University Park, Pennsylvania

Henry Berek   FISO Technologies, Rockford, Illinois

Timothy H. Bohrer   Technology and Product Development, Ivex Packaging Corporation, Lincolnshire, Illinois

Richard K. Brown*   Microwave Technology, Fort James Corporation, Sarasota, Florida

Ashim K. Datta   Department of Agricultural and Biological Engineering, Cornell University, Ithaca, New York

David Dibben   Electromagnetic Engineering Group, The Japan Research Institute, Ltd., Osaka, Japan

Carolyn Dodson   Microwave and Food Specialist, Rotonda West, Florida

Richard H. Edgar   Wastech International, Inc., Portsmouth, New Hampshire

---

* *Retired.*

Contributors

Gregory J. Fleischman National Center for Food Safety and Technology, U.S. Food and Drug Administration, Summit-Argo, Illinois

Douglas A. Little FLIR Systems, Inc., Portland, Oregon

Stuart O. Nelson Agricultural Research Service, U.S. Department of Agriculture, Athens, Georgia

John M. Osepchuk Full Spectrum Consulting, Concord, Massachusetts

Hosahalli S. Ramaswamy Department of Food Science and Agricultural Chemistry, McGill University, Ste. Anne de Bellevue, Quebec, Canada

Deborah D. Roberts Food Science and Process Research, Nestlé Research Center, Lausanne, Switzerland

Robert F. Schiffmann R. F. Schiffmann Associates, Inc., New York, New York

Triveni P. Shukla F.R.I. Enterprises, New Berlin, Wisconsin

Varoujan A. Yaylayan Department of Food Science and Agricultural Chemistry, McGill University, Ste. Anne de Bellevue, Quebec, Canada

Hua Zhang Nestlé R & D Center, Inc., New Milford, Connecticut

Handbook of
MICROWAVE TECHNOLOGY FOR FOOD APPLICATIONS

1
Electromagnetics: Fundamental Aspects and Numerical Modeling

David Dibben
*The Japan Research Institute, Ltd.
Osaka, Japan*

I. INTRODUCTION

The temperature distribution inside food heated with microwaves is determined by both the thermal properties of the food and the distribution of the absorbed microwave energy. The amount of microwave energy absorbed is, in turn, determined by the electric field inside the microwave oven or applicator. This chapter focuses on the nature of the electromagnetic fields inside a microwave oven and their mathematical descriptions as defined by Maxwell's equations. Chapter 2 will examine the influence of oven design and the influence of thermal and electromagnetic properties of food on microwave heating. Computer simulations illustrate the use of modeling techniques which are described in the present chapter.

The general solution of Maxwell's field equations for arbitrary geometries is nontrivial. Consequently we initially consider some simplified situations, such as a plane wave incident upon a slab, that provide an insight into more general behavior. We then look briefly at the difficulties of obtaining analytical solutions for the fields in a microwave applicator and follow this by a description of numerical approaches for solving Maxwell's equations. The last section of this chapter presents applications of numerical techniques for the microwave heating of food.

Microwaves occupy the portion of the electromagnetic spectrum between 300 MHz and 30 GHz, as shown in Fig. 1. However, microwave heating applications are generally limited to a discrete set of frequencies reserved for industrial, scientific, and medical use, as shown in Table 1. Outside of these ranges the extra shielding required to prevent interference with other radio systems makes the ap-

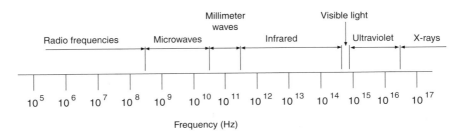

Figure 1 Electromagnetic spectrum.

plicators uneconomical for all but the most special of applications. Domestic microwave ovens operate at 2.45 GHz and industrial processing systems generally use either 2.45 GHz or 915 MHz (896 MHz in the UK).

II. FUNDAMENTAL ELECTROMAGNETICS

A sinusoidal voltage V at a frequency $\omega = 2\pi f$ applied to a pair of parallel plates, as shown in Fig. 2, will create an electric field in the space between the plates. Placing a dielectric material, such as food, between the plates will change the magnitude of the field both inside the material and in the air gap on either side of the material. The amount of energy stored in the system will also change. Inside the food there will be some dissipation of energy. Materials such as food, in which microwave power is dissipated, are generally referred to as lossy materials since some of the microwave energy is "lost" when it is converted into thermal energy inside the material.

Table 1 Some of the ISM Allocated Frequency Bands

Frequency, MHz	Frequency tolerance (+/−)	Area permitted
433.92	0.2%	Austria, Netherlands, Portugal, Germany, Switzerland, Great Britain
896	10 MHz	Great Britain
915	13 MHz	North and South America
2375	50 MHz	Albania, Bulgaria, Hungary, Romania, Czechoslovakia, Russia
2450	50 MHz	Worldwide except where 2375MHz is used

Source: Adapted from Metaxas and Meredith [10], reproduced by permission of the Institution of Electrical Engineers.

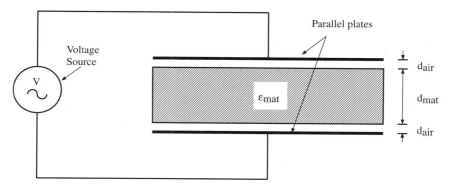

Figure 2 Parallel plate applicator.

The mechanisms of the interaction between the electric field an the material are discussed later in Chapter 3 on dielectric properties. For now, we simply assume that the interaction can be characterized by the complex permittivity ϵ and the permeability μ, which determine the storage and dissipation of electric and magnetic energy, respectively [1]. The permittivity can be written as

$$\epsilon = \epsilon_0 \epsilon_r = \epsilon_0 (\epsilon' - j\epsilon''_{\text{eff}}) \tag{1}$$

where ϵ' is the dielectric constant and ϵ''_{eff} is the effective loss factor incorporating all of the energy losses due to dielectric relaxation and ionic conduction and where $\epsilon_0 = 8.8542 \times 10^{-12}$ F/m is the permittivity of free space. For food materials the permeability is generally equal to that of free space, $\mu = \mu_0 = 4\pi \times 10^{-7}$ N/m².

A. A Parallel Plate System

If a voltage source is connected to a pair of parallel plates, a simple applicator is produced, as shown in Fig. 2. Ignoring edge effects, the electric field inside a dielectric placed between the plates will have an intensity given by [2]

$$\hat{E} = \frac{\hat{V}}{d_{\text{mat}} + 2d_{\text{air}}(\epsilon_{\text{mat}}/\epsilon_0)} \quad \text{V/m} \tag{2}$$

This electric field will cause a current to flow in the material with a density given by

$$\hat{J} = j\omega\epsilon\hat{E} \quad \text{A/m}^2 \tag{3}$$

The power density inside the material is then given by [2]

$$Q = \tfrac{1}{2}\omega\epsilon_0\epsilon''_{\text{eff}} |\hat{E}|^2 \quad \text{W/m}^3 \tag{4}$$

where $\hat{E}$ is the *peak* value of the electric field. It is clear from Eqs. (2) and (4) that the permittivity has a very strong influence upon the electric field and, in turn, upon the power dissipated in the material. This dissipated power Q is the source of microwave heating that leads to heat and moisture transfer, as discussed in Chapter 4.

B. Maxwell's Equations

Real microwave applicators used for food applications are generally more complex devices than the parallel plate applicator described above. A more complete description of the fields is needed in order to understand the behavior. This is provided by Maxwell's field equations which describe how a time-varying electric field is accompanied by a corresponding time-varying magnetic field, and vice versa. When coupled with the appropriate boundary conditions, they completely define the behavior of the electromagnetic fields inside a microwave applicator.

The first of Maxwell's equations is Ampere's law, which relates the magnetic field strength **H** to the total current density **J**$_t$ (which includes both the displacement and conduction currents.) It states that the circulation of the magnetic field intensity around a closed contour is equal to the net current passing through the surface enclosed by the contour, i.e., a current is surrounded by a magnetic field. In differential form for time harmonic fields this is written as*

$$\nabla \times \mathbf{H} = \mathbf{J}_t = j\omega\epsilon\mathbf{E} \quad \text{A/m}^2 \tag{5}$$

The second of Maxwell's equations is Faraday's law, which relates the time varying electric field **E** to the magnetic field. It states that the circulation of the electric field around a contour is determined by the rate of change of the magnetic flux through the surface enclosed by the contour. For time harmonic fields this can be written as

$$\nabla \times \mathbf{E} = -j\omega\mu\mathbf{H} \tag{6}$$

In order to completely specify the electric and magnetic fields, we must also employ Gauss's laws. These state that the net magnetic flux out of a region must be zero and that the net electric flux out of a region is related to the charge contained within it. Written in differential form, these are

---

* Equations (5) and (6) assume both time harmonic fields and that the properties ϵ and μ do not vary with time. This is true for the majority of microwave heating applications, even when the temperature dependence of the permittivity is included since the $\partial\epsilon/\partial t$ term is negligible in this context compared to the rate at which the electromagnetic fields are changing. For a more complete treatment, one of the many textbooks on electromagnetics should be consulted.

$$\nabla \cdot \mu \mathbf{H} = 0 \quad (7)$$
$$\nabla \cdot \epsilon \mathbf{E} = \rho \quad (8)$$

where ρ is the charge density.

C. Plane Waves

The coupling between the electric field and the magnetic field produced by the displacement current [Eq. (5)] and magnetic induction [Eq. (6)] provides a complete description of the propagation of electromagnetic waves. Combining the two equations to eliminate the magnetic field strength $\mathbf{H}$, we get the wave equation:

$$\nabla \times \frac{1}{\mu} \nabla \times \mathbf{E} - \omega^2 \epsilon \mathbf{E} = 0 \quad (9)$$

A similar procedure to eliminate the electric field $\mathbf{E}$ from (5) and (6) will yield an expression for $\mathbf{H}$. In a homogeneous source free medium we can use Eq. (8) to simplify the wave equation (9):

$$\nabla^2 \mathbf{E} + \omega^2 \epsilon \mu \mathbf{E} = 0 \quad (10)$$

Similarly, for the magnetic field we can obtain

$$\nabla^2 \mathbf{H} + \omega^2 \epsilon \mu \mathbf{H} = 0 \quad (11)$$

Restricting ourselves to consider only a plane wave propagating in the z direction, we can further simplify (10) to

$$\frac{d^2 E_x}{dz^2} = -\omega^2 \epsilon \mu E_x \quad (12)$$

where E_x is the electric field parallel to the x axis. Equation (12) has the solution

$$E_x(z) = A e^{+\gamma z} + B e^{-\gamma z} \quad (13)$$

where the constants A and B correspond to the magnitude of the waves propagating in the $-z$ and $+z$ directions, respectively. $\gamma = \omega \sqrt{\epsilon \mu}$ is known as the propagation constant—it is complex, and generally written in the form

$$\gamma = \alpha + j\beta \quad \text{Np/m} \quad (14)$$

where α is known as the attenuation coefficient and β as the phase constant. These are related to the material properties by the following expressions:

$$\alpha = \omega \sqrt{\frac{\mu_0 \mu_r \epsilon_0 \epsilon'}{2}} \left[\sqrt{\sqrt{1 + \left(\frac{\epsilon''_{\text{eff}}}{\epsilon'}\right)^2} - 1} \right] \quad \text{Np/m} \quad (15)$$

$$\beta = \omega\sqrt{\frac{\mu_0\mu_r\epsilon_0\epsilon'}{2}}\left[\sqrt{\sqrt{1 + \left(\frac{\epsilon''_{eff}}{\epsilon'}\right)^2} + 1}\,\right] \quad \text{Np/m} \quad (16)$$

The linkage between the electric and magnetic fields implies that the electric field of the propagating wave given by Eq. (13) will be accompanied by a normal magnetic field. Figure 3 shows a traveling electromagnetic plane wave with electric and magnetic fields at right angles to each other. When the medium in which the wave is traveling is lossy, the magnitude of the wave decays exponentially with distance.

D. Skin Depth

For a semi-infinite slab which extends from $z = 0$ to $z = \infty$ along the z axis, the constant A in (13) must be zero for the solution to remain bounded as $z \to \infty$. Setting $B = E_0$, the field strength at the surface of the slab, we obtain an equation for the field distribution inside the slab:

$$E(z) = E_0 e^{-\gamma z} \quad (17)$$

The amplitude of the field can be seen to decay exponentially (as shown in Fig. 3) in the z direction since

$$|E| = E_0 e^{-\alpha z} \quad (18)$$

The parameter α describes how far into the material the fields penetrate. This will

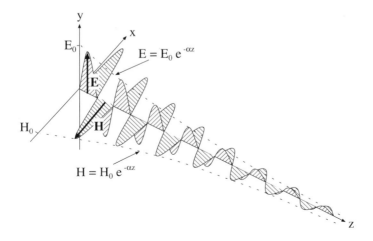

Figure 3 Traveling wave in a lossy dielectric.

depend upon both the frequency of the applied microwaves and the dielectric properties as per Eq. (15).

The skin depth δ provides a more convenient measure of how far the fields will penetrate. It is defined as the depth at which the magnitude of the field has been attenuated to $1/e$ of its value at the surface of the material. For a plane wave incident upon a semi-infinite slab, the magnitude of the field is given by Eq. (18); hence, the skin depth is simply

$$\delta = \frac{1}{\alpha} \tag{19}$$

For very lossy materials, which have a small value of δ, the field will decay very rapidly so that the heating will be confined to the surface of the material. For low loss materials, such as frozen foods, the value of δ will be much larger and the fields can penetrate much further.

For microwave heating applications, the power penetration depth δ_p is also often used. This is defined as the depth to which the *power density* reduces to $1/e$ of its value at the surface. The power, given by Eq. (4), varies as the square of the electric field,

$$Q = \tfrac{1}{2}\omega\epsilon_0\epsilon''_{\text{eff}} |\hat{E}_0|^2 e^{-2\alpha z} = Q_0 e^{-z/\delta_p} \tag{20}$$

where Q_0 is the power density or volumetric heating rate at the surface ($z = 0$) and δ_p is the power penetration depth which is given by

$$\delta_p = \frac{1}{2\alpha} = \frac{\delta}{2} \tag{21}$$

For a material subject to a plane wave, 63% of the power is dissipated within the power penetration depth and 86% within the skin depth. Food materials are generally nonmagnetic ($\mu_r = 1$), for which Eq. (15) can be simplified and substituted in Eq. (21) to obtain an expression for penetration depth as

$$\delta_p = \frac{c}{\omega\sqrt{2\epsilon'}\,(\sqrt{1 + (\epsilon''/\epsilon')^2} - 1)^{1/2}} \tag{22}$$

where the relationship

$$c = \frac{1}{\sqrt{\mu_0 \epsilon_0}} \tag{23}$$

has been used. Figure 4 shows the exponential decay of power density for beef at both 0 and 25°C (property data for the materials are shown in Table 2). Figure 3 in Chapter 4 shows ranges of penetration depths for food materials.

It must be emphasized that these skin depth calculations are valid only for materials undergoing plane wave incidence, and then only when the material is

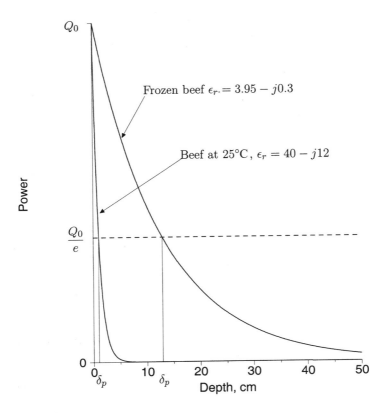

Figure 4 Power penetration depth at 2.45 GHz for beef at 0°C and 25°C. δ_p = 1.04 and 12.9 cm, respectively.

several skin depths thick so that it can be considered to be semi-infinite. For slabs less than a few skin depths in thickness resonance effects can occur inside the material which result in the field distribution not having an exponential decay from the surface. In some cases the highest field strength, and therefore power density, can actually occur in the center of the slab. This is caused by the interference of

Table 2 Example Skin Depths for Some Food Materials at 2.45 GHz

Material	Typical permittivity	Skin depth
Beef	$40 - j12$	2.08 cm
Beef (frozen)	$4 - j0.3$	26.0 cm
Mashed potato	$65 - j20$	1.59 cm
Carrots	$41 - j11$	2.23 cm

Fundamental Electromagnetics

waves reflected from the back of the slab. As discussed later in this chapter and in Chapter 2, the skin depth often does not describe accurately the microwave heating of food in a cavity.

E. Solutions for a Finite Slab

For a finite slab we must account for both constants in Eq. (13) since reflections may occur from the back wall of the slab. Figure 5 shows a comparison of the fields obtained from a one-dimensional solution of Maxwell's equations for a fi-

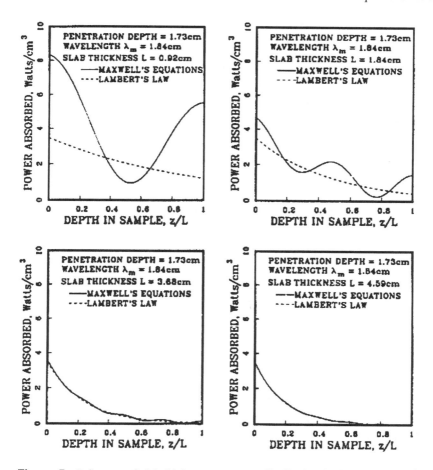

Figure 5 Influence of slab thickness on power distribution for a raw beef sample exposed to microwaves from the left at 2450 MHz. [Reprinted from Ref. 3: Chemical Engineering Science, 46 (4), Ayappa, K. G., et al., "Microwave Heating: An Evaluation of Power Formulations," Copyright © 1991, with permission from Elsevier Science.]

nite slab to a pure exponential decay for slabs of different thickness [3]. It is clear that for thin slabs the field does not decay exponentially from the surface. For a composite slab composed of more than one dielectric material, the situation is more complex [4] (as shown in Fig. 6).

Backward propagating waves in a thin material cause a highly oscillatory power absorption behavior; it is possible that small changes in thickness can produce large changes in the amount of energy absorbed. This implies that a unique value of the power penetration depth cannot be found with any certainty. Fu and Metaxas [5] proposed a new definition for the power penetration depth Δ_p, which is the depth at which the power absorbed by the material is reduced to $(1 - 1/e)$ of the total power absorbed. Figure 7 shows the normalized power penetration depth as a function of material thickness for water at 20°C. As the thickness of the slab increases, Δ_p approaches δ_p. This definition allows a unique value of Δ_p to be found for all thicknesses and also gives an indication of the validity of assuming exponential decay within the slab.

F. Boundary Conditions

At an interface between two materials with different permittivities the normal component of the electric field will be discontinuous, and similarly for the mag-

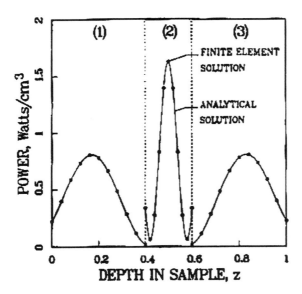

Figure 6 Field inside a composite slab. (From Ref. 4. Reproduced with permission of the American Institute of Chemical Engineers. Copyright © 1991 AICHE. All rights reserved.)

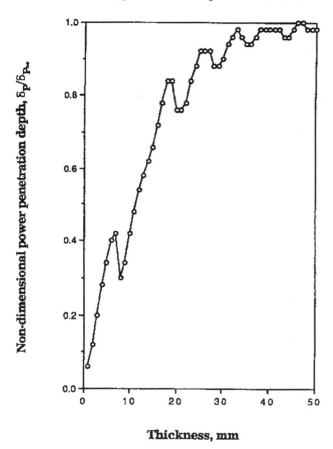

Figure 7 Power penetration depth Δ_p as a function of material thickness at 2450 MHz for water at 20°C. (After Fu and Metaxas [5], 1992, reproduced by permission of the Journal of Microwave Power and Electromagnetic Energy.)

netic field at interfaces between materials with different permeabilities. The boundary conditions at the interface between two materials are

$$\epsilon_1 \mathbf{E}_1^n = \epsilon_2 \mathbf{E}_2^n \tag{24}$$

$$\mathbf{E}_1^t = \mathbf{E}_2^t \tag{25}$$

$$\mu_1 \mathbf{H}_1^n = \mu_2 \mathbf{H}_2^n \tag{26}$$

$$\mathbf{H}_1^t = \mathbf{H}_2^t \tag{27}$$

where the subscripts 1 and 2 refer to the values in materials 1 and 2, respectively, and the superscripts t and n refer to the tangential and normal components of the field, respectively (as shown in Fig. 8). These boundary conditions have important consequences for microwave heating. An electric field parallel to the surface of the load will produce a much larger field and hence power density inside the material than a field of equal strength but normal to the surface. This has the effect that one or two resonant cavity modes which have fields parallel to the load's surface may dominate even though the cavity may support many resonant modes. In such situations the power density distribution is likely to be very nonuniform and edge overheating is likely. This is discussed further in the next chapter.

The walls of a microwave applicator are conductive and are usually represented in calculations as a perfect electrical conductor. This is a reasonable approximation since at microwave frequencies the skin depth in metal is very small ($\approx \mu m$). A perfect electrical conductor can have no electric field parallel to its surface so that the electric field tangential to the cavity walls is zero, which has the effect of reflecting all microwaves incident on the wall. Mathematically this is usually written as

$$\mathbf{E} \times \hat{\mathbf{n}} = 0 \tag{28}$$

where $\hat{\mathbf{n}}$ is the normal to the surface of the conductor. Similarly, the magnetic field will have no normal component at the perfectly conducting wall, so that

$$\mathbf{H} \cdot \hat{\mathbf{n}} = 0 \tag{29}$$

In a real applicator, the walls will not be perfect conductors and it is possible for numerical solutions to take this into account [6].

G. Reflection, Transmission, and Absorption

When a plane wave is incident upon a dielectric material, some of the energy is reflected and some transmitted into the material, as shown in Fig. 9. For a plane wave with the electric field parallel to the surface and incident at an angle θ, the transmitted wave will be refracted and the angle ϕ is given by Snell's law:

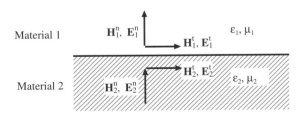

Figure 8 Boundary conditions.

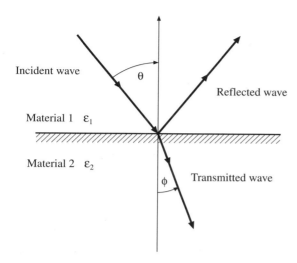

Figure 9 Reflections and transmission at a surface.

$$\sin \phi = \frac{\sqrt{\epsilon_1}}{\sqrt{\epsilon_2}} \sin \theta \tag{30}$$

The reflection coefficient Γ for the electric field is then given by Fresnel's laws [1],

$$\Gamma = \frac{E_r}{E_i} = \frac{\sqrt{\epsilon_2} \cos \theta - \sqrt{\epsilon_1} \cos \phi}{\sqrt{\epsilon_2} \cos \theta + \sqrt{\epsilon_1} \cos \phi} \tag{31}$$

For normal incidence, i.e., $\theta = 0°$, this reduces to

$$\Gamma = \frac{\sqrt{\epsilon_2} - \sqrt{\epsilon_1}}{\sqrt{\epsilon_2} + \sqrt{\epsilon_1}} \tag{32}$$

Since the power is proportional to the square of the electric field, the fraction of power transmitted into the material is given by

$$|T| = 1 - |\Gamma|^2 \tag{33}$$

For food materials which typically have permittivities in the range 40–60, the value of Γ will be about 75% [7] (T is about 50%). The transmitted wave will be refracted into the interior and for certain geometries this can lead to focusing effects which are discussed in the next chapter. Equation (33) gives the ratio of power transmitted into the food from a *single* plane wave. In a real applicator many such plane waves may be incident on the load from a variety of different directions.

Another consequence of Eq. (31) is that at a certain angle, known as the Brewster angle, the reflection from the surface will be zero. This means that all of the incident energy will be transmitted into the material—a fact that has lead to the suggestion that microwave applicators should be designed to take advantage of this.

H. Propagation Modes

The electric and magnetic fields in homogeneous media are solutions of Eqs. (10) and (11), respectively. If we consider solutions to these equations for a wave that propagates, say, along the z axis, we can find solutions of the form $f(z)g(x, y)$, where f is a function of z only and g is a function of x and y only [8]. This leads to the following standard classification of solutions:

Transverse electromagnetic (TEM): Neither the electric or magnetic fields have components in the direction of propagation, $H_z = E_z = 0$.
Transverse magnetic (TM): The magnetic field has no component in the direction of propagation, $H_z = 0$, but the electric field does, $E_z \neq 0$.
Transverse electric (TE): The electric field has no component in the direction of propagation, $E_z = 0$, but the magnetic field does, $H_z \neq 0$.

In waveguides the boundary conditions do not permit the propagation of TEM modes and the normal propagating mode for power transfer is the TE_{10} mode. The subscripts denote the number of semisinusoidal variations in the x and y directions, respectively. The TE_{10} mode has one semisinusoidal variation in the x direction and is constant in the y direction. For power transmission this is the normal mode. Inside waveguide applicators, such as the phase control applicator described in Chapter 2, other modes may be present [9].

I. Single-Mode Cavities

It is possible to design an applicator which has a single resonance near to the operating frequency. The most common applicators of this type are cylindrical applicators which have TM_{010} mode or rectangular applicators with a TE_{10n} resonant mode, as shown in Fig. 10. These applicators have a well-defined field distribution and are easy to analyze and design using analytical techniques [10]. Their use for food processing, however, is extremely limited. The reason for this is that the volume of a load with the dielectric properties typical of food has to be extremely small in order to maintain the resonance.

J. Multimode Cavities

The majority of food heating applications use a multimode resonant cavity applicator, for example, the domestic microwave oven. Inside a loaded multimode ap-

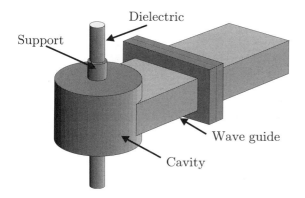

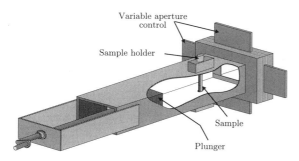

Figure 10 Single-mode resonant cavities. (a) TM_{010} single-mode resonant cavity; (b) TE_{103} single-mode resonant cavity.

plicator one can view the fields in the cavity as the superposition of multiple plane waves impinging on the load from a variety of directions. The field distribution in such cases is generally complex and beyond the scope of simple calculations; methods for its determination are the subject of the latter half of this chapter. Detailed descriptions of field distributions inside food heated in a multimode cavity are provided in Chapter 2. The assumptions used by some researchers that exponential decay is characteristic of all microwave heating simply does not hold in multimode resonant applicators. However, the skin depth can give a *qualitative* feel for the likely result; if the material is several skin depths in size the microwaves will not penetrate to the center of the material so any heating there will be by thermal conduction. If the material is less than a skin depth in thickness, the fields can penetrate to the center of the material, so more uniform heating is possible but the actual field distribution and heating pattern will be governed by many factors (as outlined in Chapter 2) and may be very nonuniform.

Inside a multimode cavity there are likely to be many resonant modes, where reflections from the walls of the cavity constructively reinforce each other to produce a stable standing wave pattern. For an *empty* rectangular cavity with dimensions $a \times b \times d$ the frequency of the resonant modes is given by the equation [10]

$$\left(\frac{l\pi}{a}\right)^2 + \left(\frac{m\pi}{b}\right)^2 + \left(\frac{n\pi}{d}\right)^2 = \left(\frac{\omega_{lmn}}{c}\right)^2 \qquad \{l, m, n\} = \{0, 1, 2, \ldots\} \quad (34)$$

where c is the speed of light and ω_{lmn} is the resonant frequency of the mode. The integer subscripts l, m, and n correspond to the number of semisinusoidal variations in the x, y, and z directions, respectively. In a similar fashion to the propagation modes in waveguides described above, the resonant cavity modes are also often classified as either TE or TM. However, there are now three subscripts corresponding to the indices l, m, and n. If all of the indices are nonzero, then the cavity can support both a TE and a TM mode [11]. However, if one of the indices is zero, then either a TE or a TM mode may be supported but not both; otherwise the boundary conditions will not be satisfied. Modes with two or more zero indices cannot satisfy the boundary conditions.

A simple calculation using Eq. (34) produces the number of modes supported by an empty cavity within a given frequency range. Table 3 gives some modes near to 2.45 GHz for a typical cavity, and Fig. 11 shows the magnitude of the field for three different modes at a plane through the center of the cavity. The total field in the cavity will be a combination of such modes.

Designing a cavity to give a large number of modes when empty is, by it-

Table 3 Some Resonant Modes for an *Empty* Rectangular Cavity

Frequency, GHz	l	m	n
2.4012	3	0	3
2.4186	4	1	2
2.4459	2	4	1
2.4602	3	1	3
2.4695	0	3	3
2.4808	0	4	2

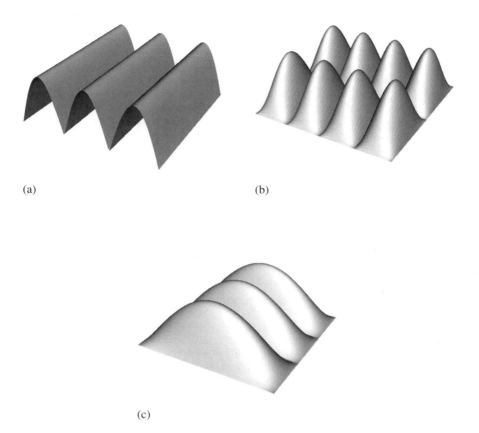

Figure 11 Example mode shapes. (a) $L = 0, m = 3$; (b) $l = 4, m = 2$; (c) $l = 3, m = 1$.

self, insufficient to ensure a good design. Many of the modes given by Eq. (34) may not actually get excited by a given feed arrangement. Secondly, introducing a load into the oven causes a very significant change in the mode pattern. Assuming the load is lossy, the sharp resonance given by Eq. (34) will become blurred and the resonant frequency will be reduced, this is shown graphically in Fig. 12. Consequently, the field distribution in the cavity will be composed of several overlapping mode distributions. Load dependent effects may also dominate the power density distribution.

In a microwave applicator there are also likely to be a few evanescent or nonpropagating modes in the region of the feed port. This may or may not have an impact on the power density produced in the load depending upon the geometry.

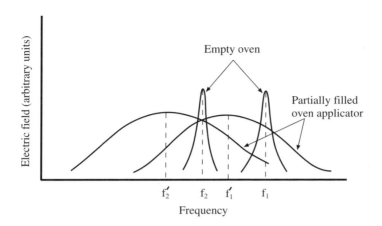

Figure 12 Oven modes. (After Metaxas and Meredith [10], 1983, reproduced by permission of the Institution of Electrical Engineers.)

III. ANALYTICAL MODELING OF MICROWAVE APPLICATORS

The simple equation given in the last section for the modes inside an empty multimode cavity provides little information on what happens in a loaded cavity. Several attempts have been made to produce analytical models for microwave heating. In single-mode cavities the fields are often easily described by a simple mathematical expression [10]. For multimode cavities the geometry generally precludes the use of analytical expressions. Results for simplified geometries, such as rectangular ovens loaded with slabs which completely fill the cross section, have been presented [12], but they do not generally reflect realistic loads.

Analytical solutions do have the great advantage, however, that the influence of the individual parameters can be easily deduced—in contrast to the numerical solutions presented below where a new simulation is required for each new parameter value.

IV. NUMERICAL MODELING OF CAVITIES

Until the late 1980s there was little research into the numerical modelling of microwave heating in multimode cavities [13]. This was partially due to the lack of availability of computers with sufficient power to solve the necessary equations. Microwave ovens are three-dimensional devices, which in general, require three-dimensional numerical solutions. Due to advances in computer technology, as well as significant advances in the simulation techniques, the methods are reaching a level of maturity where they can now be considered for use as part of the design process.

The finite element method (FEM) [14] and finite difference time domain method (FDTD) [15, 16] are the two most popular methods for simulating microwave heating. Many other methods have also been employed by researchers such as transmission line matrix (TLM) [17] and the method of moments (MoM) [13, 18]. Since FEM and FDTD are the two most common methods, they are the only two which are outlined here.

A. Finite Difference Method (FDTD)

Probably the most popular method for the simulation of microwave heating, certainly until very recently, has been the time domain finite difference method (FDTD) based on Yee's scheme [15, 19–21]. This scheme uses two complementary meshes offset from one another, one representing the electric field components and one the magnetic. The time dependent Maxwell's equations are then solved alternately. The electric field at time t is used to calculate the magnetic field at time $t + \frac{1}{2}\Delta t$ which, in turn, is used to find the electric field at time $t + \Delta t$, and so on. Δt is the time step, which due to stability constraints must be kept small. For an orthogonal grid, Δt is governed by the Courant condition which relates the time step required for stability to the spatial discretisation.

$$\frac{1}{c\,\Delta t} \geq \sqrt{\frac{1}{(\Delta x)^2} + \frac{1}{(\Delta y)^2} + \frac{1}{(\Delta z)^2}} \tag{35}$$

where c is the speed of light and Δx, Δy, and Δz the size of the grid spacing in the x, y, and z directions, respectively [16].

The advantage of FDTD is that programming is straightforward and the algorithm is extremely efficient, accurate, and easily parallelized. The main disadvantage stems from the use of a regular grid. To achieve a given accuracy with a rectangular grid when modeling curved or oblique boundaries generally requires a finer mesh than when using an equivalent boundary conforming mesh [22]. The FDTD method in its original form was restricted to a regular grid; however, this has been extended to allow variable spacing meshes to be used [23, 24]. This technique, often called subgridding, reduces the grid spacing in certain regions but still uses a grid aligned to the coordinate axis. More recent work has extended the FDTD method to use nonorthogonal grids [25, 26].

Food generally has a relatively large permittivity so that the wavelength inside the food is significantly smaller than the wavelength in the surrounding air. This means that a much finer mesh is required in the load than in the air. The subgridding techniques go some way to addressing this problem. From Eq. (35) it can be seen that a fine grid also requires a small time step. This has implications for the time needed for the calculation; the smaller the time step the more steps are required for a full solution, so the longer the calculation time.

The treatment of the material boundaries is extremely important when modeling microwave food processing. One of the most common techniques used in the FDTD scheme is to smooth the jump in dielectric properties across the interface and use a simple average of the properties for the difference equation which spans the interface. Zhao and Turner [27] found that this simple approximation leads to large errors when used for modeling microwave heating and their paper gives an alternative interface condition which leads to more accurate results.

B. Finite Element Method (FEM)

The finite element method decomposes the complex geometry to be modeled into a union of elements each of which has a simple geometry, typically quadrilaterals or triangles in two dimensions and tetrahedra or hexahedra in three dimensions [14]. Since the method does not require a regular grid, it allows a very flexible decomposition of the region. Small elements can be used in areas where the wavelength is short or high accuracy is required while larger elements can be used in the air region. The use of elements which are not aligned to the coordinate axis also means that curved boundaries can be modeled with a series of straight lines* rather than with a "staircase." This allows greater accuracy to be achieved with far fewer elements than required by the equivalent model using a purely regular grid.

This flexibility comes at a price—it is much harder to generate high-quality finite element meshes than it is to generate finite difference grids. There are now several commercial programs for doing this; their availability has increased the attractiveness of using FEM for microwave heating calculations. Some of the mesh generators can link with CAD packages, and existing geometry descriptions may be used for the analysis.

For many years FEM was plagued by "spurious modes," that is, nonphysical solutions which would corrupt the true solution rendering the simulation worthless [28]. These spurious modes are now well understood, and techniques for eliminating them are available. The most prominent of these methods uses edge elements [14, 29] instead of the more traditional node-based elements. These made their first appearance in the early 1980s but have only recently been included in commercial codes. Edge elements have additional advantages as well, such as the ability to correctly handle material discontinuities and sharp metal corners.

Traditionally, FEM has been carried out in the frequency domain as described above. However, when applied to multimode microwave cavity heating

---

* So-called "curvilinear elements" can also be used which have curved sides for better representation of curved boundaries.

problems, it was found that the system of equations became very ill-conditioned—making the solution very time consuming and prone to error [30, 31]. A possible explanation for this has been put forward by Bossavit [29, 32]. One solution to the problem is to shift to the time domain. Here we no longer have any ill-conditioned matrix problems; in fact the matrices become very well conditioned, but it becomes necessary to run the simulation for many times steps [30, 31, 33]. The time domain does have the auxiliary advantage that a broad band response can be obtained with a single calculation. This means that we can determine the response of the system to several different frequencies with a single calculation [34]. Finite element schemes employing a leap-frog time stepping scheme, similar to that used in FDTD, and complimentary meshes have also been proposed [35]. For a given number of unknowns the FDTD method is considerably faster than time domain FEM; however, the use of unstructured meshes means that FEM generally requires fewer unknowns than FDTD. Examples of FEM calculations can be seen in Chapter 2.

C. Modeling the Magnetron

In virtually all microwave systems used for the processing of food, the microwaves are generated by a magnetron which is attached to the applicator via a length of waveguide. A complete numerical simulation should therefore take into account the characteristics of the magnetron. However, the majority of numerical studies tend to ignore the magnetron, assuming that the applicator is fed from a waveguide containing a pure TE_{10} mode. This assumption greatly simplifies the calculation since accurately modeling the magnetron's output structure is non-trivial.

The standard technique for modeling the waveguide feed in the time domain (both for FDTD and TDFEM) is to place an absorbing boundary condition (ABC) at the end of the waveguide and a fictitious current sheet or an antenna [36] a short distance from it. The ABC is necessary so that any reflected energy from the cavity is absorbed; this effectively models the presence of an iso-circulator.

A FDTD study was carried out by Iwabuchi et al. [37] where the magnetron was modeled and the fields in the oven to which it was connected compared to those obtained from a model which assumed a simple TE_{10} mode excitation in the feed waveguide. They concluded that it was unnecessary to model the magnetron since for even a short length of waveguide connecting the magnetron to the applicator the predicted field distribution in the cavity remained the same.

One reason for wanting to use a more accurate model of the magnetron is to account for its nonlinear characteristics. The output frequency and power of a magnetron are dependent upon the impedance of the load to which it is connected. So if we wish to determine numerically the exact power which is delivered to the load for use in subsequent thermal calculations, then we need a model of the magnetron. Fortunately a very simple model has been proposed by Sundberg [38] which can be used with the majority of existing software. The model places a short circuit at the end of the waveguide and a dielectric block in front of that short circuit. The dielectric properties of the material are chosen according to the parameters of the equivalent circuit of the magnetron.

This model does not take into account the frequency variations with load impedance that occur with real magnetrons. However, while for lightly loaded cavities the frequency is very important [34], for heavily loaded cavities small changes in the frequency will have little effect on the power density distribution. Since the majority of food processing applications use heavily loaded cavities, modeling small frequency changes is less important.

D. Future Directions for Microwave Modeling

One of the most significant problems with using numerical simulation software as part of the microwave design process is handling the large computational resources that are required. This problem is being tackled on two fronts. As computers are becoming faster and cheaper, the computing power necessary for large simulations is increasingly affordable. It is no longer necessary to have access to a multimillion dollar supercomputer to perform realistic simulations of industrial applications. Secondly, the algorithms that are used for the calculation are being constantly improved, reducing the computing power required for a given problem. Among the areas being actively researched in relation to microwave heating are:

> *Adaptive meshing:* A solution on a coarse mesh can be used to drive the generation of finer meshes. This allows solutions of high accuracy to be obtained with a minimum number of elements, so minimizing the computational cost. This technique is available in some commercial FEM codes but is not yet routinely applied to microwave heating problems. It is, however, essential for efficient simulation.
>
> *Domain decomposition:* The region is divided into a series of subregions and the grid points internal to these regions are eliminated so that when the system is assembled the number of unknowns in the final system are greatly reduced [39, 40]. This technique has two main applications: dividing the system up so that the problem can be solved on a parallel computer and efficiently solving problems where small changes occur in some areas of the region. The majority of the work using this technique is carried out when eliminating the internal unknowns in each region. After a small change, only those regions in which a change has taken place need to be recalculated. In models which take into account temperature-dependent permittivity [41] (see also Chapter 2) or movement, the use of this technique could realize large savings in computational effort.

Multigrid techniques: Multigrid techniques, which use a hierarchical series of meshes, are used for solving a wide range of boundary value problems but have only recently been successfully applied to high-frequency electromagnetic problems [42]. Multigrid methods are optimal in the sense that the solution cost grows linearly with the problem size.

Parallel processing: Originally the use of parallel processing techniques was restricted to supercomputers. With the advent of packages, such as PVM (Parallel Virtual Machine), parallel computers can now be constructed from a heterogeneous collection of smaller machines [39, 43], e.g., networked PCs.

E. Commercial Software

There are several commercial electromagnetic simulation programs available that are capable of solving microwave heating problems, as shown in Table 4. These codes have generally not been developed with microwave heating simulations in mind. Palombizio and Yakolev [44] surveyed both the available commercial software and the attitudes of the microwave heating industry toward the software and found that there was a perception gap between what was available and what was perceived to be available. The main areas of concern for industry were cost and the user-friendliness of the software. They concluded, however, that robust, user-friendly analysis software is available, albeit at a price.

V. APPLICATION OF MODELING TO MICROWAVE HEATING

Having looked at how Maxwell's equations can be modeled, it is important to see how this may be applied to the microwave heating of food. Three separate categories can be identified where modeling can be applied:

- Domestic oven design
- Industrial oven design
- Microwaveable food packaging design

A. Domestic Oven Design

The domestic microwave oven is by far the largest application of microwave heating. Since the introduction of the microwave oven the design has been improved by the addition of turntables, mode stirrers and by better feed systems which combine to increase the uniformity of heating. However, the basic design of the oven, a rectangular box with a single microwave feed, has not changed and is unlikely to change

Table 4 Commercial EM Packages for Microwave Power Engineering

Vendor	Package	Platform
Ansoft Corp. www.ansoft.com	Ansoft HFSS, EMAS	UNIX, Windows 95/98
ANSYS Ltd. www.ansys.com	ANSYS/Emag	UNIX, Windows 95/98
CST GmbH www.cst.de	MAFIA 4	UNIX, Windows 95/98
Electro Magnetic Applications, Inc. www.sni.net/emaden	EMA 3D	UNIX, Windows NT
HP Eesof www.tmo.hp.com/tmo/hpeesof	HFFS 5.4	UNIX, Windows 95/98, NT
Infolytica Corp. www.infolytica.com	FullWave	UNIX, Windows 95/98, NT
Remcom, Inc. www.remcom.com	XFDTD 5.0	UNIX, Windows 95/98, NT
Sonnet Software, Inc. sonetusa.com	KCC Micro-strips	UNIX, Windows 95/98, NT
The Japan Research Institute www.jri.co.jp/pro-eng/jmag/E/index.html	JMAG-Works	UNIX, NT
QWED s.c. www.qwed.pl.com	QW3D 1.8	Windows 95/98, NT
Vector Fields, Inc. www.vectorfields.com	Saporano SS/EV	UNIX, Windows 95/98, NT
Weidingler Associates, Inc. www.wai.com	EMFlex	UNIX
Zeland Software, Inc. www.zeland.com	FIDELITY 2.0	Windows 95/98, NT

Source: From Ref. 44.

significantly. The oven manufacturer also has little control over how the consumer uses the microwave oven. The wide range of products which may be heated means that it is impossible to optimize for any one situation but rather it is necessary to give good performance over a range of applications. Thus the use of computer simulation to optimize domestic oven designs is somewhat limited, although the majority of published simulations have been of domestic ovens [17, 19, 20, 23, 41, 43, 45, 46]. This has been partly due to their small size compared to industrial systems and the ease with which experimental verification can be carried out.

Even though the basic design of a domestic oven is unlikely to undergo any significant changes, simulation can still play an important role in the design process.

- The position and design of the feed will have a strong impact on the field distribution that is produced in the cavity. Computer simulation can be used to study the effect of changes in feed position under a wide variety of loading conditions. This can enable designs to be optimized while building a minimum number of prototypes [45, 46]. This is also discussed in Chapter 2.
- Domestic ovens generally have fairly large manufacturing tolerances. This means that slightly different heating patterns are often produced by different ovens of the same model. Simulation allows the effect of small changes in the various oven dimensions for a variety of loading situations to be determined [47]. This form of sensitivity analysis allows the effect of the tolerances to be studied.

The use of a turntable [23] or a mode stirrer [45] inside the oven complicates the simulation. It becomes necessary to do a series of simulations with the turntable in different positions to show the effect that the rotation will have on the field distributions. Very few researchers have tackled this problem to date because of the increased computation required.

B. Industrial Oven Design

Industrial ovens, such as those used for tempering frozen products and for food pasteurization [2, 10], provide enormous scope for simulation. These ovens are generally designed to perform a very specific task with a very limited range of load configurations. Also, since only a handful of ovens of a given design may be manufactured, the design cost represents a much larger fraction of the oven's total cost than it does for a domestic oven. Consequently, if the use of numerical simulation can reduce the design cost by reducing the number of prototypes that need to be made, significant savings in the cost of the oven may be possible.

It is common when designing large industrial ovens to construct smaller pilot plant systems to test the principle and then to scale up. It is quite possible, however, to encounter problems in the scaled-up version of the oven which were not present in the prototype, especially when building multimode cavities. Numerical simulation could help to identify some of these potential problems during the design stage. Numerical simulation is also a valuable tool for "proof of principle" studies and has been used in support of patent applications for new applicator designs [9].

The main difficulty with modeling industrial ovens is their physical size. The large dimensions mean that many grid points/elements are required to accurately model the oven, which implies that a significant amount of computational resources are required. The simulation results for large industrial ovens that have been published to date have either used symmetry or special features of the oven's design to model just a small fraction of the oven [18, 21, 48] or they have used a supercomputer [49]. Alternatively, they have been restricted to using coarse grids and therefore produced results of low accuracy [50].

A second difficulty with industrial ovens is that they are often designed for use as part of a continuous system with the product moving through the oven on a conveyor. This means that simulations with the product in a variety of positions are needed to correctly characterize the system.

Even though modeling industrial ovens is still a daunting task, given the computational size of the problem, there is much that can be done. As computers get faster and algorithms improve (cf. Sec. IV.D), the simulation of industrial ovens as part of the design process rather than simply as an academic exercise will become realizable.

C. Microwaveable Food Packaging

Details of packaging techniques for microwaveable foods are covered in Chapter 12. The problems with microwave heating include variable total energy absorption, nonuniform heating, uneven distribution between the components of meals in layered foods, and surface heating intensities which are insufficient to achieve browning and crisping [51]. To overcome some of these problems, innovative packaging designs have been devised. Susceptors and surface coatings have been used to promote the browning and crisping of products such as pizzas, and metal packages have been used to shield certain areas of the product to try and prevent edge overheating, all with varying degrees of success [52]. Various matching structures have also been used which attempt to use interference effects to control the field distribution inside the product. Simpler techniques such as modifications to the geometry of the package and changing the food layout also play an important role in designing a microwaveable product.

To date, the design of microwaveable products has been based on trial and error coupled, in some cases, with some simple analytical methods. This not only requires a very extensive testing period in the design of new products, but also can fail to identify potential problems.

The computer simulation of packaging designs and product layouts holds the promise of being able to quickly try out new ideas and designs under a range of loading conditions. Furthermore, simulation can provide information which would be extremely difficult to obtain experimentally, such as the electric field distribution. These simulations can then not only provide information on how a product will behave in a given environment but also allow the effect of small changes in a single parameter to be studied. This may be difficult in an empirical environment.

The correct modeling of packaging will require better models of the packaging material. To date most numerical simulations simply ignore dielectric packaging material since it is normally thin and has a low permittivity compared to the food material. This may provide a reasonable approximation in many cases. When advanced packaging materials, such as susceptors, are used, the package must be modeled. This means including thin films in the model. The naive approach of simply using a very find grid/mesh in the package material will increase the computational complexity to an extent that the model ceases to be viable. Other techniques, such as replacing the thin film with an infinitely thin layer which is then represented by an integral term in the model, will be needed [6, 53]. It may also be possible to achieve useful results by modeling the package in isolation from the oven using a series of plane wave illuminations to get a qualitative indication of the package's performance.

VI. CONCLUSION

For a full understanding of the heating process an appreciation of the underlying electromagnetics is essential. This chapter has given a basic introduction to the electromagnetic aspects of microwave heating and the various simulation methods that can be used. With user-friendly, commercial analysis software now available, the techniques that have been outlined can be used to aid in the design of microwave food processing applications.

LIST OF SYMBOLS

c	Speed of light, 3×10^8 m/s
C_p	Specific heat capacity J/kg K
D	Electric flux density, C/m^2
E	Electric field strength, V/m
f	Frequency, Hz
H	Magnetic field strength, A/m
$\hat{\imath}, \hat{\jmath}, \hat{k}$	Cartesian unit vectors
j	$\sqrt{-1}$
J	Current density, A/m^2
J$_t$	Total current density, A/m^2
n	Normal vector to a surface
Q	Power density, W/m^3
T	Fraction of power transmitted
t	Time
Δt	Time step size
V	Voltage

Greek Symbols

α	Attenuation constant, Np/m
β	Phase constant, Np/m
γ	Propagation constant
ϵ	Permittivity, $\epsilon = \epsilon_0 \epsilon_r$, F/m
ϵ'	Relative dielectric constant, dimensionless
ϵ''	Relative dipolar loss factor, dimensionless
ϵ''_{eff}	Relative effective loss factor, dimensionless, $\epsilon''_{\text{eff}} = \epsilon'' + \dfrac{\sigma}{\omega \epsilon_0}$
ϵ_0	Permittivity of free space, 8.854×10^{-12} F/m
ϵ_r	Relative permittivity $\epsilon_r = \epsilon' - j\epsilon''_{\text{eff}}$
Γ	Reflection coefficient
λ_0	Free space wavelength, m
μ	Permeability $\mu = \mu_0 \mu_r$, H/m
μ_0	Permeability of free space, $4\pi \times 10^{-7}$ H/m
μ_r	Relative permeability, dimensionless
σ	Electrical conductivity, S/m
ω	Angular frequency, rad/s

REFERENCES

1. Arthur Von Hippel. Dielectrics and waves. Wiley, New York, 1954.
2. A. C. Metaxas. Foundations of Electroheat: A Unified Approach. John Wiley and Sons Ltd., Chichester, UK, 1996.
3. K. G. Ayappa, H. T. Davis, G. Crapiste, E. A. Davis, and J. Gordon. Microwave heating: An evaluation of power formulations. Chemical Engineering Science, 46(4):1005–1016, 1991.
4. K. G. Ayappa, H. T. Davis, E. A. Davis, and J. Gordon. Analysis of microwave heating of materials with temperature-dependent properties. AIChE Journal, 37(3):313–322, 1991.
5. W. Fu and A. C. Metaxas. A mathematical derivation of power penetration depth for thin lossy materials. Journal Microwave Power and Electromagnetic Energy, 27(4):217–222, 1992.
6. R. A. Ehlers, D. C. Dibben, and A. C. Metaxas. The effect of wall losses in the numerical simulation of microwave heating problems, 1999. Submitted to Journal Microwave Power and Electromagnetic Energy.
7. P. O. Risman and T. Ohlsson. 2450 mhz microwave heating distributions in food material slabs and cylinders. In KEMA 26–29 Sept. Arnhem, The Netherlands, 1989.
8. R. E. Collin. Foundations for Microwave Engineering. McGraw-Hill, New York, 1992.
9. J. R. Bows, M. L. Patrick, D. C. Dibben, and A. C. Metaxas. Computer simulation and experimental validation of phase controlled microwave heating. In Microwave and High Frequency Heating 1997, Fermo, Italy, pp. 23–26, 1997.
10. A. C. Metaxas and R. J. Meredith. Industrial Microwave Heating. Number 4 in IEE Power Engineering Series. Peter Peregrinus Ltd., London, 1983.

11. R.F.B. Turner, W.A.G. Voss, W. R. Tinga, and H. P. Baltes. On the counting of modes in rectangular cavities. Journal Microwave Power and Electromagnetic Energy, 19(3):199–208, 1984.
12. F. Paoloni. Caclulation of power deposition in a highly overmoded rectangular cavity with dielectric loss. Journal Microwave Power and Electromagnetic Energy, 24(1):21–32, 1989.
13. Claude Lorenson and Christine Gallerneault. Numerical methods for the modelling of microwave fields. In D. E. Clark, F. D. Gac, and W. H. Sutton, Eds., Symposium on Microwaves: Theory and Application in Materials Processing, pp. 193–200. American Ceramic Society, 1991.
14. P. Silvester and R. Ferrari. Finite Elements for Electical Engineers. Cambridge University Press, Cambridge, UK, 3rd ed., 1997.
15. K. S. Yee. Numerical solution of initial boundary value problem involving Maxwell's equations in isotropic media. IEEE Transactions on Antennas and Propagation, 14:302–307, 1966.
16. A. Taflove. Review of the formulation and applications of the finite-difference time-domain method for numerical modelling of electromagnetic wave Interactions with arbitrary structures. Wave Motion, 10:547–582, 1988.
17. R. A. Desai, A. J. Lowery, C. Christopoulos, P. Naylor, J.M.V. Blanshard, and K. Gregson. Computer modelling of microwave cooking using the transmission-line model. IEE Proceedings A, 139(1):30–38, 1992.
18. Magnus Sundberg. Moment method and FDTD analysis of industrial microwave ovens. In Microwave and High Frequency Heating 1997, Fermo, Italy, pp. 19–22, 1997.
19. F. Lui, I. Turner, and M. Bialkowski. A finite-difference time-domain simulation of power density distribution in a dielectric loaded microwave cavity. Journal Microwave Power and Electromagnetic Energy, 29(3):138–148, 1994.
20. Lizhuang Ma, Dominique-Lynda Paul, Nick Pothecary, Chris Railton, John Bows, Lawrence Barratt, Jim Mullin, and David Simons. Experimental validation of a combined electromagnetic and thermal FDTD model of a microwave heating process. IEEE Transactions on Microwave Theory and Techniques, 43(11):2565–2572, 1995.
21. M. Sundberg, P. Risman, P. S. Kildal, and T. Ohlson. Analysis and design of industrial microwave-ovens using the finite-difference time-domain method. Journal Microwave Power and Electromagnetic Energy, 31(3):142–157, 1996.
22. Richard Holland. Pitfalls of staircase meshing. IEEE Transactions on Electromagnetic Compatibility, 35(4):434–439, 1993.
23. Koji Iwabuchi, Tetsuo Kubota, Tatsuya Kashiwa, and Hiroaki Tagashira. Analysis of electromagnetic fields using the finite-difference time-domain method in a microwave oven loaded with high-loss dielectric. Electronics and Communications in Japan, 78(7):41–50, 1995. Translated from Denshi Joho Tsushin Gakkai Ronbunshi, vol. 78-C-I, no. 2, 1995.
24. K. M. Krishnaiah and C. J. Railton. A stable subgridding algorithm and its application to eigenvalue problems. IEEE Transactions on Microwave Theory and Techniques, 47(5):620–628, May 1999.
25. C. J. Railton and J. B. Schneider. An analytical and numercal analysis of several locally conformal fdtd schemes. IEEE Transactions on Microwave Theory and Techniques, 47(1):56–66, January 1999.

26. Supriyo Dey and Raj Mittra. A conformal finite-difference time-domain technique for modeling cylindrical dielectric resonators. IEEE Transactions on Microwave Theory and Techniques, 47(9):1737–1739, September 1999.
27. H. Zhao and I. W. Turner. An analysis of the finite-difference time-domain method for modelling microwave heating of dielectric materials within a three-dimensional cavity system. In Scientific and Industrial RF and Microwave Applications Conference, Melbourne, Australia. 9–10 Jul, 1996.
28. Alain Bossavit. Solving Maxwell equation in a closed cavity, and the question of spurious modes. IEEE Transactions on Magnetics, 26(2):702–705, 1990.
29. Alain Bossavit. Computational Electromagnetism: Variational Formulations, Complementarity, Edge Elemenets. Academic Press, San Diego, 1998.
30. David Dibben. Numerical and experimental modelling of microwave applicators. PhD thesis, Cambridge University, 1995.
31. David Dibben and A. C. Metaxas. Frequency domain vs. time domain finite element methods for calculation of fields in multimode cavities. IEEE Transactions on Magnetics, 33(2):1468–1471, 1997.
32. Alain Bossavit. Uniqueness of solution of Maxwell equations in the loaded microwave oven, and how it may fail to hold. In Microwave and High Frequency Heating 1995, Cambridge, England, 1995.
33. David Dibben and A. C. Metaxas. Time-domain finite-element analysis of multimode microwave applicators. IEEE Transactions on Magnetics, 32(3):942–945, 1996.
34. David Dibben and A. C. Metaxas. Finite element time domain analysis of multimode applicators using edge elements. Journal Microwave Power and Electromagnetic Energy, 29(4):242–251, 1994.
35. Mitsuo Hano and Tatsuo Itoh. Three-dimensional time-domain method for solving Maxwell's equations based in circumcenters of elements. IEEE Transactions on Magnetics, 32(3):946–949, 1996.
36. M. F. Iskander, R. L. Smith, A.O.M. Andrade, H. Kimrey, and L. M. Walsh. FDTD simulation of microwave sintering of ceramics in multimode cavities. IEEE Transactions on Microwave Theory and Techniques, 42(5):793–800, 1994.
37. Koji Iwabuchi, Tetsuo Kubota, Tatsuya Kashiwa, and Hiroaki Tagashira. Analysis of electromagnetic fields in a waveguide feed microwave oven by FD-TD method. In Japanese Electronic, Information and Communication Society Spring Meeting, vol. 2, p. 546, 1994. (In Japanese.)
38. Magnus Sundberg. Quantification of heating uniformity in multi-applicator tunnel ovens. In Microwave and High Frequency Heating 1995, St John's College Cambridge, England, September 17–21, 1995.
39. Hugo Malan and A. C. Metaxas. Domain decomposition with electromagnetic finite element analysis. In Microwave and High Frequency Heating 1997, Fermo, Italy pp. 27–30, 1997.
40. Hugo Malan and A. C. Metaxas. Domain decomposition and parallel processing in microwave applicator design. In 11th International Conference on Domain Decomposition Methods, Greenwich, UK, pp. 537–542, 1999.
41. David Dibben and A. C. Metaxas. Finite element analysis of multimode cavities with coupled electrical and thermal fields. In 29th IMPI Microwave Power Symposium, Chicago. 25–27 July, 1994.

42. Ralf Hiptmair. Multigrid method for maxwell's equations. SIAM Journal of Numerical Analysis, 36(1):204–225, 1999.
43. Seung-Woo Lee, Hong-Bae Lee, Hyun-Kyo Jung, Song-yop Hahn, PanSeok Shin, Changyul Cheon, and Jong-Chull Shon. 3D analysis of a microwave oven using vector finite element method. In CEFC 1996, Okayama, Japan, 1996.
44. Adriano Palombizio and Vadim V. Yakolev. Modeling and industry: A time to cross. Microwave World, 20(2):14–19, September 1999.
45. Akira Ahagon and Takashi Kashimoto. Electromagnetic wave analysis in developing microwave ovens. Technical Report vol. 41, no. 1, Matsushita Electric Industrial Co., Ltd., February 1995. (In Japanese.)
46. Koji Iwabuchi, Tetsuo Kubota, and Tatsuya Kashiwa. Analysis of electromagnetic fields in a mass-produced microwave oven using finite difference time domain method. Journal Microwave Power and Electromagnetic Energy, 31(3):188–196, 1997.
47. Gaetano Bellanca, Samanta Botti, Paolo Basssi, and Gabriele Falciasecca. Sensitivity of FD-TD simulations to small mesh modifications in microwave oven design. In Microwave and High Frequency Heating 1997, Fermo, Italy, pp. 60–63, 1997.
48. Magnus Sundberg. Simulation of sterilization and pasteurization in multimode applicators. In 29th IMPI Microwave Power Symposium, Chicago. July 25–27, 1994.
49. G. Bellanca, G. Golfieri, P. Basssi, and G. Falciasecca. Evaluation of dissipated power in microwave dielectric ovens by FD-TD. In Microwave and High Frequency Heating 1995, Cambridge, England, 1995.
50. D. Burfoot, C. J. Railton, A. M. Foster, and S. R. Reavell. Modelling the pasteurisation of prepared meals with microwaves at 896MHz. Journal of Food Engineering, 30:117–133, 1996.
51. Richard M. Keefer and Mel D. Ball. Improving the final quality of microwave foods. Microwave World, 13(2):14–21, 1992.
52. Harry A. Rubbright and Aaron L. Brody. Visions, realities and myths about packaging for heating food in microwave ovens. In 28th IMPI Microwave Power Symposium, Montreal. July 11–14, 1993.
53. J. M. Jin, J. L. Volakis, C. L. Yu, and A. C. Woo. Modeling of resistive sheets in finite element solutions. IEEE Transactions on Antennas and Propagation, 40(6):727–731, June 1992.

2
Electromagnetics of Microwave Heating: Magnitude and Uniformity of Energy Absorption in an Oven

Hua Zhang
Nestlé R & D Center, Inc.
New Milford, Connecticut

Ashim K. Datta
Cornell University
Ithaca, New York

I. MAJOR ELECTROMAGNETIC ISSUES IN MICROWAVE HEATING OF FOODS

A domestic microwave oven is a multimode cavity in which electromagnetic waves form a resonant pattern. When food is present inside the oven, the energy of the electromagnetic waves is transferred to the water molecules, ions, and other food components, raising the food temperature. Figure 1 shows a typical domestic microwave oven with various components. For a detailed discussion on the physical components of a microwave system, see Chapter 7 and references such as Ref. 1.

Compared with conventional heating, heat transfer is typically more difficult to study due to the complex interaction of the microwaves with the cavity and the food [2–4]. This makes generalizations difficult. The two key issues in microwave heating of food are (a) the magnitude of the energy deposited by the microwaves and (b) the uniformity of the energy deposition. The magnitude and uniformity are affected by both food and oven factors such as:

1. Strength and distribution of electromagnetic fields where the food is placed

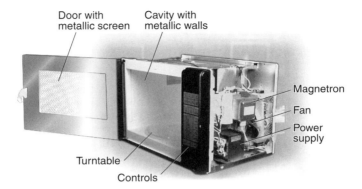

Figure 1 A domestic microwave oven showing the various components.

2. Reflection of electromagnetic waves from the food, as characterized by its property and geometry
3. Propagation of the waves inside the foods, also characterized by the food properties and geometry

The following sections are organized according to the oven and food factors as they affect the magnitude and uniformity of energy absorption. The sections will attempt to provide an understanding based on numerical calculations and experimental measurements, as available.

II. ELECTROMAGNETIC FIELDS INSIDE A DOMESTIC MICROWAVE OVEN

Spatial variation of electromagnetic fields are difficult to measure (see Chapter 8 on measurement). Computational models serve as excellent tools for visualization of the electromagnetic fields and how they are affected by food and oven factors. Some of the mathematical details of the modeling process are discussed in Chapter 1. A brief outline of the governing equations and boundary conditions are provided in the Appendix. Schematics of two of the ovens for which electromagnetic fields are computed are shown in Fig. 2.

Computed electromagnetic fields for the cavity in Fig. 2b containing a cylindrical food (ham) following the work of Ref. 5 are presented in Fig. 3 (see color insert). The feeding port is located asymmetrically in Fig. 3 where the incoming microwaves are in transverse electric (TE) mode with some evanescence modes due to the junction with the cavity. Figure 3 also shows the electric field distribution in the cylindrical ham inside the oven. Such computed contours provide a comprehensive picture of the electromagnetic field (and power absorption) pattern expected for a food and oven system. As mentioned earlier, power ab-

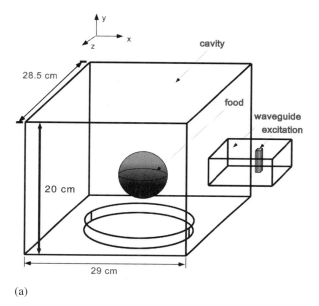

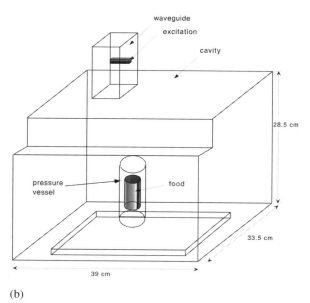

Figure 2 Schematic of two microwave oven systems used in some of the computational studies presented here: (a) General Electric, Inc., Louisville, KY, (b) MDS 2000 Microwave Digestion System, CEM Corporation, Matthews, NC. The rated power of the GE oven is 635 W and for the CEM oven is 850 W.

sorption patterns can be characterized by magnitude and uniformity. Their dependence on food and oven factors are discussed in Secs. III and IV.

III. MAGNITUDE AND UNIFORMITY OF ENERGY ABSORPTION: FOOD FACTORS

Food factors, such as volume, surface area, and dielectric properties, are important in determining the magnitude and uniformity of power absorption. The effects of shape, size, aspect ratio, and properties on the magnitude of power absorption are discussed.

A. Total Energy Absorption as Affected by Load Volume and Surface Area

It is important to compare the effects of load volume and surface area in conventional heating before discussing microwave heating. In many food heating situations, the ideal heating scenario would be to have the same volumetric heating, i.e., a constant power per unit volume. In conventional hot air heating, instantaneous absorbed power increases with volume only to the extent the surface area increases, as given by the surface convective heat transfer equation:

$$q_{total} = hA(T_{amb} - T_{surface}) \tag{1}$$

where q_{total} is the total heat transfer between the solid and the fluid, h is the convective heat transfer coefficient, A is the surface area exposed to the fluid, $T_{surface}$ is the solid surface temperature at any time, and T_{amb} is the hot air temperature. This equation can be written in terms of power per unit volume as

$$\frac{q_{total}}{V} = h\frac{A}{V}(T_{amb} - T_{surface}) \tag{2}$$

For a spherical shaped food, $A/V = 3/r$ and power per unit volume would decrease as the size r increases. Note, however, that q_{total} would keep increasing with radius r.

In microwave heating, like conventional heating, the total power absorbed also increases with volume, eventually leveling off at a power primarily dependent on the magnetron power level, and to a lesser extent on permittivity and geometry. Thus, total power absorption with load volume is typically described by a curve similar to that shown in Fig. 4. The curves shown in this figure are obtained using experimental measurements and numerical simulation of heating water in a cylindrical container [6]. This curve is sometimes described by an empirical equation of the form

$$q_{total} = a(1 - e^{-bV}) \tag{3}$$

which can also be written as

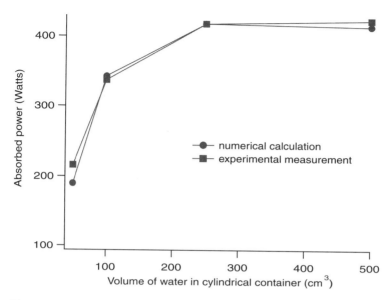

Figure 4 Magnitude of power absorbed in different volumes of water loads, obtained from experiment and electromagnetic simulations for the oven in Fig. 2a. (From Ref. 13.)

$$\frac{q_{\text{total}}}{V} = \frac{a}{V}(1 - e^{-bV}) \qquad (4)$$

where a and b are empirical constants. Thus, power absorption per unit volume decreases as load volume increases, reducing some of the benefits of the volumetric heating of microwaves. If we look at power absorption per unit surface area for microwave heating (Fig. 5), both from experimental data and computational results, it decreases and is in sharp contrast with conventional heating. In conventional heating, power absorption per unit surface area is approximately a constant (changes with time as T_{surface} changes), as given by Eq. (1).

1. Load Properties and Volume

The microwave power output of an oven for a given load is generally different from the rated power by the oven manufacturer. Usually a testing method recommended by IEC is employed [7] to check the output power of the oven. Before the IEC standard, there were many methods [8] used that produced varying results. According to Gerling [8], the Japanese method typically gave 10 to 15% more output power than the American method because they used 2000 g of water instead of 1000 g, based on typical relationship such as the one shown in Fig. 4.

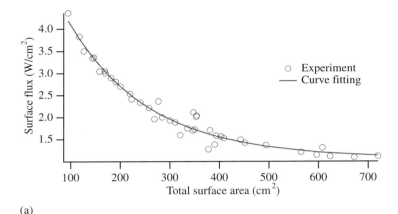

(a)

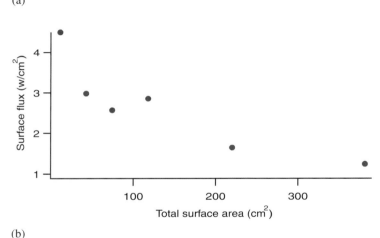

(b)

Figure 5 Surface flux (obtained by dividing the absorbed energy equally over the entire surface area) decreases with increase in surface area, obtained from (a) experiments and (b) electromagnetic simulation. (The experimental data is from Ref. 75 and electromagnetic computation is from Ref. 13.)

To compare power absorption in two food loads of different shape, a relative absorption cross section (RACS) is used [9]. RACS is the total power absorption per unit effective shadow area that is normal to the wave propagation direction and is defined as

$$\text{RACS} = \frac{P_{\text{total}}}{S} \quad (5)$$

where S is the geometric cross section (πR^2 for a sphere). Although RACS is defined for plane waves, it is still applicable for cavity heating.

For plane wave irradiation on spherical materials, the total power absorption was theoretically calculated in terms of RACS [10] as a function of the quantity $2\pi R/\lambda$ [9, 11, 12], where R is the radius and λ is the wavelength in the air. As shown in Fig. 6a, from Ref. 9, the value of the RACS under plane wave irradiation montonically increases for $2\pi R/\lambda < 0.3$ and reduces asymptotically for $2\pi R/\lambda > 1$. For $0.3 < 2\pi R/\lambda < 1$, the RACS has its maximum value but fluctu-

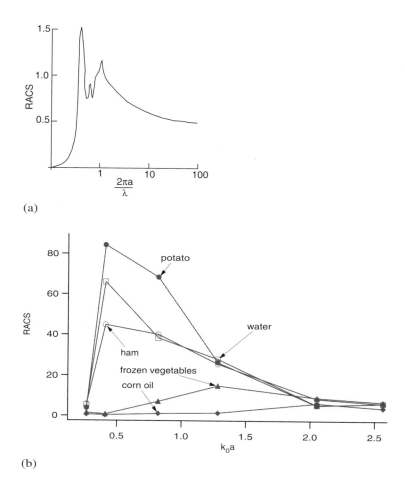

Figure 6 Power absorption per unit cross-sectional area (RACS) for a sphere heated by (a) plane waves and (b) in a cavity. The plane wave results in part (a) show that RACS increases for very small values of radius and decreases asymptotically for large values, with resonance complicating the relationship at intermediate radius. The cavity results seem to follow the same overall trends at small and large volumes. [Part (a) from Ref. 9; part (b) from Ref. 13.]

ates due to resonances. Variation in RACS for cavity heating is shown in Fig. 6b. The trends at the low and high values of $2\pi R/\lambda$ are the same as in plane waves, with maximum occurring somewhere in between. Thus, it can be concluded that the efficiency of microwave power absorption is small for very small and very large volumes, the exact variation being dependent on the food properties and the oven system for cavity heating.

In cavity heating, the efficiency of microwave energy absorption also varies significantly with food composition [6]. In Fig. 7, the RACS in a sphere is plotted against the parameter $\epsilon''/|k|$ for cavity heating. The parameter $\epsilon''/|k|$ is designed to combine the two factors affecting the power absorption [13], ϵ'' representing power absorption (see Chapter 1) and $|k| \alpha (\epsilon'^2 + \epsilon''^2)^{1/4}$ representing the reciprocal of field strength inside foods. As shown in Table 1, corn oil has the smallest value of $\epsilon''/|k|$, and ham has the largest value. This figures illustrates that the efficiency of power absorption increases with $\epsilon''/|k|$. Similar trend was also seen for different shapes [6].

2. Load Shape

Electromagnetic fields inside the food in a microwave cavity vary with shape. As shown in Fig. 8, the power absorbed for heating low loss materials [14] is different for the three different shapes—spherical, cylindrical, and rectangular. For all three shapes, the overall trend shows an increase in absorbed power with volume, with

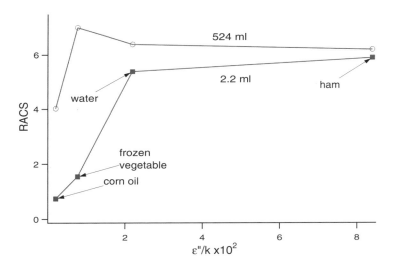

Figure 7 Efficiency of power absorption as related to the dielectric properties, for spheres of 2.2 and 524 ml volume. Computations are for the oven shown in Fig. 2a. (From Ref. 13.)

Table 1 Dielectric Properties of Materials

Material	$\epsilon' - j\epsilon''$	λ, cm	L, cm	$|k|$	Source
Corn oil	2.0–j0.15	8.65	36.8	72.6	[67]
Frozen vegetables	5.1–j0.9	5.4	4.9	116.7	[67]
Meat	17.6–j5.7	2.88	1.45	220.6	[68]
Ham	38–j30	1.86	0.48	356.9	[69]
Potato	51.5–j16.3	1.68	0.87	377.1	[70]
Water	78–j10	1.38	1.72	454.9	[71]

some local variations at small volumes consistent with the discussion in the previous section on volume effect. Figure 8 shows that the power absorption for one particular shape relative to the others change with volume, and no definite conclusion can be made over the entire volume range. Thus, it is difficult to conclude on the best shape for total power absorption, as has been claimed for a sphere [15].

3. Aspect Ratio

Power absorption can vary significantly with aspect ratio, defined as surface area/volume (cm^2/cm^3). Thawing time for frozen tylose (a food analog) blocks is found significantly shorter for flatter shapes of the same volume [16]. Computed power absorptions in Fig. 9 for a high loss food (ham) also show that the absorbed

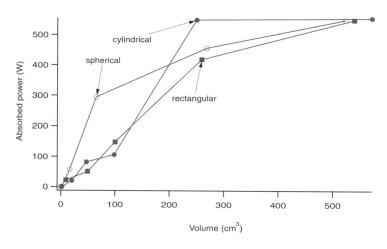

Figure 8 Effect of shape on the magnitude of power absorption at different volumes. Computations are for the oven shown in Fig. 2a and properties of frozen vegetables. (From Ref. 13.)

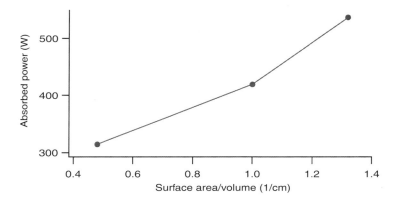

Figure 9 Effect of aspect ratio on the magnitude of power absorption for rectangular foods (volume = 260 cm^3) for the oven shown in Fig. 2a. (From Ref. 13.)

power increases with aspect ratio [13]. The significant increase in power absorption with aspect ratio in Fig. 9 is mainly due to surface heating of ham that has a small penetration depth and this trend may not be true for low loss materials [16].

B. Uniformity of Energy Absorption as Affected by Load Shape, Size, and Properties

Like the magnitude of power absorption, the uniformity of power absorption is also related to the food geometry and dielectric properties. When a small food material is heated in a cavity, the uniformity of power absorption is mainly related to food itself. However, for a typical food size, such as a potato, a frozen meal, or a cup of water, the uniformity is also related to the power density distribution inside an empty oven. The food parameters contributing to heating uniformity are now discussed.

1. Corner and Edge Overheating

One of the major problems in microwave heating is the high intensity of electromagnetic fields at the edges and corners of the food. This is illustrated in Fig. 10 (see color insert), which shows the computed electric fields in a rectangular block in a domestic oven. A common intuitive explanation [3, 17, 18] is that microwave power impinges in several directions at edges and corners, whereas it comes only in one direction at the center area of the foods.

For heating in plane waves, this edge and corner effect can be simply explained using Fig. 11. The refracted angles of the incoming waves are smaller than

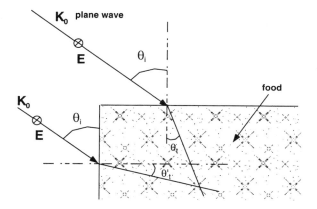

Figure 11 A simple explanation of heating concentration at edges and corners using refraction of plane waves.

the incident angles. At an edge this results in the energy concentration. The refracted angle θ_t is given by

$$\cos \theta_t = \left(1 - \frac{k_0}{k} \sin^2 \theta_i \right)^{1/2} \tag{6}$$

where the wave number k is given by

$$k = \frac{2\pi}{\lambda} \sqrt{\epsilon' - j\epsilon''} \tag{7}$$

Thus, foods with smaller values of dielectric properties (small $|k|$), such as frozen products, will have insignificant edge and corner effect.

In cavity heating, as shown in Fig. 10, the fields can be thought of as superposed by a number of plane waves with various directions and amplitudes [19]. Though complicated, the same mechanism as in plane waves should still be applicable. Absence of strong incoming plane waves may be the reason that some of the edges and corners are not overheated in Fig. 10.

Transverse magnetic (TM^z) modes have been considered [20, 21] to reduce the edge effect in some slab food loads since the dominant component of the electric field (in z direction) is perpendicular to the largest surfaces of the load. Although, the precise mechanism for the reduction in the edge effect was not provided [20, 21]. A cylindrical cavity that allows only two TM^z modes is claimed to reduce the edge and corner effect of microwave heating [22] using mode selection technique. For sharp edges and corners that have angles less than 90°, the overheating has been considered to be more serious [21] and is attributed to concentrations of electric charges (caused by geometric effect) that create localized heating.

2. Focusing

Focusing or internal heating concentration is one of the most significant features of microwave heating as compared with conventional heating. An illustration of focusing is shown in Fig. 12, where the internal heating is higher than that near the surface. Focusing can change the temperature distribution in a food profoundly, and is related to the wavelength and penetration depth of the microwaves in the material and the size of the material being heated. Focusing of plane waves incident on a sphere has been studied at length [11, 23–27] in the context of accidental exposure of a human head to microwaves. For plane waves incident on a sphere, the distribution of fields has been analyzed [24] to show that internal heating is expected for a *small radius* when

$$\lambda < 2.73R \tag{8}$$

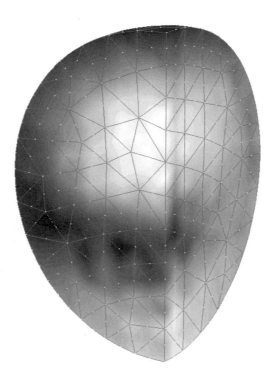

Figure 12 Example of focusing effect with enhanced interior heating (lighter shade) due to a curved surface. Computations are for a sphere of radius 3 cm, properties of egg white, $\epsilon = 70 - j15.8$ in an oven rated at 1000W (Model No. R-5H06, Sharp Inc., New Jersey) with dimensions $W \times D \times H = 40 \times 40 \times 23.5$ cm. (From Ref. 76.)

is satisfied. For larger radius, the wavelength as well as the penetration depth (representing the lossiness) in the material are factors in deciding focusing. Combining these two factors in the expression $QOI = 7.27\pi e^{-R/\delta_P}(R/\lambda)^2$, a criterion for focusing in *larger radius* was defined as [24]

$$QOI > 1 \qquad (9)$$

Thus, foods with a smaller wavelength and a longer penetration depth are likely to have focusing (since QOI value will be higher). This criteria can be successfully used to predict experimentally observed focusing, as illustrated in Table 4. Possible implications of these plane wave results [Eqs. (8) and (9)] on food systems are

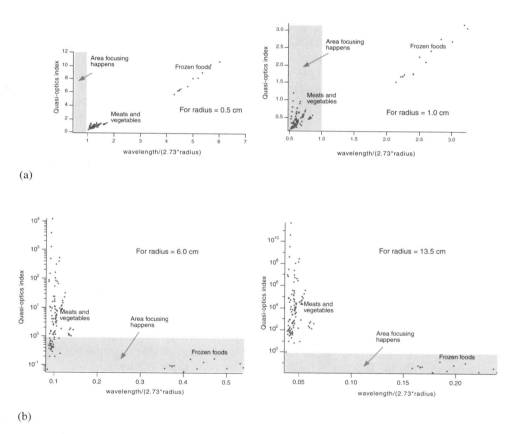

Figure 13 Implications of plane wave analysis in predicting focusing for various food groups, with (a) using criteria for small size [Eq. (8)] and (b) using criteria for large size [Eq. (9)]. (From Ref. 6.)

illustrated in Fig. 13. Figure 13a shows that for very small size, using criteria for small radius [Eq. (8)], most of the food materials are not expected to have heating concentration for plane waves. When radius increases to 1 cm, most meats and vegetables have focusing, but frozen foods still have surface heating. For a much larger radius of 6 cm (Fig. 13b), using criteria for larger radius [Eq. (9)], most of the meats and some of the vegetables do not have focusing. At the largest radius of 13.5 cm shown, only the frozen foods are expected to have focusing. For cavity heating, criterion for focusing is hard to obtain from analytical solutions. Extensive numerical solutions show that the *general trends* mentioned for plane wave heating are likely to be valid for cavity heating.

Figure 14 shows computed radial distributions of heating potentials for various sizes of spheres heated in a domestic oven [14]. The property values for the materials are shown in Table 1. For a small radius (Fig. 14a) the heating potential along the radial direction has its maximum close to the surface for most of the materials, regardless of penetration depths. The inward shifting of the maximum value of heating potential follows the order of wavelength. For the longest wavelength (e.g., frozen vegetables), the maximum heating potential occurs at the very surface. For the smallest wavelength (e.g., water), the maximum potential happens at the most interior place. It is evident from Fig. 14a that for a small size, the distribution of heating potential along radial direction is primarily a function of wavelength and is not affected much by penetration depth.

As the radius increases, penetration depths start to play an important role, as shown in Fig. 14b. There is strong focusing at the larger radius for most of the materials because of their relatively longer penetration depths. For the smallest penetration depth (ham), there is no focusing and the maximum heating potential is at the surface for this radius as is commonly identified in the literature [7, 28, 29]. As the radius increases further (Fig. 14c), for materials with intermediate penetration depth, focusing begins to diminish as seen for potato and meat.

Heating concentration index, an index representing the intensity and extent of focusing (defined in Ref 14), is shown for spherical foods of different radii in Fig. 15. It shows that a material with a small wavelength (water) will have a higher intensity of focusing over a larger range of radii. Focusing is less intense for short penetration depths, as illustrated by the data for ham. Foods with longer wavelength and penetration depth, such as frozen vegetables, have a large range of radius where the maximum heating potential is inside, but the focusing is not as intense.

Focusing in cavity heating has been reported experimentally by numerous researchers [3, 30–34]. An example [14] of experimentally observed focusing is shown in Fig. 16, including comparisons with model predictions. Experimental measurements used an infrared camera and the computed temperature contours are from a coupled thermal-electromagnetic model. This figure also serves to illustrate the effect of variation in material properties on focusing. The dielectric and thermal properties of tylose with different salt contents used in this figure are

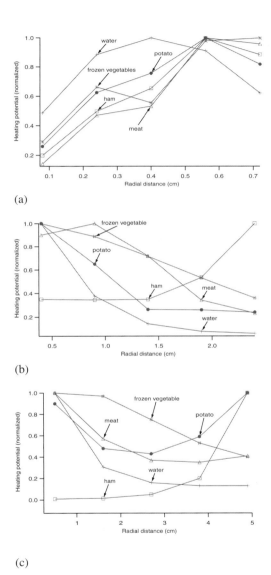

Figure 14 Heating potential distributions along radial direction for different spherical foods of radii (a) 0.8 cm, (b) 2.5 cm, and (c) 5.0 cm placed in a cavity (Fig. 2a), obtained from electromagnetic simulations. (From Ref. 14.)

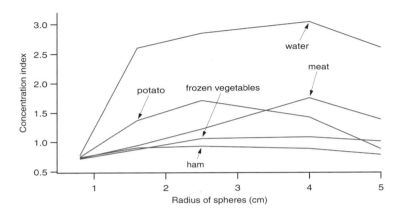

Figure 15 Concentration index (a measure of intensity and extent of focusing) as a function of radius for different spherical food materials heated in a cavity (Fig. 2a), obtained from electromagnetic simulations. (From Ref. 14.)

listed in Table 2. Figure 16a shows strong focusing effect for a low loss material (tylose with no salt) at the particular size. For more lossy materials (tylose with 2 and 4% salt), numerical computations and experimental data show lack of focusing, as seen in Fig. 16b,c. Interestingly, the focusing effects shown here are also predicted by the criterion developed for plane waves [Eq. (9)] since in Fig. 16a, $QOI = 3.08 > 1$ for tylose with 0% salt, predicting focusing. For tylose with 2 and 4% salt, $QOI = 0.39$ and 0.078, respectively, both predicting no focusing (since $QOI < 1$).

In summary, focusing for small sizes depends on the wavelength in the material, while for large sizes it depends on both the wavelength and the penetration

Table 2 Properties of Tylose at Room Temperature

Parameter	0% salt	2% salt	4% salt	Source
ϵ'	58	54.8	50.2	[72]
ϵ''	17.5	29.6	38.4	[72]
λ, cm	1.59	1.6	1.62	
L, cm	0.86	0.504	0.382	
$|k|$	399	404.8	407.8	
ρ, kg/m^3	1060	1060	1060	[73]
c_p, J/kgK	3700	3700	3700	[73]
k_t, W/mK	0.49	0.49	0.49	[73]

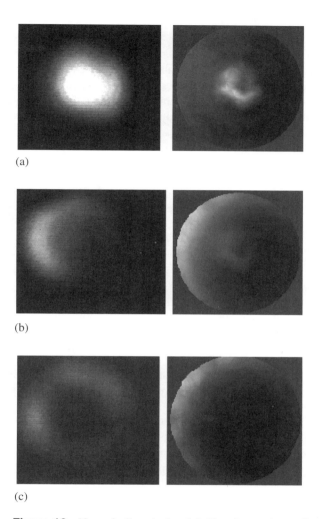

Figure 16 Numerically calculated (left) and experimentally measured (right) temperature distributions for tylose spheres of 2.5 cm radius for (a) 0%, (b) 2%, and (c) 4% salt contents placed at the center and 4 cm above the floor in the oven. Computations are for the cavity described in Fig. 2a. (From Ref. 14.)

depth. Compared with spherical shapes, cylindrical shapes have a lower intensity and a lower upper limit of radius for focusing since they have a curved surface along one dimension [14]. For a spherical food, it was also shown that the relative distribution of heating potential along the radial direction varied little with location inside the oven where the sphere is placed [14].

3. Resonance

A food material with a certain size and geometry can itself be a dielectric resonator (DR) if its dielectric constant is high, such as for water or a water rich vegetable. During such resonance, the food itself behaves like a cavity. In microwave communication literature, DR is widely used [35, 36]. For low-loss foods in cylindrical or rectangular shapes, it is possible to predict resonance with exact dimensions and properties. During resonance, most energy can be transmitted into foods. The heating is predictable but nonuniform.

In the theoretical study of resonance for cylinders and slabs under plane waves, Ayappa et al. [37] concluded that TE^z polarization has a higher power absorption peak than that in the TM^z case, and the resonance is generally related to the material's dimension and wavelength. They developed a criteria for resonance to occur in a cylinder as

$$\frac{D}{\lambda} = 0.5n - 0.257 \qquad (10)$$

where D is the diameter of the cylinder, λ is the wavelength of radiation in the cylinder material and $n = 1, 2, \ldots$ correspond, respectively, to the first, second resonances and so on. For a slab of thickness L, the equivalent relationship is

$$\frac{L}{\lambda} = 0.5n + 0.001 \qquad (11)$$

Experiments by Barringer et al. [38] verified the criteria [Eq. (10)] for cylinders in a specially designed oven, but these resonances are not observed in a common domestic oven.

The dielectric loss factor is important for the occurrence of resonance, as well as for the amplitude of resonance. As shown in Fig. 17, for a cylindrical food load in a cavity, the occurrence of resonance is mainly a function of dielectric constant, while the peak value of absorbed power is damped as the dielectric loss factor increases [13].

C. Differential Heating in Multicomponent Foods

Since different foods absorb energy and their temperatures rise at different rates due to their dielectric and thermal properties, it is expected that if they are placed together, they would also heat at different rates. This is a significant problem in heating a multicomponent ready meal, such as a frozen dinner. Differential heating of such food components have been studied [39–42]. In particular, the effect of food component layout was shown to significantly vary the heating rates of various components in frozen ready meals [40]. Such differences are primarily due to electromagnetic as opposed to thermal effect. As shown in Fig. 18 from electromagnetic calculations (see color insert) [42], the energy absorption rate itself

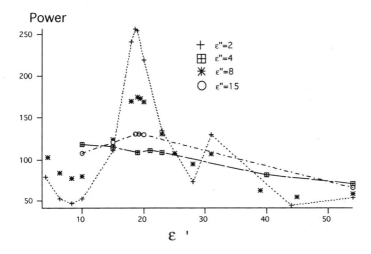

Figure 17 Variation of absorbed power with dielectric properties for a cylindrical food of radius 2.5 cm and a height of 4 cm. Computations are for the Sharp oven described in Fig. 12. (From Ref. 76.)

varies with different components in a multicomponent meal. Note that the temperature rise would additionally be a function of thermal properties, particularly the specific heat. Measured temperatures in studies such as Refs. 40, 41 show the tremendous differences during the heating process.

The heating rates and uniformity in multicomponent meals are affected by the relative placement of the components, their composition, and geometry [39], as well as the interaction of these factors. For example, mashed potato, which heats the slowest, is recommended for placement on the sides of the tray [39]. They [39] also noted little change in heating pattern with product formulation changes. In the three types of trays (1, 2, and 3 compartments) that were approximately equal in overall size and made of CPET (crystallized polyethylene terephthalate) plastic, they found significant differences in heating rates.

In order to better understand the simultaneous heating process, simultaneous heating of two components was studied using electromagnetic simulation [6]. Figure 19 from this study shows that for water and corn oil, water absorbs significantly more power than corn oil at small volume. As volume increases, power absorption in water relative to oil reduces considerably, until at about 400 cm^3, water absorbs only twice as much power as oil. It is interesting to note that at this higher volume, the oil temperature will rise faster than the water temperature, as has been reported in the literature, due to a lower specific heat and density of oil.

Simultaneous heating of several combinations of high- and low-loss food materials are shown in Fig. 20. Between a high and a low loss material, the

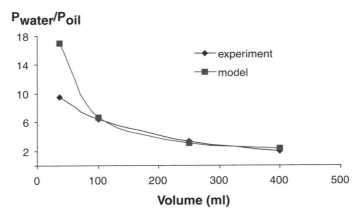

Figure 19 Ratio of power absorption in water to that in corn oil placed in rectangular containers. Experiments and computations are for the oven described in Fig. 2a and container dimensions [volume in ml ($W \times D \times H$ cm)] of 36 (4.4 × 3.75 × 2.2), 100 (6.0 × 7.0 × 2.4), 250 (9.5 × 8.1 × 3.25) and 400 (10.0 × 8.0 × 5.0), respectively. (From Ref. 6.)

low-loss material absorbs more power at larger volume. In the small volume range this trend is reversed, as shown in Fig. 20a, b. Between two materials that are both lossy, as shown in Fig. 20c, although the relative power absorption changes with volume, there is insufficient information to conclude on the general trend. These results are consistent with how two materials would absorb power when they are heated one at a time. That is, if one material (of certain geometry and size) absorbs more power than another material when heated separately, that material will absorb relatively more power when heated simultaneously.

D. Effect of Temperature and Moisture Changes in the Food During Processing

Changes in dielectric properties due to changes in temperature and moisture during heating couples the electromagnetics with heat transfer and moisture transfer. This is discussed in detail in Chapter 4. As an example, Fig. 21 shows how the estimated electric field inside a grape changes as it loses moisture during a drying process [43].

E. Modifications Using Active Packaging

Active packaging, where the package interacts with the microwaves, has been used to successfully emulate conventional surface heating or to modify the elec-

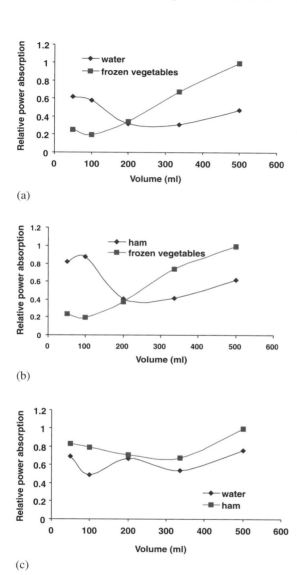

Figure 20 Relative power absorption in simultaneous heating of two different materials as rectangular blocks: (a) water and frozen vegetables, (b) ham and vegetables, (c) water and ham. Computations are for the oven described in Fig. 2a and container dimensions [volume in ml ($W \times D \times H$ cm)] are 50 ($4 \times 5 \times 2.5$), 100 ($5.7 \times 7.0 \times 2.5$), 200 ($12 \times 7 \times 2.4$), 336 ($12 \times 7 \times 4$) and 500 ($16.2 \times 9.5 \times 3.25$), respectively. (From Ref. 6.)

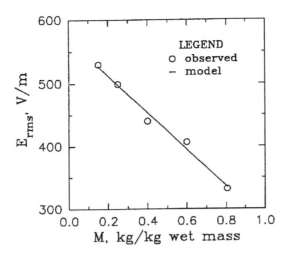

Figure 21 Increase in magnitude of electric field in a material during processing due to decrease in moisture level. (From Ref. 43.)

tromagnetic field pattern around the food to obtain a more desirable heating pattern. Calculations of electromagnetic field modifications in a package containing aluminum foil strips were performed using commercial software [44], although the results were not made public. An example of modification of electric field using aluminum foil on the sides of a rectangular block of whey protein gel is shown in Fig. 22 (see color insert). This figure shows how the edge effect can be reduced by the use of the foil. For a detailed discussion of various types of active packaging, see Chapter 12.

IV. MAGNITUDE AND UNIFORMITY OF ENERGY ABSORPTION: OVEN FACTORS

The power distribution in a single-mode cavity is simple and usually has the maxima at the center plane. In a multimode cavity, such as the domestic oven, power distribution is not so obvious. Power distribution in a cavity, mapped using a container with 64 water cells [45], showed that it is highly nonuniform, varying in all three directions. Computation of electromagnetic fields for an empty cavity also shows nonuniform electric fields (Fig. 3). Jia [45] found that power distribution is generally not symmetric even when the power entry port is symmetrically located. Usually this nonuniformity of power distribution in an empty cavity is the major determinant of the magnitude and uniformity of microwave heating of foods.

The magnitude of energy absorption in foods is obviously related to the power output of the magnetron. For a given magnetron, the energy absorption is primarily a function of the impedance of the oven system [46]. The overall impedance includes the geometry and placement of the food, size and geometry of the cavity and waveguide, and locations and geometry of the feed ports. These factors are discussed in detail in this section. Usually the size of the food is much smaller than the cavity, thus having little impact on the overall impedance of the system. Therefore, the oven system without the loads can be tuned to minimize the impedence in the system design stage for higher heating efficiency [47].

A. Effect of Placement Inside the Oven

For a given microwave system, a load placed in different locations inside the oven absorbs microwaves differently. Figure 23 shows the temperature patterns in a cylindrical potato due to placement at three different locations (Table 3) in an oven [48]. It is obvious from the figure that heating patterns are significantly different for the three locations.

Measurements of power distribution using water [45, 49, 50] assume the maximum power absorption of water cells occur at the highest electromagnetic fields inside the cavity. This common sense is challenged by Jackson and Barmatz [19] as they observed reversed situations where the absorbed microwave power in a loaded cavity can attain its maximum value at the location where the electrical field is a minimum for an empty cavity. This perhaps indicates the significant influence of the load to the overall power distribution inside the oven. Location and dielectric properties of loads may influence the modal pattern inside a microwave oven [51] and a small change in them can result in a different heating pattern.

B. Effect of Oven Size and Geometry

The fields in an oven (a resonant cavity), which are often referred to as modes, are partially determined by the size of the oven. The larger the oven size, the more modes it has [47]. Usually more cavity modes lead to more uniform heating [17]. The effect of oven size on the efficiency of power absorption by load is studied by Perch and Schubert [52]. By reducing the cavity size of an oven system, the power absorption data were collected for different volumes of water loads. Then a single efficiency index is estimated for the different size cavities. They concluded that the smaller size cavity is less efficient, as shown in Fig. 24. Lower efficiency in a smaller cavity can be attributed to increased reflection of the magnetron power. A large output power of magnetron may not significantly increase the heating rate if a small cavity is used. A smaller cavity, however, is not necessarily less efficient overall, since reflection of magnetron power also depends on the waveguide, location of magnetron and the coupling method.

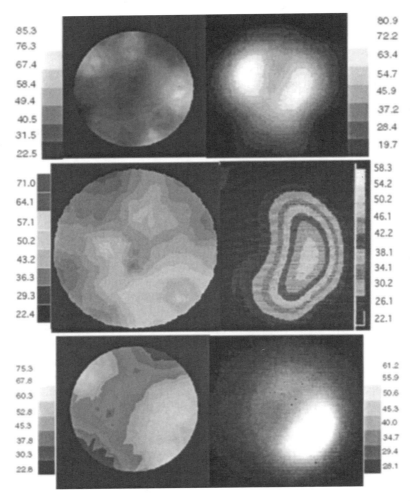

Figure 23 Calculated (left) and experimental (right) temperature profiles at 15 s for three locations as noted in Table 2. Shown is the bottom view. Dimensions for the Sharp oven used here are described in Fig. 12. (From Ref. 48.)

C. Effect of Turntables

The turntable inside the domestic microwave oven (see Fig. 1) is one of the most effective and less expensive approaches to improve uniformity of heating. The effects of the spatial variation in field, as shown in Fig. 3, can be reduced by moving the food continuously during heating through high and low electric fields, as is done by

Table 3 Three Locations for Which Electromagnetic Field Patterns Are Provided in Fig. 23

Location	Distance from bottom, cm	Distance from center, cm
1	4	0
2	4	4
3	6	0

the turntable. Electromagnetic fields in the food on a rotating turntable have not been reported in the literature, perhaps due to the computational complexities.

D. Effect of Mode Stirrers

A mode stirrer in a microwave oven is a rotating metallic blade assembly near the feedport that is intended to improve the uniformity of heating. Mode stirrers are more common in commercial ovens. Typically, ovens do not include both a mode stirrer and a turntable [7]. Mode stirrer is said to disperse the microwave irradiation, leading to uniform heating. The metal blades rotating inside the oven will change and smooth its power density distribution patterns, as claimed in some of the invention patents [53–55]. However, this change may not necessarily translate into a more uniform power distribution overall. Literature is inconclusive in terms of the precise way the mode stirrers work and the extent of uniformity that they contribute. Perhaps electromagnetic simulation of a cavity with a mode stirrer can

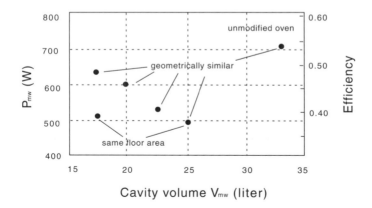

Figure 24 Absorbed microwave power P_{mw} and efficiencies as a function of cavity volumes. (From Ref. 52.)

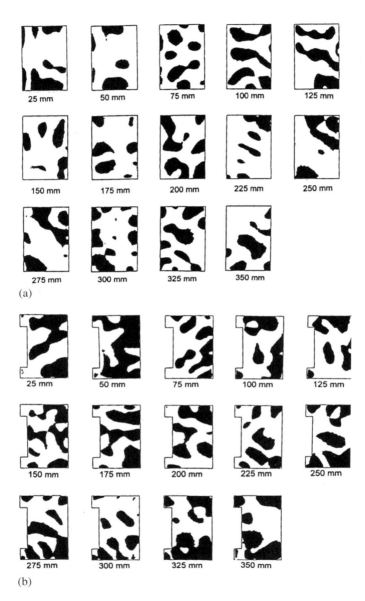

Figure 25 Power distribution in the absence (a) and presence (b) of mode stirrer in an empty cavity, from a qualitative study using thermal paper. Heights above the oven floor are indicated in mm. The oven used is a Daewoo domestic oven rated at 850 W with dimensions $W \times D \times H = 26 \times 36 \times 35$ cm. (From Ref. 56.)

provide more insight into the process, but this has not been performed to date. Qualitative evaluation of the effectiveness of a mode stirrer in improving heating uniformity has been performed [56]. They used thermal paper that changes with temperature and the black and white images of uniformity obtained this way are shown in Fig. 25 for an empty oven, with or without a mode stirrer. Although some improvement in the heating uniformity can be seen in this figure, it is not obvious as to how significant is the improvement.

E. Effect of Feed Location

In microwave ovens, the locations of power feedports have many possibilities—symmetrically or asymmetrically placed on five of the six sides of the cavity (other than the front door). Even though the location and orientation of the feeding port have significant influence on the heating pattern in a loaded cavity [57], there is no reported study on the optimization of the feedport location for microwave ovens in general.

Multiple feeding ports have been claimed [22, 58–60] to improve uniformity of heating. The principle of multiple feeding is to increase the number of resonant modes as many as possible. When a large number of modes are present in a cavity, the electromagnetic fields tend to be more uniform. Each feeding port located at a specific place on the cavity wall excites certain modes corresponding to the excitation source. When another feeding port is used, a different set of modes is added into the cavity, thus, more uniform heating may be achieved.

It is, however, generally difficult to predict the extent of uniformity that may be achieved by arbitrary addition of feedports. Electromagnetic simulation is an effective tool to optimize such feedport location design [57], although this is often achieved by trial and error. In one example [22] of an innovative design using feedport location, the two feedports are located in a cylindrical cavity at an angle of 90° to each other and the excitation from the two ports are phase-shifted by 90°. This results in a rotating TM_{11}^z mode that is claimed to provide more uniform heating due to circumferential averaging of the electromagnetic fields.

V. FUTURE CONCEPTS AND DEVELOPMENTS IN OVEN DESIGN

Efforts are being made to improve the heating uniformity and selectivity by making dramatic changes in oven design. Some of these changes are phase control, variable frequency, fringe field, and the combination of microwave with other means of heating. Some details of these systems are provided below. It is important to note that other than combination microwave and conventional heating, they are mostly in concept stages for food processing applications.

A. Phase Control Heating

Direct control of the microwave field has been utilized by researchers outside the food area to achieve targeted heating to overcome the undesirable or dominant heating patterns. Such arrangements can also have advantages for food applications. An example is the use of constructive waveguide interference techniques [61]. A schematic of the arrangement from their study is shown in Fig. 26. Here two mutually exclusive coherent signals are constructively interfered in a practical waveguide arrangement, where the path length, and hence phase of one signal can be varied relative to the other. Figure 27 (see color insert) compares the thermal images from a phase control experiment with those from a domestic oven. The heating patterns from the domestic oven were not only complex but also changed significantly during heating. However, the phase control heating shows two entirely different heating patterns that were found to remain constant with longer heating. The heating patterns are also simpler, and it is foreseeable how selection of appropriate phase conditions could be combined to achieve more uniform heating. It is reasonable to expect differences attributable solely to the different frequencies, but the phase control heating patterns are claimed to be reproducible and can be directly controlled to account for factors such as temperature-dependent permittivity [61].

B. Variable Frequency Oven

Variable frequency microwave (VFM) technique was developed (http://www.microcure.com) specifically to solve fixed-frequency microwave heating uniformity

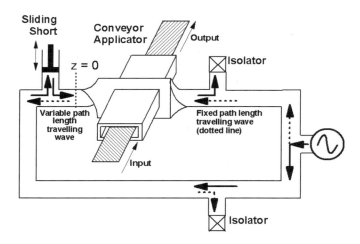

Figure 26 Schematic of a phase control system, showing conditions for constructive interference: coherent sources, mutually exclusive signals, and means of controlling their relative phase. (From Ref. 61.)

problems that have limited the use of microwave in nonfood related applications. The VFM processing technique is geared toward advanced materials processing and chemical synthesis. VFM uses preselected bandwidth sweeping around a central frequency employing frequency agile sources, such as traveling wave tubes as the microwave power amplifier. The use of this technology overcomes the nonuniform energy distribution, hot spots, and thermal runaway of fixed frequency microwave heating of advanced materials in multimode cavities. By sweeping over a range of frequencies, a series of resonant and off-resonant states can be traversed. Also, dielectric properties of some materials vary significantly with frequency, which can be advantageous.

A detailed study of variable frequency microwave ovens using mathematical modeling can be seen in Ref. 2 in the context of polymer processing which showed significant improvements in heating uniformity. The variable frequency ovens are 2 to 3 orders of magnitude more expensive than the domestic microwave oven. The only study of variable frequency microwave ovens for food applications was by Ref. 62. Figure 28 (see color insert) shows the heating patterns obtained using a variable-frequency oven, but processing at the discrete frequencies mentioned. Results for sweeping a range of frequencies during heating, with different durations for each frequency, were also shown [62]. One of the advantages mentioned for variable-frequency microwave heating was more controlled heating of foodstuffs whose geometries impose a dominant heating pattern, such as spheres and cylinders. For example, in Fig. 28 for spheres, center heating occurs in domestic microwave ovens, but the pattern changes with frequency in a variable-frequency oven. Development of the optimum heating procedure requires experimentation to choose frequency values and the duration of heating at each frequency.

C. Fringing Field Applicators

Most microwave heating applications require some sort of cavity to completely surround the product to be heated. Fringing field applicators (sometimes called surface wave applicators or surfatrons) are types of slow wave devices [63] that provide an intense electric field at the surface of the applicator. They do not require that the applicator completely surround the product being processed because they are nonradiating. They are therefore suited to heating large surface areas and surface layers of lossy products [64]. They only radiate when a product is adjacent (typically of the order of millimeters to centimeters) to the applicator surface. In the context of food processing, fringing field applicators have been used only at the research stage, such as for drying [65] and setting batter on frozen food products [64].

D. Combination Microwave and Other Heating

To overcome the uneven heating, excessive moisture migration toward the surface and other unwanted aspects of microwave heating, microwave ovens have been combined with hot air, infrared heat source, or both. Addition of these two sources are not expected to modify the microwave fields in a direct way, although the modified temperature (and moisture) patterns due to these sources can influence the electric fields. Since the effects of adding these are primarily felt in the heat and moisture transport, they are discussed in detail in Chapter 4. In one instance of a domestic microwave oven, microwaves were combined with hot air, infrared and induction heating [66]. Another new development is in so-called "speed cookers," some of which combine microwave with jet air impingement or halogen lamps. See, for example, the website http://www.geappliances.com/advantium/home_fr.htm. The "speed cookers" claim to cook with conventional oven quality but in a fraction of the time.

APPENDIX

A very brief description is provided here of the governing equations and boundary conditions used to compute the electromagnetic fields in the cavity presented in this paper. For description of the symbols below, see Chapter 1. Further generic details of electromagnetic modeling can be seen in many of the papers and textbooks cited in this chapter. Microwave reflection and transmission inside the cavity (such as in Fig. 2) and the foods are described by the Maxwell's equations:

$$\nabla \times \mathbf{E} = j\omega\mu\mathbf{H} \tag{12}$$

$$\nabla \times \mathbf{H} = -j\omega\epsilon_0\epsilon^*\mathbf{E} \tag{13}$$

$$\nabla \cdot (\epsilon \mathbf{E}) = 0 \tag{14}$$

$$\nabla \cdot \mathbf{H} = 0 \tag{15}$$

where $\bar{\mathbf{E}} = \mathbf{E}e^{-j\omega t}$ and $\bar{\mathbf{H}} = \mathbf{H}e^{-j\omega t}$ are the harmonic electric and magnetic fields, respectively. The complex permittivity ϵ^* is given by

$$\epsilon^* = \epsilon' + j\epsilon''_{\text{eff}} \tag{16}$$

Boundary conditions for the electromagnetic modeling of a cavity are set on the walls of the cavity which are considered perfect conductors. The entire cavity interior is treated as a dielectric with appropriate dielectric properties of air and food in the regions that they occupy. Note that in modeling of the entire cavity, the food-air interface does not have to be treated in any special way. In the interior of a perfect electrical conductor, the electric field is zero. This condition, together with the Maxwell's equations, lead to the boundary condition at the air-metallic wall interface as

$$E_{t,\text{air}} = 0 \tag{17}$$

$$B_{n,\text{air}} = 0 \tag{18}$$

Table 4 Diameters for Heating Concentration in the Dielectric Spheres or Cylinders

Material	$\epsilon' - j\epsilon''$	λ, cm	L, cm	Diameter for heating concentration, cm	QOI	Source
Phantom food	$52 - j14$	1.01	1.68	1.8–3.5	2.96–4.83	[74]
Agar gel	$70 - j15$	1.09	1.45	10	3.04	[31]
Meat	$41.6 - j11.3$	0.98	1.87	6	4.12	[32]
Water	$78 - j10$	1.38	1.72	5	4.19	[33]

Here the subscripts t and n stand for tangential and normal directions, respectively. The excitation of the microwave power is treated as sinusoidal at 2.45 GHz (domestic oven frequency) given by

$$I = I_0 \sin(\omega t) \qquad (19)$$

The microwave power absorption or heating potential in food, $P(\mathbf{r}, T)$, is related to the electric fields by

$$P(\mathbf{r},T) = \tfrac{1}{2}\omega\epsilon_0\epsilon''_{\text{eff}} \mathbf{E}^2 \qquad (20)$$

This absorbed energy causes the food temperature to rise.

REFERENCES

1. R. V. Decareau and R. A. Peterson, Microwave Processing and Engineering. Chichester, England: Ellis Horwood Ltd., 1986.
2. C. Saltiel and A. K. Datta, Heat and mass transfer in microwave processing, Advances in Heat Transfer, vol. 32, 1999.
3. P. O. Risman, T. Ohlsson, and B. Wass, Principles and models of power density distribution in microwave oven loads, Journal of Microwave Power and Electromagnetic Energy, pp. 193–198, 1987.
4. R. Mudgett, Microwave food processing, Technologist, pp. 4–10, 12–13, 22, 1989.
5. H. Zhang and A. K. Datta, Experimental and numerical investigation of microwave sterilization of solid foods, American Institute of Chemical Engineers Journal, 2000.
6. H. Zhang and A. K. Datta, Heating concentrations of microwaves in spherical foods. Part one: In plane waves, Submitted to Journal of Journal of Microwave Power and Electromagnetic Energy, 2000.
7. C. R. Buffler, Microwave Cooking and Processing. Van Nostrand Reinhold, New York, 1993.
8. J. E. Gerling, Microwave oven power: A technical review, Journal of Microwave Power and Electromagnetic Energy, vol. 22, no. 4, pp. 199–207, 1987.
9. C. M. Weil, Absorption characteristics of multilayered sphere models exposed to

uhf/microwave radiation, IEEE Transactions on Biomedical Engineering, vol. 22, no. 6, pp. 468–476, 1975.
10. A. Anne, Scattering and absorption of microwaves by dissipative dielectric objects: The biological significance and Hazards to mankind. PhD thesis, University of Pennsylvania, Philadelphia, 1963.
11. J. Lin, A. Guy, and C. C. Johnson, Power deposition in a spherical model of man exposed to 1–20 mhz electromagnetic fields, IEEE Trans. on Microwave Theory and Techniques, vol. 21, no. 12, pp. 791–797, 1973.
12. J. M. Osepchuk, A history of microwave heating application, IEEE Transactions on Microwave Theory and Techniques, vol. MTT-32, no. 9, pp. 1200–1224, 1984.
13. H. Zhang and A. K. Datta, Microwave power absorption in single and multi-compartment foods, Submitted to Inst. of Chemical Engineers, 2000.
14. H. Zhang and A. K. Datta, Heating concentrations of microwaves in spherical and cylindrical foods. Part Two: In a cavity, Submitted to Journal of Microwave Power and Electromagnetic Energy, 1999.
15. J. Giese, Advances in microwave food processing, Food Technology, pp. 118–123, September 1992.
16. M. Chamchong and A. K. Datta, Thawing of foods in a microwave oven: Effect of load geometry and dielectric properties, Journal of Microwave Power and Electromagnetic Energy, vol. 34, no. 1, pp. 22–32, 1999.
17. R. Mudgett, Electromagnetic energy and food processing, Journal of Microwave Power and Electromagnetic Energy, vol. 234, no. 4, pp. 225–230, 1988.
18. H. A. Rubbright, Packaging for microwavable foods, Cereal Foods World, September 1990.
19. H. W. Jackson and M. Barmatz, Microwave absorption by a lossy dielectric sphere in a rectangular cavity, Journal of Applied Physics, vol. 70, no. 15, pp. 5193–5204, 1991.
20. M. Sundburg, P. O. Risman, and T. Ohlsson, Analysis and design of industrial microwave ovens using the finite-difference time-domain method, Journal of Microwave Power and Electromagnetic Energy, vol. 31, no. 3, pp. 142–157, 1996.
21. P. O. Risman and T. Ohlsson, Metal in the microwave oven, Microwave World, vol. 13, no. 1, pp. 28–33, 1992.
22. P. O. Risman and C. Buffler, Cylindrical microwave heating applicator with only two modes, US Patent 5632921, 1996.
23. G. W. Ford and S. A. Werner, Scattering and absorption of electromagnetic waves by a gyrotropic sphere, Physics Review B, vol. 18, pp. 6752–6759, 1978.
24. H. N. Kritikos and H. Schwan, The distribution of heating potential inside lossy spheres, IEEE Transactions on Biomedical Engineering, vol. 22, no. 6, pp. 457–465, 1975.
25. H. N. Kritikos and H. Schwan, Hot spots generated in conducting spheres by electromagnetic waves and biological implications, IEEE Transactions on Biomedical Engineering, vol. 19, no. 1, 53–58, 1971.
26. D. Livesay and K.-M. Chen, Electromagnetic field induced inside arbitrarily shaped biological bodies, IEEE Transactions on Microwave Theory and Techniques, vol. MTT-22, 12, pp. 1273–1280, 1974.
27. A. R. Shapiro, R. Lutomirski, and H. T. Yura, Induced fields and heating within a cra-

nial structure irradiated by an electromagnetic plane wave, IEEE Transactions on Microwave Theory and Techniques, vol. 19, no. 2, pp. 187–196, 1971.
28. F. Peyre, A. K. Datta, and C. E. Seyler, Influence of the dielectric property on microwave oven heating patterns: Application to food materials, Journal of Microwave Power and Electromagnetic Energy, vol. 32, no. 1, pp. 3–15, 1997.
29. G. Roussy and J. Pearce, Foundations and Industrial Applications of Microwaves and Radio Frequency Fields. John Wiley, New York, 1995.
30. T. Ohlsson, Some microwave and heat transfer fundamentals of microwave cooking, Microwave Power Symposium, pp. 1–2, 1980.
31. S. Swami, Microwave heating characteristics of simulated high moisture foods, Master's thesis, University of Massachusetts, Amherst, 1982.
32. W. E. Nykvist and R. V. Decareau, Microwave meat roasting, Journal of Microwave Power, vol. 11, no. 1, pp. 3–24, 1976.
33. H. Prosetya and A. K. Datta, Batch microwave heating of liquids: An experimental study, Journal of Microwave Power and Electromagnetic Energy, vol. 26, no. 3, pp. 215–226, 1991.
34. L. Lu, J. Tang, and L. Liang, Moisture distribution in spherical foods in microwave drying, Drying Technology, 16, pp. 503–524, 1998.
35. D. Kajfez, Dielectric rod waveguides, in Dielectric Resonators (D. Kajfez and P. Guillon, eds.), Artech House, Inc., Norwood, MA, 1986.
36. M. Postieszalski, Cylindrical dielectric resonators and their applications in tem line microwave circuits, IEEE Transactions on Microwave and Techniques, vol. 27, no. 3, pp. 233–238, 1979.
37. K. G. Ayappa, H. T. Davis, and S. A. Barringer, Resonant microwave power absorption in slabs and cylinders, AIChE Journal, vol. 43, pp. 615–624, 1997.
38. S. Barringer, K. G. Ayappa, H. T. Davis, E. A. Davis, and J. Gordon, Effect of sample size on the microwave heating rate oil vs water, AIChE Journal, vol. 40, no. 9, pp. 1433–1439, 1994.
39. S. Ryynanen and T. Ohlsson, Microwave heating uniformity of ready meals as affected by placement, composition and geometry, Journal of Food Science, vol. 61, no. 3, pp. 620–624, 1996.
40. J. R. Bows and P. S. Richardson, Effect of component configuration and packaging materials on microwave reheating of a frozen three-component meal, International Journal of Food Science and Technology, vol. 25, pp. 538–550, 1990.
41. H. Ni, A. K. Datta, and R. Parmeswar, Microwave heating characteristics of multicomponent foods, in Presented at the IFT Annual Meeting (Anaheim, CA), 1995.
42. A. K. Datta, Understanding and improvements of microwave oven heating, in Proceedings of the Appliance Manufacturers Conference and Exposition (Nashville, TN), pp. 35–45, October 1998.
43. T. N. Tulasidas, C. Ratti, and G. S. V. Raghavan, Modelling of microwave drying of grapes, Canadian Journal of Agricultural Engineering, vol. 39, no. 1, pp. 57–67, 1996.
44. C. Lorenson, The why's and how's of mathematical modelling for microwave heating, Microwave World, vol. 11, no. 1, pp. 13–23, 1990.
45. X. Jia, Experimental and numerical study of microwave power distributions in a microwave heating application, Journal of Microwave Power and Electromagnetic Energy, 28, no. 1, pp. 11–21, 1993.

46. R. E. Collin, Foundations of Microwave Engineering. McGraw-Hill, New York, 1998.
47. D. Pozar, Microwave Engineering. Addison-Wesley, New York, 1993.
48. H. Zhang and A. K. Datta, Coupled electromagnetic and thermal modeling of microwave oven heating of foods, Journal of Microwave Power and Electromagnetic Energy, 2000.
49. J. R. White, Measuring the strength of microwave field in a cavity, Journal of Microwave Power and Electromagnetic Energy, vol. 5, no. 2, pp. 145–147, 1970.
50. C. S. MacLatchy and R. M. Clements, A simple technique for measuring high microwave electric field strengths, Journal of Microwave Power and Electromagnetic Energy, vol. 15, no. 1, pp. 7–14, 1980.
51. E.-D. M. El-Sayed and S. S. Farghaly, Influence of load location in mode tuning of microwave ovens, Journal of Microwave Power, vol. 18, no. 2, pp. 197–207, 1983.
52. C. Persch and H. Schubert, Characterization of household microwave ovens by their efficiency and quality parameters, in The 5th International Conference on Microwave and High Frequency Heating (Cambridge University, UK), 1995.
53. R. P. Corcorn, A. E. Hirtler, and R. F. Mittel, Including a semiresonant slotted mode stirrer, US Patent 3872276, 1975.
54. W. W. Teich, Radiating mode stirrer heating system, US Patent 4342896, 1982.
55. M. S. Miller, Drive arrangement for microwave oven mode stirrer, US Patent 4296297, 1981.
56. S. Bradshaw, S. Delport, and E. van Wyk, Qualitative measurement of heating uniformity in a multimode microwave cavity, Journal of Microwave Power and Electromagnetic Energy, vol. 32, no. 2, pp. 87–95, 1997.
57. W. Fu and A. Metaxas, Numerical prediction of three-dimensional power density distributions in a multi-mode cavity, Journal of Microwave Power and Electromagnetic Energy, vol. 29, no. 2, pp. 67–75, 1994.
58. J. S. Claesson and P. O. G. Risman, Feeding arrangement for a microwave oven, US Patent 4695693, 1987.
59. L. E. Berg and P. O. Risman, Microwave oven, a method for excitation of the cavity of a microwave oven, and a wave guide device for carrying out the method, US Patent 5237139, 1993.
60. J. E. Staats and L. H. Fitzmayer, Triangular antenna array for microwave oven, US Patent 4695693, 1987.
61. J. R. Bows, M. L. Patrick, R. Janes, A. C. Metaxas, and D. C. Dibben, Microwave phase control heating, International Journal of Food Science and Technology, vol. 34, pp. 295–304, 1999.
62. J. R. Bows, Variable frequency microwave heating of food, Journal of Microwave Power and Electromagnetic Energy, vol. 34, no. 4, pp. 227–238, 1999.
63. A. C. Metaxas, Industrial Microwave Heating. Peter Peregrinus, 1983. Stevenage, Herts, Great Britain.
64. J. Bows, Applications of microwave fringing field devices, in Proceedings of the 5th Microwave and High Frequency Heating Conference (Cambridge University, UK), 1995.

65. U. S. Shivhare, G.S.V. Raghavan, and R. G. Bosisio, Drying of corn using variable microwave power with a surface wave applicator, Journal of Microwave Power and Electromagnetic Energy, vol. 26, no. 1, pp. 38–44, 1991.
66. W. G. Phillips, The multitalented microwave, Popular Science, vol. 251, no. 6, p. 30, 1997.
67. T. Ohlsson, N. E. Bengtsson, and P. O. Risman, The frequency and temperature dependence of dielectric food data as determined by a cavity perturbation technique, Journal of Microwave Power, vol. 9, no. 2, pp. 129–145, 1974.
68. T. Ohlsson, M. Henriques, and N. E. Bengtsson, Dielectric properties of model meat emulsions at 900 and 2800 MHz in relation to their composition, Journal of Food Science, vol. 39, pp. 1153–1156, 1974.
69. A. Romine and S. A. Barringer, Dielectric properties of ham as a function of temperature, moisture, and salt, in Annual Meeting of Institute of Food Technologists (Atlanta, GA), 1998.
70. S. O. Nelson, J. W. R. Forbus, and K. C. Lawrence, Microwave permittivities of fresh fruits and vegetables from 0.2 to 20 GHz, Transactions of the ASAE, vol. 37, no. 1, pp. 183–189, 1994.
71. U. Kaatze, Complex permittivity of water as a function of frequency and temperature, Journal of Chemical Engineering Data, vol. 34, pp. 371–374, 1989.
72. M. Chamchong, Microwave Thawing of Foods. PhD thesis, Cornell University, Ithaca, NY, 1997.
73. J. Mannapperuma and R. P. Singh, A computer-aided method for the prediction of properties and freezing/thawing times of foods, Journal of Food Engineering, vol. 9, pp. 275–304, 1989.
74. T. Ohlsson and P. Risman, Temperature distribution of microwave heating—spheres and cylinders, Journal of Microwave Power, vol. 13, no. 4, pp. 303–313, 1978.
75. H. Ni, A. K. Datta, and R. Parmeswar, Moisture loss as related to heating uniformity in microwave processing of solid foods, Journal of Food Process Engineering, vol. 22, no. 5, pp. 367–382, 1999.
76. H. Zhang, Electromagnetic and thermal studies of microwave processing of foods, PhD thesis, Cornell University, 2000.
77. A. Yang, A. Taub, H. Zhang, and A. K. Datta, "Effectiveness of metallic shielding on the uniformity of microwave heating," in Presented at the 33rd Microwave Power Symposium (Chicago, IL), July 1998.

3
Dielectric Properties of Food Materials and Electric Field Interactions

Stuart O. Nelson
U.S. Department of Agriculture
Athens, Georgia

Ashim K. Datta
Cornell University
Ithaca, New York

I. INTRODUCTION

The electrical properties of materials known as dielectric properties (see Chapter 1) are of critical importance in understanding the interaction of microwave electromagnetic energy with those materials. These properties, along with thermal and other physical properties, and the characteristics of the microwave electromagnetic fields determine the absorption of microwave energy and consequent heating behavior of food materials in microwave heating and processing applications.

In this chapter, we provide a basic background for understanding the interaction of microwave electric fields with foods as dielectric materials and illustrate the way in which the dielectric properties of food materials vary with important variables. Various techniques for measuring the dielectric properties are discussed, and some of the advantages and limitations of the methods are pointed out. The dielectric properties of granular or particulate materials and their relationships to the bulk density or packing of such food materials are discussed. Also included is a discussion of the influence of food moisture content and composition on the dielectric properties of the materials.

Finally, sources of data on the dielectric properties of food materials are identified where representative data for various food materials may be obtained for the commonly used frequencies for microwave food processing.

II. DEFINITION OF TERMS AND BASIC PRINCIPLES

The fundamental electromagnetic characteristics of materials have been defined in detail in terms of electromagnetic field concepts [1] and in terms of parallel-equivalent circuit concepts [2], some details of which are provided in Chapter 1. Dielectric properties, introduced in Chapter 1, are the relevant material properties for explaining interactions with electric fields. The dielectric properties of usual interest are the dielectric constant ϵ' and the dielectric loss factor ϵ'', which are the real and imaginary parts, respectively, of the relative complex permittivity ϵ_r given by

$$\epsilon_r = \epsilon' - j\epsilon'' \tag{1}$$

or

$$\epsilon_r = |\epsilon_r| e^{-j\delta} \tag{2}$$

where δ is the loss angle of the dielectric. Hereafter in this chapter, "permittivity" is understood to represent the relative complex permittivity, i.e., the permittivity relative to free space, or the absolute permittivity divided by the permittivity of free space ϵ_0 (see Chapter 1). For convenience, the subscript r in ϵ_r will be dropped from now on. Often, the loss tangent, tan δ, or dissipation factor, defined as

$$\tan \delta = \frac{\epsilon''}{\epsilon'} \tag{3}$$

is also used as a descriptive dielectric parameter, and sometimes the power factor (tan $\delta/\sqrt{1 + \tan_2 \delta}$) is used. The conductivity of the dielectric σ is related to the dielectric loss factor, and is given by

$$\sigma = \omega \epsilon_0 \epsilon'' \tag{4}$$

where σ is in S/m, $\omega = 2\pi f$ is the angular frequency in rad/s, with frequency f in Hz. In this chapter, ϵ'' (and σ) is interpreted to include the energy losses in the dielectric due to all operating dielectric relaxation mechanisms and ionic conduction.

The dielectric properties of materials dictate, to a large extent, the behavior of the materials when subjected to radio-frequency (RF) or microwave fields for purposes of heating or drying the materials. The power dissipated per unit volume P (denoted as Q in other chapters of this book), in the dielectric can be expressed as

$$P = \sigma E^2 = 55.63 f\epsilon'' E^2 \times 10^{-12} \text{ W/m}^3 \tag{5}$$

where E represents the rms electric field intensity in V/m. The time rate of temperature increase depends not only on the absorbed power but also on the specific heat and density of the material and the vaporization of any water that may take place in the material. Temperature and moisture changes due to heating are discussed in detail in Chapter 4.

From Eq. (5), it is obvious that the power dissipation is directly related to the dielectric loss factor ϵ''; however, it may also be dependent on the dielectric constant in that, depending on the geometry and the electric field configuration, the electric field intensity may be a function of ϵ'. A simple example illustrates this influence of ϵ' values. A spherical inclusion of dielectric constant ϵ'_2 embedded in an infinite medium of dielectric constant ϵ'_1 will have an electric field intensity in the spherical material E_2 given by

$$E_2 = E_1 \left(\frac{3\epsilon'_1}{2\epsilon'_1 + \epsilon'_2} \right) \tag{6}$$

[4], where E_1 is the electric field intensity in the infinite medium. Thus, the dielectric constant can have an important influence on the electric field intensity in the material to be heated and, therefore, affects the power dissipation in that material [Eq. (5)]. Power dissipation in the material, therefore, can depend on both the dielectric loss factor and the dielectric constant. Also, from Eq. (5), power dissipation depends on the frequency of the wave and on the square of the electric field intensity.

The dielectric properties of the materials are very important in evaluating the penetration of energy that can be achieved. For a plane wave propagating in the z direction in a thick (semi-infinite) slab, the magnitude of the electric field intensity $|E|$ was shown in Chapter 1 to be given by

$$|E| = E_0 e^{-\alpha z} \tag{7}$$

where E_0 is the rms electric field intensity at the surface of the slab in which the wave is traveling, and z is the distance in the direction of travel. The attenuation constant α is related to the dielectric properties of the medium as given in Chapter 1. It can be simplified for a nonmagnetic material as

$$\alpha = \frac{2\pi}{\lambda_0} \sqrt{\frac{\epsilon'}{2} (\sqrt{1 + \tan^2 \delta} - 1)} \tag{8}$$

Equation (7) shows that the magnitude of the electric field intensity E decreases as the wave propagates into the material. Since power dissipated in the material P is proportional to the square of the electric field intensity [Eq. (5)], the decay of the power dissipation can be written as

$$P = P_0 e^{-2\alpha z} \tag{9}$$

The penetration depth D_p is defined as the distance at which the power drops to $1/2.7183 = 37\%$ of its value at the surface of the material. Thus,

$$D_p = \frac{1}{2\alpha} \tag{10}$$

which, by Eq. (8), can be written as

$$D_p = \frac{\lambda_0}{2\pi\sqrt{2\epsilon'}} \left[\sqrt{1 + \left(\frac{\epsilon''}{\epsilon'}\right)^2} - 1 \right]^{-1/2} \tag{11}$$

Note that the penetration depth in Eq. (11) is formulated for a plane wave traveling in a semi-infinite region. Thus, in general, penetration depth may not properly characterize the decay of power in a material of finite dimensions. For example, in microwave oven heating as discussed in Chapter 2, penetration depth is not an appropriate descriptor of the spatial variation in power dissipation. For low-loss dielectrics with $\epsilon''/\epsilon' \ll 1$, Eq. (11) reduces to

$$D_p = \frac{\lambda_0 \sqrt{\epsilon'}}{\epsilon''} \tag{12}$$

For practical reasons, attenuation of the wave is often expressed in decibels (dB). In terms of power densities and electric field intensity values, attenuation can be expressed [5] as

$$10 \log_{10}\left[\frac{P_0}{P(z)}\right] = 20 \log_{10}\left[\frac{E_0}{E(z)}\right] \tag{13}$$

$$= 8.686\alpha z$$

where P_0 is the power level at a point of reference and $P(z)$ is the power level at distance z from the reference point. The unit of attenuation [left-hand side of Eq. (13)] is dB. Whereas the attenuation constant α is usually expressed in nepers/m, it may also be expressed in dB/m, with 1 neper = 8.686 dB. If attenuation is high in the material, the dielectric heating will taper off quickly as the wave penetrates the material. Attenuation and phase measurements are often used in measuring dielectric properties.

III. VARIATION OF DIELECTRIC PROPERTIES

The dielectric properties of most materials vary with several different factors. In hygroscopic materials such as foods, the amount of water in the material is generally a dominant factor. The dielectric properties also depend on the frequency of the applied alternating electric field, the temperature of the material, and on the density, composition, and structure of the material. In granular or particulate materials, the bulk density of the air-particle mixture is another factor that influences the dielectric properties. Of course, the dielectric properties of materials are dependent on the chemical composition and especially on the presence of mobile ions and the permanent dipole moments associated with water and any other molecules making up the material of interest.

A. Frequency Dependence

With the exception of some extremely low-loss materials, i.e., materials that absorb essentially no energy from RF and microwave fields, the dielectric properties of most materials vary considerably with the frequency of the applied electric fields. This frequency dependence has been discussed previously [1, 3]. An important phenomenon contributing to the frequency dependence of the dielectric properties is the polarization, arising from the orientation with the imposed electric field, of molecules which have permanent dipole moments. The mathematical formulation developed by Debye [6] to describe this process for pure polar materials can be expressed as

$$\epsilon = \epsilon_\infty + \frac{\epsilon_s - \epsilon_\infty}{1 + j\omega\tau} \tag{14}$$

where ϵ_∞ represents the dielectric constant at frequencies so high that molecular orientation does not have time to contribute to the polarization, ϵ_s represents the static dielectric constant, i.e., the value at zero frequency (dc value), and τ is the relaxation time, the period associated with the time for the dipoles to revert to random orientation when the electric field is removed. Separation of Eq. (14) into its real and imaginary parts yields

$$\epsilon' = \epsilon_\infty + \frac{\epsilon_s - \epsilon_\infty}{1 + \omega^2\tau^2} \tag{15}$$

$$\epsilon'' = \frac{(\epsilon_s - \epsilon_\infty)\omega\tau}{1 + \omega^2\tau^2} \tag{16}$$

The relationships defined by these equations are illustrated in Fig. 1. Thus, at frequencies very low and very high with respect to the molecular relaxation process, the dielectric constant has constant values ϵ_s and ϵ_∞, respectively, and the losses are zero. At intermediate frequencies, the dielectric constant undergoes a dispersion, and dielectric losses occur with the peak loss at the relaxation frequency, $\omega_o = 1/\tau$.

The Debye equation can be represented graphically in the complex ϵ'' versus ϵ' plane as a semicircle with locus of points ranging from ($\epsilon' = \epsilon_s$, $\epsilon'' = 0$) at

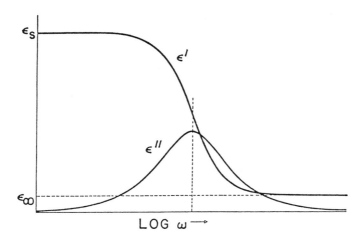

Figure 1 Graphical representation of the Debye dielectric relaxation for polar molecules with a single relaxation time.

the low-frequency limit to ($\epsilon' = \epsilon_\infty$, $\epsilon'' = 0$) at the high-frequency limit (Fig. 2). Such a representation is known as a Cole-Cole diagram [7].

Since few materials of interest consist of pure polar materials with a single relaxation time, many other equations have been developed to better describe the frequency-dependent behavior of materials with more relaxation times or a distribution of relaxation times [8]. One such equation is the Cole-Cole equation [7],

$$\epsilon = \frac{\epsilon_s - \epsilon_\infty}{1 + (j\omega\tau)^{1-\alpha}} \tag{17}$$

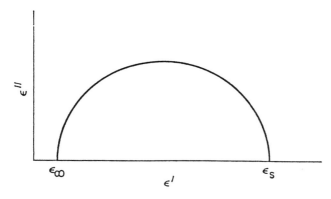

Figure 2 Cole-Cole plot for a polar substance with a single relaxation time.

Table 1 Microwave Dielectric Properties of Water at Indicated Temperatures

Frequency, GHz	20°C		50°C	
	ϵ'	ϵ''	ϵ'	ϵ''
0.6	80.3	2.75	69.9	1.25
1.7	79.2	7.9	69.7	3.6
3.0	77.4	13.0	68.4	5.8
4.6	74.0	18.8	68.5	9.4
7.7	67.4	28.2	67.2	14.5
9.1	63.0	31.5	65.5	16.5
12.5	53.6	35.5	61.5	21.4
17.4	42.0	37.1	56.3	27.2
26.8	26.5	33.9	44.2	32.0
36.4	17.6	28.8	34.3	32.6

Source: Refs. 9, 10.

where α denotes the spread of relaxation times, and this empirical relaxation-time distribution parameter takes on values between 0 and 1.

Water in its liquid state is a good example of a polar dielectric. The microwave dielectric properties of liquid water are listed in Table 1 for several frequencies at temperatures of 20 and 50°C as selected from data listed by Hasted [9] and Kaatze [10]. Although earlier work indicated a better fit of experimental data with the Cole-Cole equation (17) with an α of about 0.012 than with the Debye equation (14), Kaatze has shown that the dielectric spectra for pure water can be sufficiently well represented by the Debye equation when using the relaxation parameters given in Table 2. The relaxation frequency, $(2\pi\tau)^{-1}$, is provided in Table 2 along with the static and high-frequency values of the dielectric constant, ϵ_s and

Table 2 Debye Dielectric Relaxation Parameters for Water

Temperature, °C	ϵ_s	ϵ_∞	τ, ps	Relaxation frequency, GHz
0	87.9	5.7	17.67	9.007
10	83.9	5.5	12.68	12.552
20	80.2	5.6	9.36	17.004
30	76.6	5.2	7.28	21.862
40	73.2	3.9	5.82	27.346
50	69.9	4.0	4.75	33.506
60	66.7	4.2	4.01	39.690

Source: Ref. 10.

ϵ_∞, for water at temperatures from 0 to 60°C. Thus Eqs. (15) and (16), together with relaxation parameters listed in Table 2, can be used to provide close estimates for the dielectric properties of water over a wide range of frequencies and temperatures.

However, water in its pure liquid state appears in food products very rarely. Most often it has dissolved constituents, is physically absorbed in material capillaries or cavities, or is chemically bound to other molecules of the material. Dielectric relaxations of absorbed water take place at lower frequencies than the relaxation of free water [9], which occurs at about 19.5 GHz for water at 25°C. Depending upon the material structure, there may be various forms of bound water, differing in energy of binding and in dielectric properties. Moist material, in practice, is usually an inhomogeneous mixture, often containing more than one substance with unknown dielectric properties. Thus, it is difficult to understand and predict the dielectric behavior of such material at different frequencies, temperatures, and hydration levels.

B. Temperature Dependence

The dielectric properties of materials are also temperature dependent, and the nature of that dependence is a function of the dielectric relaxation processes operating under the particular conditions existing and the frequency being used. As temperature increases, the relaxation time decreases, and the loss-factor peak illustrated in Fig. 1 will shift to higher frequencies. Thus, in a region of dispersion, the dielectric constant will tend to increase with increasing temperature as a result of dielectric relaxation, whereas the loss factor may either increase or decrease, depending on whether the operating frequency is higher or lower than the relaxation frequency. However, for complex dielectrics such as food materials, other mechanisms may mask or dominate the dielectric relaxation effects. The temperature dependence of ϵ_∞ is relatively small [8], and while that of ϵ_s is larger, its influence is minor in a region of dispersion. Below and above the dispersion region, the dielectric constant tends to decrease with increasing temperature. Distribution functions can be useful in expressing the temperature dependence of dielectric properties [8], but the frequency and temperature-dependent behavior of the dielectric properties of most materials is complicated and can perhaps best be determined by measurement at the frequencies and under the other conditions of interest.

C. Density Dependence

Since the influence of a dielectric depends on the amount of mass interacting with the electromagnetic fields, the mass per unit volume, or density, will have an effect on the dielectric properties. This is especially notable with particulate dielectrics such as pulverized or granular materials. In understanding the nature of the density dependence of the dielectric properties of particulate materials, relationships between the dielectric properties of solid materials and those of air-particle mixtures, such as granular or pulverized samples of such solids, are useful.

In some instances, the dielectric properties of a solid may be needed when particulate samples are the only available form of the material. This was true for cereal grains, where kernels were too small for the dielectric sample holders used for measurements [11, 12] and in the case of pure minerals that had to be pulverized for purification [13]. For some materials, fabrication of samples to exact dimensions required for dielectric properties measurement is difficult, and measurements on pulverized materials are more easily performed. In such instances, proven relationships for converting dielectric properties of particulate samples to those for the solid material are important. Several well-known dielectric mixture equations have been considered for this purpose [14–16].

The notation used here applies to two-component mixtures, where ϵ represents the effective permittivity of the mixture, ϵ_1 is the permittivity of the medium in which particles of permittivity ϵ_2 are dispersed, and v_1 and v_2 are the volume fractions of the respective components, where $v_1 + v_2 = 1$. Two of the mixture equations found particularly useful for cereal grains were the complex refractive index mixture equation

$$(\epsilon)^{1/2} = v_1(\epsilon_1)^{1/2} + v_2(\epsilon_2)^{1/2} \tag{18}$$

and the Landau and Lifshitz, Looyenga equation

$$(\epsilon)^{1/3} = v_1(\epsilon_1)^{1/3} + v_2(\epsilon_2)^{1/3} \tag{19}$$

To use these equations to determine ϵ_2, one needs to know the dielectric properties (permittivity) of the pulverized sample at its bulk density (air-particle mixture density), ρ, and the specific gravity or density of the solid material, ρ_2. The fractional part of the total volume of the mixture occupied by the particles (volume fraction), v_2, is then given by ρ/ρ_2. Solving Eq. (18) and (19), respectively, for the complex permittivity of the solid material and substituting $1 - j0$ for ϵ_1 (the permittivity of air), the permittivity of the solid materials can be calculated as

$$\epsilon_2 = \left(\frac{\epsilon^{1/2} + v_2 - 1}{v_2}\right)^2 \tag{20}$$

$$\epsilon_2 = \left(\frac{\epsilon^{1/3} + v_2 - 1}{v_2}\right)^3 \tag{21}$$

It has been noted that Eqs. (18) and (19) imply the linearity of $\epsilon^{1/2}$ and $\epsilon^{1/3}$, respectively, with the bulk density of the mixture ρ [16]. The complex refractive index and Landau and Lifshitz, Looyenga relationships thus provide a relatively reliable method for adjusting the dielectric properties of granular and powdered materials with characteristics like grain products from known values at one bulk density to corresponding values for a different bulk density. It follows from Eq. (18) that for an air-particle mixture, where $\epsilon_1 = 1 - j0$, and since $v_1 = 1 - v_2$ and $v_2 = \rho/\rho_2$, that

$$(\epsilon_x)^{1/2} = \left[\frac{(\epsilon_2)^{1/2} - 1}{\rho_2}\right]\rho_x + 1 \tag{22}$$

for a mixture of density ρ_x. Similarly,

$$(\epsilon_y)^{1/2} = \left[\frac{(\epsilon_2)^{1/2} - 1}{\rho_2}\right]\rho_y + 1 \tag{23}$$

for the same mixture of density ρ_y. Equating the slopes of these two lines (the terms in brackets in the last two equations) and solving for ϵ_x gives the following:

$$\epsilon_x = \left[(\epsilon_y^{1/2} - 1)\frac{\rho_x}{\rho_y} + 1\right]^2 \tag{24}$$

which provides an expression for the permittivity of the mixture at any given density ρ_x when the permittivity ϵ_y is known at density ρ_y. In an analogous way, it follows from Eq. (19) that

$$\epsilon_x = \left[(\epsilon_y^{1/3} - 1)\frac{\rho_x}{\rho_y} + 1\right]^3 \tag{25}$$

Either Eq. (24) or (25) should provide reliable conversions of permittivity from one mixture density to another, but the Landau and Lifshitz, Looyenga relationship [Eq. 19)] provided somewhat closer estimates within the range of measured densities in work with whole kernel wheat, ground wheat, and finely pulverized coal [17]; so Eq. (25) is preferred.

IV. MEASUREMENT PRINCIPLES AND TECHNIQUES

Techniques for the measurement of dielectric properties of materials are many and varied. At microwave frequencies, generally about 1 GHz and higher, transmission-line, resonant cavity, and free-space techniques have been useful. Principles and techniques of microwave dielectric properties measurements have been discussed in reviews of such methods [18–22]. Microwave dielectric properties measurement techniques can be classified as reflection or transmission measurements using resonant or nonresonant systems, with open or closed structures for the sensing of the properties of material samples [23]. Closed-structure methods include waveguide and coaxial-line transmission measurements and short-circuited waveguide or coaxial-line reflection measurements. Open-structure techniques include free-space transmission measurements and open-ended coaxial-line or open-ended waveguide measurements. Resonant structures can include either closed resonant cavities or open resonant structures operated as two-port devices for transmission measurements or as one-port devices for reflection measurements.

With the development of suitable equipment for time-domain measurements, techniques were developed for measurement of dielectric properties of materials over wide ranges of frequency [24–29]. Since modern microwave network analyzers have become available, the methods of obtaining dielectric properties over wide frequency ranges have become even more efficient. Extensive reviews have included methods for both frequency-domain and time-domain techniques [30, 31].

Dielectric sample holder design for the specific materials is an important aspect of the measurement technique. The Roberts and von Hippel [32] short-circuited line technique for dielectric properties measurements provides a suitable method for many materials. For this method, the sample holder can be simply a short section of coaxial-line or rectangular waveguide with a shorting plate or other short-circuit termination at the end of the line against which the sample rests. This is convenient for particulate samples, because the sample holder, and also the slotted line or slotted section to which the sample holder is connected can be mounted in a vertical orientation so the top surface of the sample can be maintained perpendicular to the axis of wave propagation as required for the measurement. The vertical orientation of the sample holder is also convenient for liquid or particulate materials when the measurements are taken with a network analyzer instead of a slotted line.

Dielectric properties of cereal grains, seed, and powdered or pulverized materials have been taken with various microwave measurement systems assembled for such measurements. 21 mm, 50 ohm coaxial-line systems were used for these measurements at frequencies from 1 to 5.5 GHz [33–35]. A 54 mm, 50 ohm coaxial sample holder, designed for minimal reflections from the transition, was used with this same system for measurements on larger-kernel cereals such as corn [36]. A rectangular-waveguide X-band system was used to determine dielectric properties of grain and seed samples at 8–12 GHz [37, 38]. A rectangular waveguide K-band system [39] was used for measurements on ground and pulverized materials for measurements at 22 GHz [11–16, 35]. The Roberts and von Hippel method [32] requires measurements to determine the standing-wave ratios (SWRs) in the line with and without the sample inserted at the short-circuited end of the line. From the shift of the standing-wave node and changes in node widths related to SWRs, sample length, and waveguide dimensions, etc., ϵ' and ϵ'' can be calculated with suitable computer programs [40, 41]. Similar determinations can be made with a vector network analyzer or other instrumentation by measurement of the complex reflection coefficient of the empty and filled sample holder.

Computer control of impedance analyzers [42] and network analyzers [43] has facilitated the automatic measurement of dielectric properties over wide frequency ranges. Special calibration methods have also been developed to eliminate errors caused by unknown reflections in the coaxial-line systems [42, 44].

Microwave dielectric properties of wheat and corn have been measured at several frequencies by free-space measurements with a network analyzer and dielectric sample holders with rectangular cross-sections between horn antennas and other types of radiating elements [45–48]. Measurement of the complex transmission coefficient, the components of which are attenuation and phase shift, permits the calculation of ϵ' and ϵ''. For free-space permittivity measurements, it is important that an attenuation of about 10 dB through the sample layer be maintained to avoid disturbances resulting from multiple reflections within the sample and between the sample interfaces and the antennas. The sample size, laterally, must be sufficiently large to avoid problems caused by diffraction at the edges of the sample [49]. Focused-beam horn/lens antennas are helpful in avoiding such problems with samples of reasonable size [50].

Open-ended coaxial-line probes have been used successfully for convenient broadband permittivity measurements [51, 52] on liquid and semisolid materials of relatively high loss, which includes most food materials. This technique has been used for permittivity measurements on fresh fruits and vegetables [53–56]. The technique is subject to errors if there are significant density variations in material or if there are undetected air gaps or air bubbles between the end of the coaxial probe and the sample, and the technique is not suitable for determining permittivities of very low-loss materials [57], but it has been used to provide broad-band information on granular and pulverized materials when sample bulk densities were established by auxiliary permittivity measurements [58]. For measurements at temperatures other than the temperature at which the calibration is performed, extreme caution is advisable with checking to verify reliability of the permittivity measurements.

The choices of measurement technique, equipment, and sample holder design depend upon the dielectric materials to be measured and the frequency or frequency range of interest. Vector network analyzers are very versatile and useful if studies are extensive. Scalar network analyzers and impedance analyzers are simpler and can be appropriate for certain programs. For limited studies, more commonly available microwave laboratory measurement equipment can suffice if suitable sample holders are constructed. When data are required at only one microwave frequency, a resonant cavity technique may be the logical choice [59, 60]. Such cavities can be easily constructed with rectangular waveguide sections or from waveguide flanges and waveguide stock [61]. Construction of a cylindrical cavity [60] may be advantageous, depending on the needs. For temperature-dependent studies, a cavity with provision for alternate dielectric properties measurement and microwave heating of the sample for temperature control may be advantageous [62, 63]. At common microwave frequencies, a measurement system can often be assembled from microwave laboratory components using a short waveguide section with a shorting plate as the sample holder [37] and an available general computer program [40, 41] for calculation of the dielectric properties.

V. DIELECTRIC BEHAVIOR OF FOOD MATERIALS

Measured dielectric properties data, along with some of the models for these properties, are presented here for several representative groups of food products. This will illustrate the nature of the variation of these properties with respect to the variables already discussed. Sources of additional data on these and other materials will be identified. Prediction of dielectric properties based on composition [64–67] have generally been unsuccessful for large groups of food materials, primarily because of the complex interactions of the components. An example of such an interaction has been noted in relatively simple cases of alcohol and sugar solutions where the dielectric loss of the mixture is greater than the loss of either the solute or the solvent alone [68]. For grains, the modeling of dielectric properties has been more successful [69–78].

A. Dielectric Properties of Salt Solutions

Many foods contain water as the major constituent. Also, salt ions can affect the dielectric properties significantly, especially the dielectric loss factor. Thus, it is useful to study the dielectric properties of salt solutions [67, 79] which are relatively simple systems and they may suggest trends in the dielectric behavior of some food materials. The dielectric properties of saline solutions have been studied in detail, and equations for calculating these properties as functions of temperature and salinity have been developed by Stogryn [80] and tested against experimental data from the literature. Plots of the 2450 MHz dielectric constant and loss factor of aqueous NaCl solutions of between 0 and 2% salinity (concentration by weight) are shown in Fig. 3 as a function of temperature [80]. Stogryn's equations (corrected for typographic errors in the published paper) provide parameters for the Debye model which adequately represent the behavior of aqueous NaCl solutions. Also, Stogryn's equations provide dielectric properties for pure water that agree closely with those reported by Kaatze from -4.1 to $60°C$ [10]. Measurements on salt solutions closely approximating the values shown in Fig. 3, especially at the lower temperatures, were reported by Sun et al. [67]. The dielectric constants of the salt solutions decrease with an increase in salt concentration, whereas the dielectric loss factor increases with increasing salt concentration. Di-

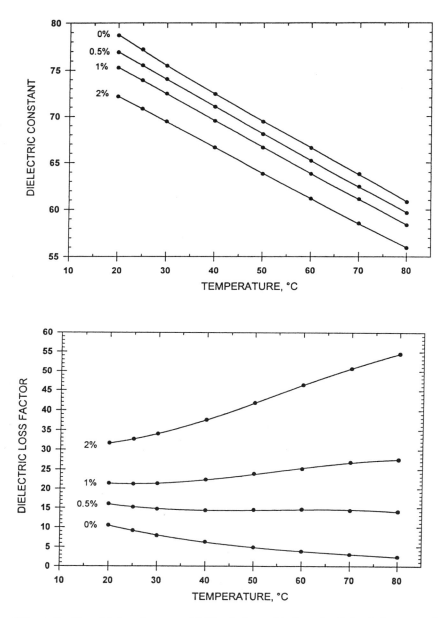

Figure 3 Permittivity of aqueous NaCl solutions of indicated salinity (salt content, by weight) as a function of temperature at 2450 MHz. Points are values calculated from Stogryn's equations [80].

electric constants of salt solutions decrease as temperature rises. However, the sign of the temperature coefficient of the dielectric loss factor depends on the concentration for the range of temperatures and concentrations shown in Fig. 3. At salt concentrations between about 0.5 and 1%, this temperature coefficient changes from negative to positive for temperatures above 25°C at 2450 MHz.

B. Compositional Effects on Dielectric Properties

When dielectric properties data for many different types of foods (meats, fruits, and vegetables) were considered together, they showed very little correlation with composition [67]. This was attributed to variability in sample composition and measurement errors, and the general unavailability of detailed composition data. Correlations of data at 2450 MHz from a restricted set containing about 30 data points [81], for raw beef, beef juice, raw turkey, and turkey juice, with composition data taken from the USDA Handbook [82] revealed significant relationships between both the dielectric constant and loss factor and the water and ash contents and temperature [67]. The addition of terms for components such as protein, carbohydrates, and fat, did not improve the correlation significantly. Above a critical moisture content of about 35 to 40%, most of the additional water in foods is thought to be in a free form in the capillaries [83, 84]. Also, de Loor and Meijboom [85] found that high-water-content foods, such as potato, agar gels, and milk, had similar relaxation times, which they attributed to the availability of free water in the foods. Thus, since the moisture content of all the meats and meat products used in the correlation was greater than 60%, the free water should be the dominant component governing the overall dielectric behavior of these foods.

The ash content was taken to be a good indicator of the total salts in these foods. Increased ash content served to elevate the dielectric constant in the foods, which agreed with the experimental data reported by Bengtsson and Risman [86], in contrast to the behavior of aqueous salt solutions which show a reduction in the dielectric constant with increasing salt concentration (Fig. 3). In the range of temperatures studied [67], the ash content elevated the dielectric loss, indicating that increased salt adds conductive charge carriers that increase the loss of the system as a result of charge migration.

The dielectric loss of aqueous solutions at microwave frequencies is attributable to effects of dipole rotation and ionic charge migration, with ionic conduction having greater influence at lower frequencies and ionic conduction losses decreasing with increasing frequency. The dipole loss also decreases as temperature increases [87, 88]. At a given frequency, the loss resulting from ionic conduction increases with increasing temperature. Results of Sun et al. showed that for salt contents greater than 2%, the predicted dielectric loss increased with temperature [67], which is in accord with the behavior of aqueous salt solutions (Fig. 3).

Water in foods can be free or bound. There is among food materials a continuum from largely free to largely bound water with intermediate states often being termed loosely bound or tightly bound. Because the degree of binding varies with material, moisture level, temperature, and composition, it has been difficult to predict the dielectric properties of materials with bound water. Bound water contributes less to the dielectric constant than free water in the microwave region, because the irrotationally bound water is unable to respond to the alternating electric field at such frequencies [9]. For low-moisture foods, the dielectric constant and loss factor increase with moisture content. For high-moisture foods, in which bound water plays an insignificant role, the properties are influenced not only by the water content but also by dissolved constituents and ionic conduction, although the contribution of charge carriers to the dielectric loss may be very small at frequencies above 2 or 3 GHz.

There is little dielectric properties data available for many foods covering a wide range of moisture contents for the frequencies of interest in microwave heating. Data of Mudgett et al. [83] illustrates some of the trends over a wide moisture range for a food material as shown in Fig. 4. At very low moisture contents

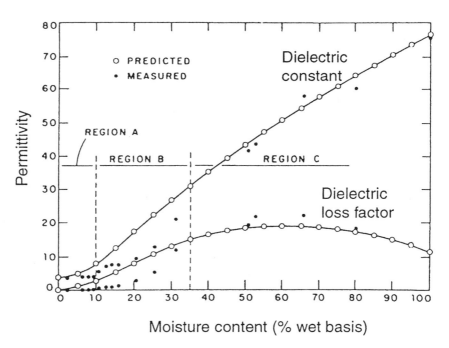

Figure 4 Dielectric constant and loss factor as a function of added moisture in freeze-dried potato at 3 GHz and 25°C. (Adapted from Mudgett et al. [83].)

(<10%), both the dielectric constant and loss factor are very low, as noted in region A of the graph. In region B, they increase rapidly with moisture content. This can be partially attributed to ionization of bound salts associated with increased availability of loosely bound or free water. In region C, at moisture contents above 35%, the further availability of water appears to dilute the dissolved salts [65].

C. Dielectric Properties at Freezing and Sterilizing Temperatures

Accurate knowledge of dielectric properties in partially frozen material is critical to determining the rates and uniformity of heating in microwave thawing. As the ice in the material melts, absorption of microwave energy increases tremendously. Thus, the portions of material that thaw first, absorb significantly more energy and heat at increasing rates that can lead to localized boiling temperatures while other areas are still frozen. Dielectric properties of frozen food materials have been reported in the literature [86, 89, 90], and these properties can be heavily influenced by composition, particularly total water and salt content. Salt affects the situation through freezing point depression, leaving more water unfrozen at a given temperature in this range. Salt also increases the ionic content and consequently the interaction with the microwave fields.

Chamchong [91] reported 2.45 GHz dielectric properties of tylose, a food analog, at temperatures including the frozen range as shown in Fig. 5. By measuring the apparent specific heat with a differential scanning calorimeter, the fraction of water frozen at selected temperatures was measured directly as shown in Fig. 6. Distinctly different frozen fractions were observed for the three salt concentrations in tylose, with higher salt concentration producing less frozen water at a given temperature. Measured properties above freezing were extrapolated to the initial freezing point. For the partially frozen region, the dielectric constant and loss factor were predicted as linear functions of the unfrozen water determined experimentally.

Both the dielectric constant and loss factor decrease significantly as more water freezes. Since the fraction of unfrozen water is a nonlinear function of temperature, the increase in the dielectric properties of the partially frozen material is also nonlinear with temperature. Above the freezing range, the dielectric constant of tylose decreased linearly with temperature while the dielectric loss factor of tylose increased linearly with temperature. With the addition of salt, the dielectric constant decreased, while the dielectric loss factor increased. Thus, changes above the freezing range followed those exhibited by salt solutions (Fig. 3).

Dielectric properties at high temperatures, useful for microwave pasteurization and sterilization, have been scarce. In addition to temperature effects per se, physical and chemical changes such as gelatinization of starch [92] and denaturation of protein, causing release of water and shrinkage [93] at higher temperatures,

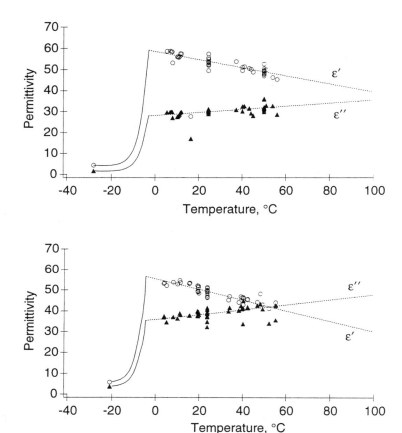

Figure 5 Permittivities of tylose at 2.45 GHz with 2% salt (upper graph) and 4% salt (lower graph). (From Ref. 91.)

can significantly change dielectric properties. Such effects are also strongly dependent on moisture content.

The dielectric properties of several powdered hydrocolloids, including potato starch, locust bean gum, gum arabic, carrageenan, and carboxymethylcellulose, were measured at 2.45 GHz as functions of temperature (20–100°C) and moisture content between 0 and 30%, wet basis [94, 95]. The dielectric constants and loss factors of all hydrocolloids increased with moisture content and with temperature indicating a relaxation frequency below 2.45 GHz at all moisture contents and temperatures. The degree of temperature dependence of all five hydro-

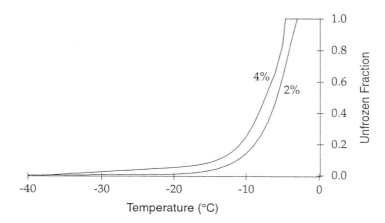

Figure 6 Unfrozen fraction of water in tylose (a food analog) as a function of temperature for two different salt contents.

colloids increased as moisture content increased but to a lesser extent in potato starch and a much lesser extent in locust bean gum.

D. Dielectric Properties of Cereal Grains

For hard red winter wheat, the dielectric constant and loss factor over the frequency range from 250 Hz to 12 GHz are shown in Fig. 7. The dielectric constant always decreases with increasing frequency, but the loss factor may increase or decrease as frequency increases, depending upon the frequency and the moisture content of the wheat. The loss factor for very dry wheat (Fig. 7) indicates a dielectric relaxation, probably attributable to bound water, in the range between 1 and 100 MHz, but this is not very evident in the curves for the dielectric constant. The variables affecting the dielectric properties of grain have been studied in some detail [96–99]. Reference data have been compiled for the dielectric properties of grain and seed [100] as they are influenced by frequency and moisture content. For soft red winter wheat, additional temperature- and frequency-dependent data are available for the 0.1–100 MHz frequency range [101, 102].

The frequency and moisture dependence of the dielectric properties of shelled yellow-dent field corn are shown in Fig. 8. The regularity of the relationship between the dielectric constant and both moisture and frequency is evident in all these curves, but, as already noted, the behavior of the loss factor is much less predictable. The variation of the dielectric constant with temperature is linear at low moisture contents but becomes nonlinear at higher moisture levels as illustrated in Fig. 9 for three frequencies and a range of moisture contents. The varia-

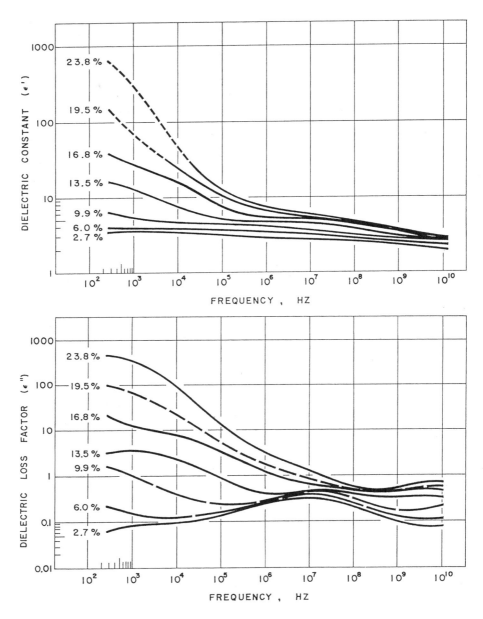

Figure 7 Frequency dependence of the dielectric properties of hard red winter wheat, *Triticum asestivum* L., at 24°C and indicated moisture contents. (From Ref. 96.)

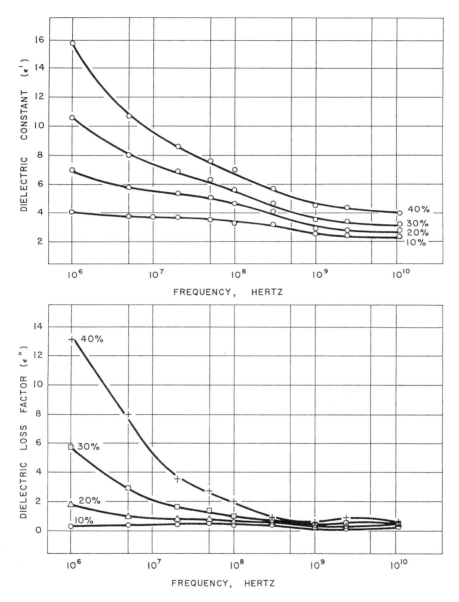

Figure 8 Frequency dependence of the dielectric properties of shelled yellow-dent field corn, *Zea mays* L., at 24°C and indicated moisture contents. (From Ref. 97.)

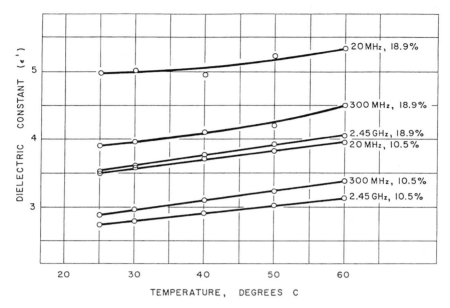

Figure 9 Temperature dependence of the dielectric constant of shelled yellow-dent field corn at indicated frequencies and moisture contents. (From Ref. 36.)

tion of the dielectric constant over a range of natural bulk densities appears linear as shown in Fig. 10, but this dependence is also nonlinear over wider ranges of bulk density [71, 103]. Data of the type illustrated in Figs. 9 and 10 were used in developing models for the dielectric constant of shelled corn at 20, 300, and 2450 MHz. Based on linear relationships between the dielectric constant ϵ' and moisture content M, temperature T, and bulk density ρ, the following equation was obtained for the dielectric constant at 2450 MHz [36]:

$$\epsilon'_{2450} = 2.48 + 0.099(M - 10) \\ + (\rho - \bar{\rho})(0.387M - 3.22) + 0.013(T - 24) \quad (26)$$

where M is in percent, wet basis, ρ is in g/cm^3, T is in degrees C, and $\bar{\rho}$ is the density corresponding to the mean normal test weight for the 21 yellow-dent field corn lots used in developing the model. This mean test weight $\bar{\rho}$ was expressed by a cubic equation [36]:

$$\bar{\rho} = 0.6829 + 0.01422M - 0.000979M^2 + 0.0000153M^3 \quad (27)$$

Models were later developed for the dielectric constants of corn [71], wheat [72], barley [73, 74], oats [73], rye [75], and soybeans [76] as functions of frequency, moisture content, and bulk density over wide ranges of each variable [75, 100].

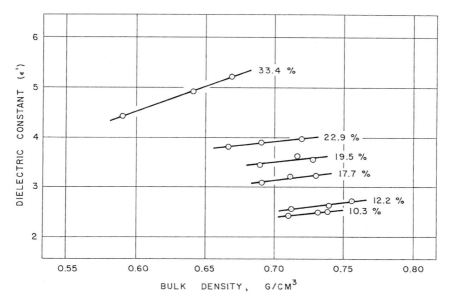

Figure 10 Dependence of the dielectric properties of shelled-yellow-dent field corn on bulk density at indicated moisture contents at 2.45 GHz and 24°C. (From Ref. 36.)

A composite model for cereal grain permittivity was developed as follows from data on all of these cereal grains [104]:

$$\epsilon' = \left[1 + \frac{0.504M\rho}{\sqrt{M} + \log f}\right]^2 \quad (28)$$

$$\epsilon'' = 0.146\rho^2 + 0.00461M^2\rho^2 (0.32 \log f + 1.743/\log f - 1) \quad (29)$$

where the variables have the same units as those for Eq. (26).

For rough rice, the frequency and moisture dependence of the dielectric properties [105] are illustrated in Fig. 11. The relationship between dielectric constant and moisture content is nearly linear for the three frequencies 20 MHz, 300 MHz, and 2.45 GHz. In accordance with Eqs. (24) and (25), the square root and cube root, respectively, of the dielectric constants are expected to be linear with bulk density as shown in Fig. 12 for ground rough rice. These relationships were useful in modeling the dielectric constants of rice [106]. These principles were also used in determining grain kernel and soybean seed dielectric constants by linear extrapolations to kernel density [11, 12] and by complex computations based on dielectric mixture equations [3].

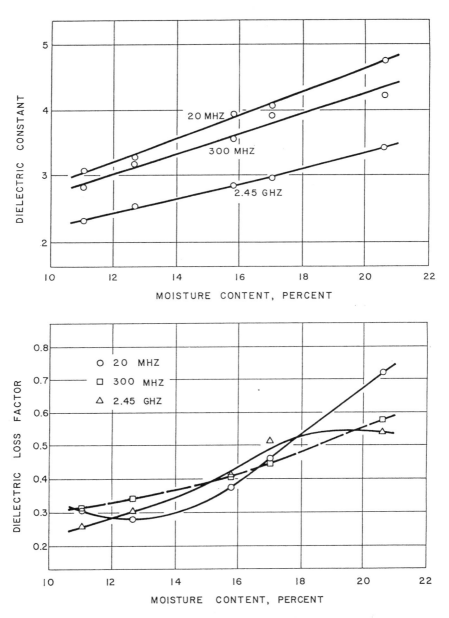

Figure 11 Moisture dependence of the dielectric properties of rough rice, *Oryza sativa* L., at 24°C and indicated frequencies. (From Ref. 105.)

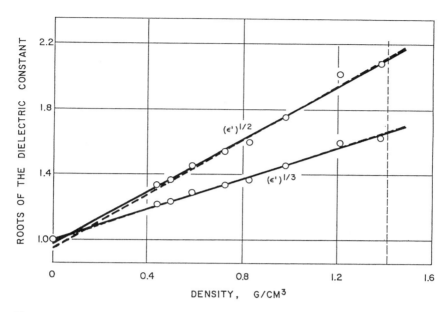

Figure 12 Linear regressions of the square roots and cube roots of the dielectric constants of ground rough rice of 11.5% moisture content at 24°C and 11.0 GHz on the bulk density of the sample. (From Ref. 11.)

E. Dielectric Properties of Fresh Fruits and Vegetables

Dielectric constants and loss factors of fresh fruits and vegetables have been explored by several researchers [53–56, 107, 108]. Examples of the frequency dependence observed for 24 different fruits and vegetables are shown in Fig. 13 for potato and apple. The dielectric constant exhibits the expected monotonic decrease in value with frequency in the 0.2–20 GHz range. In all the fruits and vegetables measured, the loss factor decreases as frequency increases from 0.2 GHz, reaches a broad minimum in the region between 1 and 3 GHz, and then increases as frequency approaches 20 GHz. This behavior is dominated by ionic conductivity at lower frequencies, by bound water relaxation, and by the free water relaxation near the top of the frequency range. The dielectric properties of these fresh fruits and vegetables at 915 MHz and 2.45 GHz are given in Table 3 along with other descriptive information.

F. Dielectric Properties of Nuts

Dielectric properties data on nuts are quite limited. Dielectric constants and loss factors for peanuts have been reported for the 1–50 MHz frequency range [1,

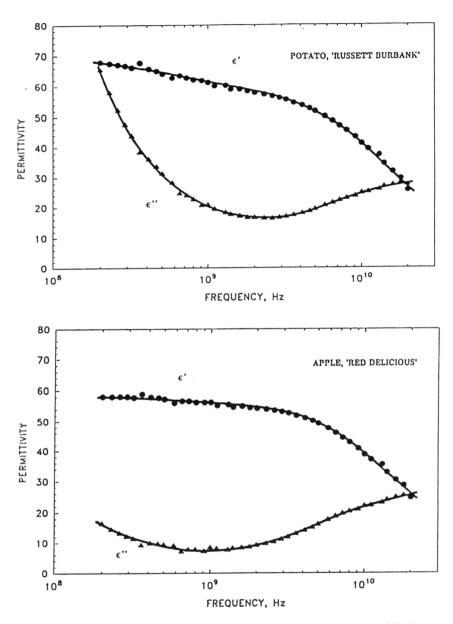

Figure 13 Frequency dependence of the microwave dielectric properties of fresh potato and apple tissue at 23°C. (From Refs. 54, 55.)

Table 3 Permittivities of Fresh Fruits and Vegetables at 23 °C

Fruit or vegetable	Moisture content %, wet basis	Tissue density g/cm^3	915 MHz ϵ'	915 MHz ϵ''	2.45 GHz ϵ'	2.45 GHz ϵ''
Apple	88	0.76	57	8	54	10
Avocado	71	0.99	47	16	45	12
Banana	78	0.94	64	19	60	18
Cantaloupe	92	0.93	68	14	66	13
Carrot	87	0.99	59	18	56	15
Cucumber	97	0.85	71	11	69	12
Grape	82	1.10	69	15	65	17
Grapefruit	91	0.83	75	14	73	15
Honeydew	89	0.95	72	18	69	17
Kiwi fruit	87	0.99	70	18	66	17
Lemon	91	0.88	73	15	71	14
Lime	90	0.97	72	18	70	15
Mango	86	0.96	64	13	61	14
Onion	92	0.97	61	12	64	14
Orange	87	0.92	73	14	69	16
Papaya	88	0.96	69	10	67	14
Peach	90	0.92	70	12	67	14
Pear	84	0.94	67	11	64	13
Potato	79	1.03	62	22	57	17
Radish	96	0.76	68	20	67	15
Squash	95	0.70	63	15	62	13
Strawberry	92	0.76	73	14	71	14
Sweet potato	80	0.95	55	16	52	14
Turnip	92	0.89	63	13	61	12

Permittivity, $\epsilon = \epsilon' - j\epsilon''$

Source: Refs. 55, 56.

109] showing the expected variation with frequency and moisture content. More detailed information is available for chopped pecans [110], and these data are summarized for the frequency range from 50 kHz to 12 GHz over the moisture range from 3 to 9%, wet basis, at 22°C in Fig. 14. The temperature dependence at 0–40°C of the dielectric constant and loss factor were determined for the frequency range from 100 kHz to 110 MHz over a similar range of moisture contents [111]. Both the dielectric constant and loss factor increased regularly with moisture content at all frequencies and decreased as frequency increased. At low moisture contents, the temperature dependence was minimal, but both the dielectric constant and loss factor increased rapidly with temperature at high moisture levels.

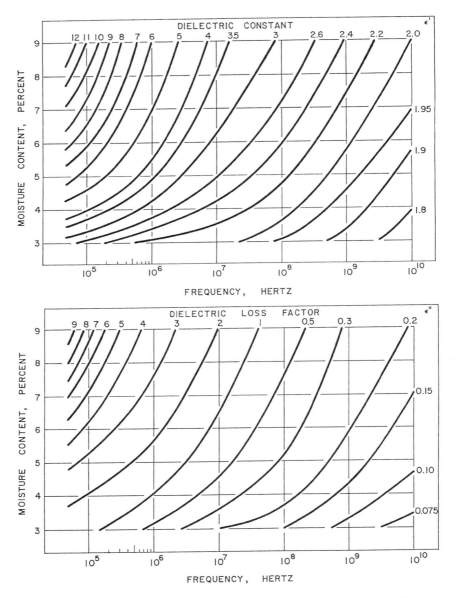

Figure 14 Contour plots of the dielectric properties of chopped pecan, *Carya illinoensis* (Wangenh.) Koch, nuts at 22°C. (From Ref. 110.)

G. Dielectric Properties of Dairy Products

Dielectric properties of dairy products are relatively scarce. Properties of whey and skim milk powders have been measured by Rzepecka and Pereira [112]. Those of aqueous solutions of nonfat dried milk were modeled by Mudgett et al. [113, 114]. Representative dielectric properties of milk and its constituents at 2.45 GHz are shown in Table 4 from the work of Kudra et al. [115]. For butter, measurements between 30 Hz and 5 MHz were reported by Sone et al. [116]. The dielectric constant and loss factor of butter at 2.45 GHz as functions of temperature were measured by Rzepecka and Pereira [112], as shown in Fig. 15. The small positive slope in the temperature range between 0 and 30°C, which is less pronounced than the slope below 0°C, was explained by compensating effects of the free and bound water with the effect of bound water being predominant. The permittivity of bound water at 2.45 GHz has a positive temperature coefficient, while the permittivity of free water decreases with temperature. At temperatures below the freezing point, free water crystallizes, and its permittivity decreases rapidly to very small values, so the behavior appears to result from the influence of bound water. The rapid increase in the permittivity of butter above 30°C is caused by the disintegration of the emulsion. Over a smaller range of moisture content, Parkash and Armstrong [117] showed a linear increase in dielectric constant with moisture content. Seasonal variation in dielectric properties of butter [118] showed no significant changes. In butter, mixing is important, because the size of water droplets

Table 4 Dielectric Properties of Milk and Its Constituents at 2.45 GHz

Description	Fat %	Protein %	Lactose %	Moisture %	ϵ'	ϵ''
1% Milk	0.94	3.31	4.93	90.11	70.6	17.6
3.25% Milk	3.17	3.25	4.79	88.13	68.0	17.6
Water+Lactose I[a]	0	0	4.0	96.0	78.2	13.8
Water+Lactose II	0	0	7.0	93.00	77.3	14.4
Water+Lactose III	0	0	10.0	90.00	76.3	14.9
Water+Sodium Caseinate I	0	3.33	0	96.67	74.6	15.5
Water+Sodium Caseinate II	0	6.48	0	93.62	73.0	15.7
Water+Sodium Caseinate III	0	8.71	0	91.29	71.4	15.9
Lactose (solid)	0	0	100	0	1.9	0.0
Sodium Caseinate (solid)	0	100	0	0	1.6	0.0
Milk fat (solid)	100	0	0	0	2.6	0.2
Water (distilled)	0	0	0	100	78.0	13.4

[a] Level of concentration.
Source: Ref. 115.

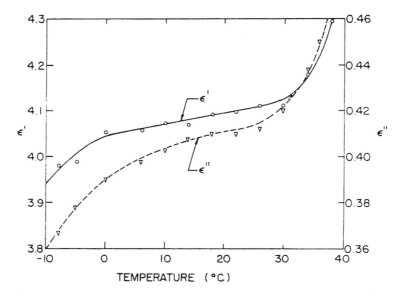

Figure 15 Permittivity of butter (16.5% moisture) at 2450 MHz as a function of temperature. (From Ref. 112.)

can influence the dielectric properties. Large water droplets behave as free water, while very small water inclusions exhibit a bound water character.

Measured dielectric properties of cheese include those for cheddar cheese [119], processed cheese [120], and mozzarella cheese [120]. Using rectangular waveguide sample holders, Green [119] measured the dielectric constant and loss factor of cheddar cheese at 20°C in five frequency bands from 750 MHz to 12.4 GHz. Figure 16 shows the dielectric constant and loss factor for a cheddar cheese which fitted the following equation with errors of less than 1% for the dielectric constant and 4% for the loss factor:

$$\epsilon = 3.22 + \frac{15.30}{1 + j\omega(1148.76)} + \frac{5.16}{1 + j\omega(51.43)} \\ + \frac{9.63}{1 + j\omega(7.94)} + \frac{0.5318}{j\omega\epsilon_0} \quad (30)$$

where the numerators of the first three fractional terms represent the dielectric increments at relaxation frequencies of 139 MHz, 3.1 GHz, and 20.1 GHz, and the corresponding relaxation times appear in the denominator for each of these terms. The static conductivity in the numerator of the last term was measured by using a conductivity cell. Variation of the dielectric properties with moisture content was

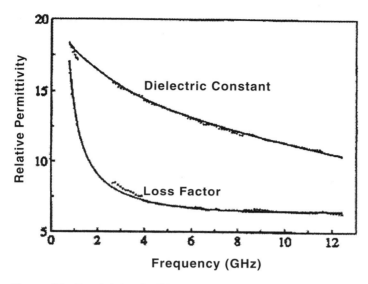

Figure 16 Permittivity of solid cheddar cheese at 20°C as a function of frequency. (From Ref. 119.)

also reported, as shown in Fig. 17 for grated and chopped samples with bulk densities of about 0.38 g/cm^3 at 3.95 GHz.

Dielectric constant and loss factor for processed cheese of four different compositions are shown in Table 5 for temperatures of 20 and 70°C [120]. At higher moisture and lower fat contents, the loss factor increases somewhat with temperature. However, in general, the dielectric constant and loss factor of processed cheese are not very temperature dependent.

During storage, dielectric properties of cheese can change significantly. For example, Fig. 18 shows that the dielectric constant and loss factor for mozzarella cheese decreased as much as 30–40% during wrapped storage [120]. Such changes are likely due to compositional changes as a result of strong proteolysis (fragmentation of protein molecules by addition of water to peptide bonds) during storage.

H. Dielectric Properties of Fish and Seafood

Dielectric properties of codfish were measured by Bengtsson et al. at frequencies from 10 to 200 MHz at temperatures from −25 to 10°C [89]. An abrupt change in the dielectric constant was noted in the region of freezing temperatures which became slightly more gradual as frequency decreased. At temperatures below the freezing point, dielectric constants increased slightly from values below 10 to val-

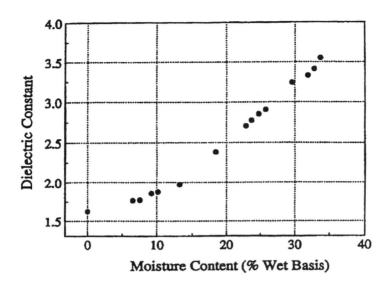

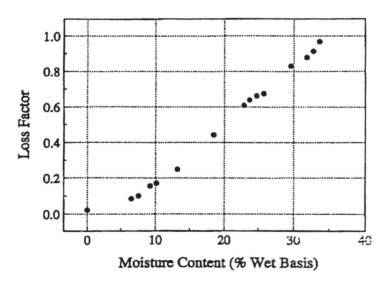

Figure 17 Permittivity of grated and chopped cheddar cheese with a bulk density of 0.39 g/cm^3 at 3.95 GHz and 20°C as a function of moisture content. (From Ref. 119.)

Table 5 Dielectric Properties of Processed Cheese at 2.45 GHz as Related to Composition

Composition		Temperature, °C			
		20°C		70°C	
% fat	% moisture	Dielectric constant	Loss factor	Dielectric constant	Loss factor
0	67	43	29	43	37
12	55	30	21	32	23
24	43	20	14	22	17
36	31	14	8	13	9

Source: Ref. 120.

ues between 10 and 20 before the abrupt increase, on thawing, to values between 60 to 90, depending on the frequency. The 10 GHz dielectric properties of white fish meal were studied by Kent [121, 122] as functions of temperature and moisture content, and some resulting data are shown in Fig. 19. Both the dielectric constant and loss factor increased nonlinearly with moisture content. They also increased with temperature in a relatively more linear fashion. Kent also examined the complex permittivity of fish meal in relation to changes in temperature, den-

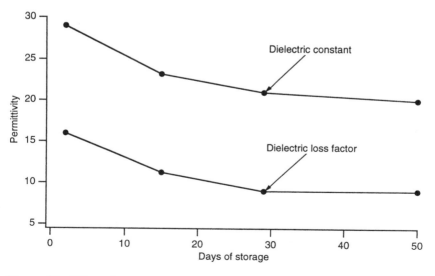

Figure 18 Effect of storage on the dielectric properties of mozzarella cheese at 2.45 GHz and 24°C. (From Ref. 120.)

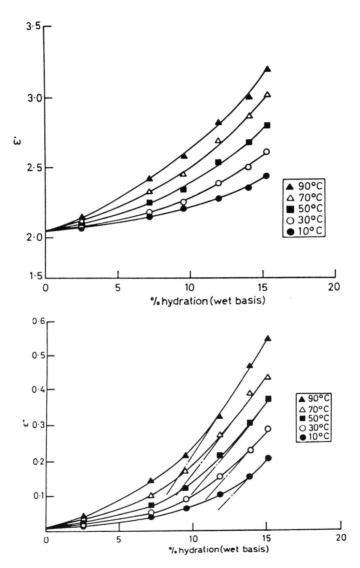

Figure 19 Permittivity of white fish meal at 10 GHz as a function of moisture content at indicated temperatures. Bulk density, 0.67 g/cm³. (From Ref. 122.)

sity, and moisture content [123], and concluded that at high moisture contents and temperatures the dependence of the permittivity on density was similar in nature to its dependence on temperature and moisture content.

Kent and Anderson [124] have reported the ability to distinguish differences in scallops from those soaked in water and also those soaked in polyphosphate solutions by a microwave spectra of the dielectric properties taken between 200 MHz and 12 GHz with data subsequently subjected to principal component analysis.

Zheng et al. [125] recently reported dielectric properties data at 915 and 2450 MHz on raw nonmarinated and marinated catfish and shrimp at temperatures from about 10 to 90°C. Measurements showed that marination increased both the dielectric constant and loss factor. The dielectric constant generally decreased with increasing temperature, whereas the loss factor increased with temperature.

I. Dielectric Properties of Meats

Bengtsson and Risman [86] reported dielectric properties data for a large variety of foods, measured at 2.8 GHz with a resonant cavity as a function of temperature, including both raw and cooked meats. Some of their results are shown in Fig. 20. In general, both the dielectric constant and loss factor increased with increasing moisture content. When temperature increased through the freezing point, sharp increases were noted in both the dielectric constant and loss factor. Both raw beef and pork and cooked beef exhibited a decrease in dielectric constant and loss factor as temperature increased above freezing to temperatures of 60°C. Brined ham, however, showed definite increases in loss factor with increasing temperature as a result of added salt. The dielectric constant also increased with temperature, which is just the opposite behavior that might be expected from the temperature dependence of salt solutions.

J. Dielectric Properties of Baked Food Products

Microwave and radio-frequency heating have also been used for baking processes [126–129]. Microwave permittivities of bread dough were measured over the frequency range from 600 MHz to 2.4 GHz by Zuercher et al. [130] as a function of water-flour composition, proofing time, and baking time. The dielectric constant and the loss factor both decreased as the water content was reduced. Both also tended to decrease with baking time. Microwave permittivities of cracker dough, starch and gluten were measured over the 0.2–20 GHz frequency range by Haynes and Locke who also studied the dielectric relaxation in this frequency range [131]. They identified two relaxation regions, one for doughs below 35% moisture associated with bound water, and another for moisture contents above that level associated with free water.

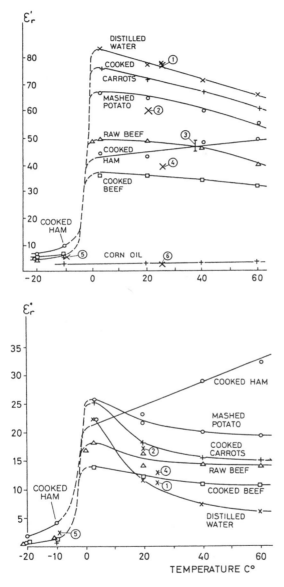

Figure 20 Temperature dependence of the dielectric properties of selected foods at 2.8 GHz. (From Ref. 86.)

The dielectric constant and loss factor of baked dough, as a function of moisture and temperature, were measured at 27 MHz by Kim et al. [132] as shown in Fig. 21. The dielectric constant gradually increased with moisture content and temperature. The loss factor showed a sudden exponential increase with moisture content beyond a certain moisture content in the range from 15% to 20%. Tem-

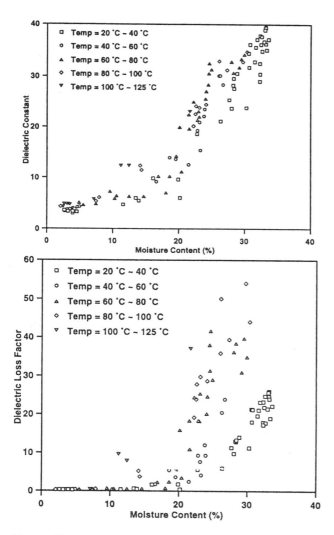

Figure 21 Permittivity of biscuit dough at different temperatures as a function of moisture content at 27 MHz. (From Ref. 130.)

perature also affected the dielectric loss factor beyond this point. The ionic conductivity and the bound water relaxation are considered the dominant loss mechanisms in the baked dough at this frequency. Temperature, in combination with water mobility, affects the ionic conductivity.

VI. DIELECTRIC PROPERTIES: DATA COMPILATIONS

Dielectric properties data for food and other materials are widely dispersed in the technical literature. Significant compilations of such data have been prepared, probably the most comprehensive effort to date being that of von Hippel [133]. Reasonably comprehensive tabulations of dielectric properties data are available for agricultural products [1], biological substances [134], and various materials for microwave processing [135]. Useful compilations of food dielectric properties data are also available in printed form [136, 137] and in electronic form [138]. A global representation showing the range of the dielectric properties of food materials at 2450 MHz is shown in Fig. 22. The various tabulations can provide guidelines, but variability of composition of food products, and other specific conditions for particular applications, often require carefully conducted measurements when dielectric properties data are critically needed.

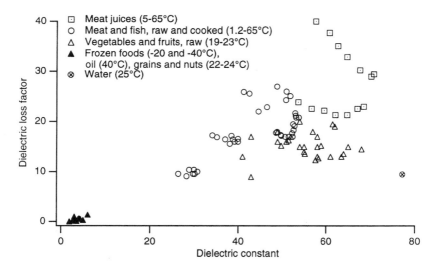

Figure 22 Global representation of the dielectric properties of food materials at about 2450 MHz for indicated temperature ranges.

ACKNOWLEDGMENTS

The authors are indebted to Andrzej W. Kraszewski for supplying the computations for aqueous NaCl solutions plotted in Fig. 3 and to Drs. Kraszewski, Mike Kent, Allen Edison, Samir Trabelsi, and Kurt Lawrence for reading the manuscript and for helpful discussions and suggestions.

LIST OF SYMBOLS

α	Attenuation constant
β	Phase constant
δ	Loss angle of the dielectric
ϵ	Relative complex permittivity
ϵ'	Dielectric constant
ϵ''	Dielectric loss factor
ϵ_0	Permittivity of free space
γ	Propagation constant
λ	Wavelength
λ_0	Free-space wavelength
π	Mathematical constant
ρ	Specific gravity or density (g/cm^3)
σ	Electrical conductivity
τ	Relaxation time
f	Frequency
t	Time
D_p	Penetration depth
E	Electric field intensity
P	Power
V	Potential difference
ω	Angular frequency

REFERENCES

1. S. O. Nelson, Electrical properties of agricultural products—a critical review, Trans. ASAE 16(2):384–400 (1973).
2. S. O. Nelson, Dielectric properties of grain and seed in the 1 to 50-mc range, Trans. ASAE 8(1):38–48 (1965).
3. S. O. Nelson, Dielectric properties of agricultural products—measurements and applications, Digest of Literature on Dielectrics (A. de Reggie, ed.), IEEE Trans. Electr. Insul. 26(5):845–869 (1991).
4. J. A. Stratton, Electromagnetic Theory, McGraw Hill, New York (1941).
5. A. von Hippel, Dielectrics and Waves, John Wiley, New York (1954).

6. P. Debye, Polar Molecules, The Chemical Catalog Co., New York (1929).
7. K. S. Cole and R. H. Cole, Dispersion and absorption in dielectrics. I. Alternating current characteristics, J. Chem. Phys. 9:341–351 (1941).
8. C.J.F. Böttcher and P. Bordewijk, Theory of Electric Polarization II, Elsevier Science Publ. Co., Amsterdam (1978).
9. J. B. Hasted, Aqueous Dielectrics, Chapman and Hall, London (1973).
10. U. Kaatze, Complex permittivity of water as a function of frequency and temperature, J. Chem. Eng. Data 34:371–374 (1989).
11. T.-S. You and S. O. Nelson, Microwave dielectric properties of rice kernels, J. Microwave Power 23(3):150–159 (1988).
12. S. O. Nelson and T.-S. You, Microwave dielectric properties of corn and wheat kernels and soybeans, Trans. ASAE 32(1):242–249 (1989).
13. S. O. Nelson, D. P. Lindroth, and R. L. Blake, Dielectric properties of selected minerals at 1 to 22 GHz, Geophys. 10:1344–1349 (1989).
14. S. O. Nelson and T.-S. You, Relationships between microwave permittivities of solid and pulverised plastics, J. Phys. D: Appl. Phys. 23:346–353 (1990).
15. S. O. Nelson and T.-S. You, Use of dielectric mixture equations for estimating permittivities of solids from data on pulverized samples, in Physical Phenomena in Granular Materials (G. D. Cody, T. H. Geballe, and P. Sheng, eds.), Materials Res. Soc. Symp. Proc., vol. 195, pp. 295–300, Materials Research Society, Pittsburgh, PA (1990).
16. S. O. Nelson, Estimation of permittivities of solids from measurements on pulverized or granular materials, chap. 6, in Dielectric Properties of Heterogeneous Materials (A. Priou, ed.), vol. 6, Progress in Electromagnetics Research, (J. A. Kong, ed.) Elsevier, New York (1992).
17. S. O. Nelson, Density dependence of the dielectric properties of particulate materials, Trans. ASAE 26(6):1823–1825 (1983).
18. W. B. Westphal, 2. Distributed Circuits, A. Permittivity, II. Dielectric measuring techniques, in Dielectric Materials and Applications, (A. von Hippel, ed.), John Wiley, New York (1954).
19. H. M. Altschuler, Dielectric constant, chap. IX, pp. 495–546, in Handbook of Microwave Measurements (M. Sucher and J. Fox, eds.), Polytechnic Press of the Polytechnic Inst. Brooklyn, New York (1963).
20. R. M. Redheffer, The measurement of dielectric constants. In Technique of Microwave Measurements (C. G. Montgomery, ed.), vol. 11, MIT Radiation Laboratory Series. (L. N. Ridenour, ed.-in-chief) Boston Technical Publishers, Inc. (1964).
21. H. E. Bussey, Measurement of RF properties of materials—a survey, Proc. IEEE 55(6):1046–1053 (1967).
22. G. Franceschetti, A complete analysis of the reflection and transmission methods for measuring the complex permittivity of materials at microwaves, Alta Frequenza 36(8):757–764 (1967).
23. A. Kraszewski, Microwave aquametry—a review, J. Microwave Power 15(4):209–220 (1980).
24. H. Fellner-Feldegg, The measurement of dielectrics in the time domain. J. Phys. Chem. 73:616–623 (1969).
25. A. M. Nicolson and G. R. Ross, Measurement of the intrinsic properties of materials by time-domain techniques. IEEE Trans. Instrum. Meas. 19(4):377–382 (1970).

26. M.J.C. van Gemert, High-frequency time-domain methods in dielectric spectroscopy, Philips Res. Repts. 28:530–572 (1973).
27. M. Kent, Time domain measurements of the dielectric properties of frozen fish, J. Microwave Power 10(1):37–48 (1975).
28. B. P. Kwok, S. O. Nelson, and E. Bahar, Time-domain measurements for determination of dielectric properties of agricultural materials, IEEE Trans. Instrum. Meas. 28(20):109–112 (1979).
29. W. L. Bellamy, S. O. Nelson, and R. G. Leffler, Development of a time-domain reflectometry system for dielectric properties measurements. Trans. ASAE, 28(4):1313–1318 (1985).
30. U. Kaatze and K. Giese, Dielectric relaxation spectroscopy of liquids: frequency domain and time domain experimental methods, J. Physics E: Sci. Intrum. 13:133–141 (1980).
31. M. N. Afsar, J. R. Birch, R. N. Clarke, and G. W. Chantry, The measurement of the properties of materials, Proc. IEEE 74(1):183–199 (1986).
32. S. Roberts and A. von Hippel, A new method for measuring dielectric constant and loss in the range of centimeter waves. J. Appl. Phys. 17(7), 610–616 (1946).
33. S. O. Nelson, Microwave dielectric properties of grain and seed, Trans. ASAE 16(5):902–905 (1973).
34. S. O. Nelson, G. E. Fanslow, and D. D. Bluhm, Frequency dependence of the dielectric properties of coal, J. Microwave Power 15(4):277–282 (1980).
35. S. O. Nelson, D. P. Lindroth, and R. L. Blake, Dielectric properties of selected and purified minerals at 1 to 22 GHz, J. Microwave Power Electromagn. Energy 24(4):213–220 (1989).
36. S. O. Nelson, RF and microwave dielectric properties of shelled, yellow-dent field corn, Trans. ASAE 22(6):1451–1457 (1979).
37. S. O. Nelson, A system for measuring dielectric properties at frequencies from 8.2 to 12.4 GHz, Trans. ASAE 15(6):1094–1098 (1972).
38. S. O. Nelson and L. E. Stetson, 250-Hz to 12-GHz dielectric properties of grain and seed, Trans. ASAE 18(4):714–718, (1975).
39. S. O. Nelson, Dielectric properties of some fresh fruits and vegetables at frequencies of 2.45 to 22 GHz, Trans ASAE 26(2):613–616 (1983).
40. S. O. Nelson, C. W. Schlaphoff, and L. E. Stetson, Computer program for calculating dielectric properties of low- or high-loss materials from short-circuited waveguide measurements. ARS-NC-4, Agric. Res. Ser., U.S. Dept. Agric. (1972).
41. S. O. Nelson, L. E. Stetson, and C. W. Schlaphoff, A general computer program for precise calculation of dielectric properties from short-circuited waveguide measurements. IEEE Trans. Instrum. Meas. 23(4), 455–460 (1974).
42. K. C. Lawrence, S. O. Nelson, and A. W. Kraszewski, Automatic system for dielectric properties measurements from 100 kHz to 1 GHz, Trans. ASAE 32(1):304–308 (1989).
43. D. G. Waters and M. E. Brodwin, Automatic material characterization at microwave frequencies, IEEE Trans. Instrum. Meas. 37(2):280–284 (1988).
44. A. Kraszewski, M. A. Stuchly, and S. S. Stuchly, ANA calibration method for measurements of dielectric properties, IEEE Trans. Intrum. Meas. 32(2):385–386 (1983).
45. A. Kraszewski and S. Nelson, Study on grain permittivity measurements in free space, J. Microwave Power Electromagn. Energy 25(4):202–210 (1990).

46. A. W. Kraszewski, S. Trabelsi, and S. O. Nelson, Wheat permittivity measurements in free space, J. Microwave Power Electromagn. Energy 31(3):135–139 (1996).
47. A. W. Kraszewski, S. Trabelsi, and S. O. Nelson, Moisture content determination in grain by measuring microwave parameters, Meas. Sci. Technol. 8:857–863 (1997).
48. S. Trabelsi, A. W. Kraszewski, and S. O. Nelson, Microwave dielectric properties of shelled, yellow-dent field corn, J. Microwave Power Electromagn. Energy 32(2):188–194 (1997).
49. S. Trabelsi, A. W. Kraszewski, and S. O. Nelson, Nondestructive microwave characterization for determining bulk density and moisture content of shelled corn, Meas. Sci. Technol. 9(9):1548–1556 (1998).
50. S. Trabelsi, A. W. Kraszewski, and S. O. Nelson, Use of horn/lens antennas for free-space microwave permittivity measurements, IEEE Antennas and Propagation Soc. Int. Symp. Digest 4:2006–2009 (1998).
51. J. P. Grant, R. N. Clarke, G. T. Symm, and N. M. Spyrou, A critical study of the open-ended coaxial line sensor technique for RF and microwave complex permittivity measurements. J. Phys. E: Sci. Instrum. 22:757–770 (1989).
52. D. V. Blackham and R. D. Pollard, An improved technique for permittivity measurements using a coaxial probe, IEEE Trans. Instrum. Meas. 46(5):1093–1099 (1997).
53. V. N. Tran, S. S. Stuchly, and A. Kraszewski, Dielectric properties of selected vegetables and fruits 0.1–10.0 GHz, J. Microwave Power 19(4):251–258 (1984).
54. S. O. Nelson, W. R. Forbus, Jr., and K. C. Lawrence, Microwave permittivities of fresh fruits and vegetables from 0.2 to 20 GHz, Trans. ASAE 37(1):183–189 (1994).
55. S. O. Nelson, W. R. Forbus, Jr., and K. C. Lawrence, Permittivities of fresh fruits and vegetables at 0.2 to 20 GHz, J. Microwave Power Electromagn. Energy 29(2):81–93 (1994).
56. S. O. Nelson, W. R. Forbus, Jr., and K. C. Lawrence, Assessment of microwave permittivity for sensing peach maturity, Trans. ASAE 38(2):579–585 (1995).
57. S. O. Nelson and P. G. Bartley, Jr., Open-ended coaxial probe permittivity measurements on pulverized materials, IEEE Inst. Meas. Technol. Conf. Proc. 1:653–657 (1997).
58. S. O. Nelson, P. G. Bartley, Jr., and K. C. Lawrence, Measuring RF and microwave permittivities of adult rice weevils, IEEE Trans. Instrum. Meas. 46(4):941–946 (1997).
59. M. A. Rzepecka, A cavity perturbation method for routine permittivity measurement. J. Microwave Power 8(1):3–11 (1973).
60. P. O. Risman and N. E. Bengtsson, Dielectric properties of foods at 3 GHz as determined by cavity perturbation technique, I. Measuring technique, J. Microwave Power 6(2):101–106 (1971).
61. A. W. Kraszewski, and S. O. Nelson, Resonant cavity perturbation—some new applications of an old measuring technique, J. Microwave Power Electromagn. Energy 31(3):178–187 (1996).
62. C. Akyel, R. G. Bosisio, R. Chahine, and T. K. Bose, Measurement of the complex permittivity of food products during microwave power heating cycles, J. Microwave Power 18(4):355–365 (1983).
63. R. G. Bosisio, C. Akyel, R. C. Labelle, and W. Wang, Computer-aided permittivity measurements in strong RF fields (Part II), IEEE Trans. Instrum. Meas. 35(4):606–611 (1986).

64. R. E. Mudgett, S. A. Goldblith, D. I. C. Wang, and W. B. Westphal, Prediction of dielectric properties in solid foods of high moisture content at ultrahigh and microwave frequencies. J. Food Proc. Preserv. 1:119–151 (1977).
65. R. E. Mudgett, Electrical properties of foods, chap. 8 in Engineering Properties of Foods (M. A. Rao and S. S. H. Rizvi, eds.), Marcel Dekker, New York, 1995, pp. 389–455.
66. R. K. Callay, M. Newborough, D. Probert and P. S. Calay, Predictive equations for the dielectric properties of foods, Int. J. Food Sci. Technol. 29:699–713 (1995).
67. E. Sun, A. K. Datta, and S. Lobo, Composition-based prediction of dielectric properties of foods. J. Microwave Power Electromagn. Energy 30(4):205–212 (1995).
68. B. D. Roebuck and S. A. Goldblith, Dielectric properties of carbohydrate-water mixtures at microwave frequencies. J. Food Sci. 37:199–204 (1972).
69. A. Kraszewski, A model for the dielectric properties of wheat at 9.4 GHz, J. Microwave Power 13(4):293–296 (1978).
70. S. D. Ptitsyn, Y. P. Sekanov, and M. Y. Batalin, Modeling the dielectric properties of grain (Russian), Mekhanizatsiya i Elektrifikatsiya Sel'skogo Khozyaistva 12:47–49 (1982).
71. S. O. Nelson, Moisture, frequency, and density dependence of the dielectric constant of shelled, yellow-dent field corn, Trans. ASAE 27(5):1573–1578, 1585 (1984).
72. S. O. Nelson, A mathematical model for estimating the dielectric constant of hard red winter wheat, Trans. ASAE 28(1):234–238 (1985).
73. S. O. Nelson, Mathematical models for the dielectric constants of spring barley and oats, Trans. ASAE 29(2):607–610, 615 (1986).
74. S. O. Nelson, Models for estimating the dielectric constant of winter barley, International Agrophysics 2(3):189–200 (1986).
75. S. O. Nelson, Models for the dielectric constants of cereal grains and soybeans, J. Microwave Power 22(1):35–39 (1987).
76. S. O. Nelson, A model for estimating the dielectric constant of soybeans, Trans. ASAE 28(6):2047–2050 (1985).
77. K. C. Lawrence, S. O. Nelson, and A. W. Kraszewski, Temperature-dependent model for the dielectric constant of soft red winter wheat, Trans. ASAE 34(5):2091–2093 (1991).
78. A. Kraszewski and S. O. Nelson, Composite model of the complex permittivity of cereal grain, J. Agric. Eng. Res. 43:211–219 (1989).
79. J. B. Hasted, D. M. Ritson, and C. H. Collie, Dielectric properties of aqueous ionic solutions. Parts I and II. J. Chemical Physics 16:1–21 (1948).
80. A. Stogryn, Equations for calculating the dielectric constant of saline water, IEEE Trans. Microwave Theory Techn. 19:733–736 (1971).
81. E. C. To, R. E. Mudgett, D. I. C. Wang, and S. A. Goldblith, Dielectric properties of food materials. J. Microwave Power 9:303–315 (1974).
82. USDA, Composition of foods: Raw, processed, prepared. Agricultural Handbook No. 8, Agricultural Research Service, United States Dept. Agric. (1976).
83. R. E. Mudgett, S. A. Goldblith, D. I. C. Wang, and W. B. Westphal, Dielectric behavior of a semi-solid food at low, intermediate and high moisture content, J. Microwave Power 15(1):27–36 (1980).
84. V. N. Tran and S. S. Stuchly, Dielectric properties of beef, beef liver, chicken and

salmon at frequencies from 100 to 2,500 MHz. J. Microwave Power 22:29–33 (1987).
85. G. P. deLoor, and F. W. Meijboom, The dielectric constant of foods and other materials with high water contents at microwave frequencies, J. Food Technol. 1:313–322 (1966).
86. N. E. Bengtsson and P. O. Risman, Dielectric properties of foods at 3 GHz as determined by a cavity perturbation technique, II. Measurements on food materials, J. Microwave Power 6(2):107–123 (1971).
87. R. E. Mudgett, Dielectric properties of food, in Microwaves in the Food Processing Industry (R. V. Decareau, ed.) Academic Press, New York (1985).
88. T. Ohlsson, M. Henriques, and N. E. Bengtsson, Dielectric properties of model meat emulsions at 900 and 2800 MHz in relation to their composition. J. Food Sci. 39:1153–1156 (1974).
89. N. E. Bengtsson, J. Melin, K. Remi, and S. Soderlind, Measurements of the dielectric properties of frozen and defrosted meat and fish in the frequency range 10–200 MHz, J. Sci. Food Agric. 14:592–604 (1963).
90. M. Kent and A. C. Jason, Dielectric properties of foods in relation to interactions between water and the substrate, in Water Relations of Foods (R. B. Duckworth, ed.), Academic Press, London, 1975, pp. 211–231.
91. M. Chamchong, Microwave thawing of foods: effect of power levels, dielectric properties, and sample geometry. PhD dissertation, Cornell University, Ithaca, NY (1997).
92. L. A. Miller, J. Gordon, and E. A. Davis, Dielectric and thermal transition properties of chemically modified starches during heating, Cereal Chem. 68(5):441–448 (1991).
93. A. Li and S. A. Barringer, Effect of salt on the dielectric properties of ham at sterilization temperatures, IFT Annual Meeting Abstracts, 55–5 (1997).
94. A. Prakash, S. O. Nelson, M. E. Mangino, and P. M. T. Hansen, Variation of microwave dielectric properties of hydrocolloids with moisture content, temperature and stoichiometric charge, Food Hydrocolloids 6(3):315–322 (1992).
95. S. Nelson, A. Prakash, and K. Lawrence, Moisture and temperature dependence of the permittivities of some hydrocolloids at 2.45 GHz, J. Microwave Power Electromagn. Energy 26(3):178–185 (1991).
96. S. O. Nelson and L. E. Stetson, Frequency and moisture dependence of the dielectric properties of hard red winter wheat, J. Agric. Eng. Res. 21:181–192 (1976).
97. S. O. Nelson, Frequency and moisture dependence of the dielectric properties of high-moisture corn, J. Microwave Power 13(2):213–218 (1978).
98. S. O. Nelson, Review of factors influencing the dielectric properties of cereal grains, Cereal Chem. 58(8):487–492 (1981).
99. S. O. Nelson, Factors affecting the dielectric properties of grain, Trans. ASAE 25(4):1045–1049 (1982).
100. ASAE D293.2, Dielectric properties of grain and seed, ASAE Standards 1999, 46th ed., pp. 553–562 (1999).
101. K. C. Lawrence, S. O. Nelson, and A. W. Kraszewski, Temperature dependence of the dielectric properties of wheat, Trans. ASAE 33(2):535–540 (1990).
102. K. C. Lawrence, S. O. Nelson, and A. W. Kraszewski, Temperature-dependent

model for the dielectric constant of soft red winter wheat, Trans. ASAE 34(5):2091–2093 (1991).
103. S. O. Nelson, Microwave dielectric properties of insects and grain kernels, J. Microwave Power 11(4):299–303 (1976).
104. A. Kraszewski and S. O. Nelson, Composite model of the complex permittivity of cereal grain, J. Agric. Eng. Res. 43:211–219 (1989).
105. S. H. Noh and S. O. Nelson, Dielectric properties of rice at frequencies from 50 Hz to 12 GHz, Trans. ASAE 32(3):991–998 (1989).
106. S. O. Nelson and S. H. Noh, Mathematical models for the dielectric constants of rice, Trans. ASAE 35(5):1533–1536 (1992).
107. S. O. Nelson, Microwave dielectric properties of fresh fruits and vegetables, Trans. ASAE 23(5):1314–1317 (1980).
108. R. Seaman and J. Seals, Fruit pulp and skin dielectric properties for 150 MHz to 6400 MHz. J. Microwave Power and Electromagn. Energy 26(2):72–81 (1991).
109. J. D. Whitney and J. G. Porterfield, Dielectric properties of peanuts, Trans. ASAE 10(1):38–42 (1967).
110. S. O. Nelson, Frequency and moisture dependence of the dielectric properties of chopped pecans, Trans. ASAE 24(6):1573–1576 (1981).
111. K. C. Lawrence, S. O. Nelson, and A. W. Kraszewski, Temperature dependence of the dielectric properties of pecans, Trans. ASAE 35(1):251–255 (1992).
112. M. A. Rzepecka and R. R. Pereira, Permittivity of some dairy products at 2450 MHz. J. Microwave Power 9(4):277–288 (1974).
113. R. E. Mudgett, A. C. Smith, D.I.C. Wang, and S. A. Goldblith, Prediction of relative dielectric loss factor in aqueous solutions of non-fat dried milk through chemical simulation. J. Food Sci. 26:915–918 (1971).
114. R. E. Mudgett, A. C. Smith, D.I.C. Wang, and S. A. Goldblith, Prediction of dielectric properties in nonfat milk at frequencies and temperatures of interest in microwave processing, J. Food Sci. 39:52–54 (1974).
115. T. Kudra, V. Raghavan, C. Akyel, R. Bosisio, and F. van de Voort, Electromagnetic properties of milk and its constituents at 2.45 GHz, J. Microwave Power Electromagn. Energy 27:199–204 (1992).
116. T. Sone, S. Taneya, and M. Handa, Dielectric properties of butter and their application for measuring moisture content during continuous processing, 18th Int. Dairy Congr. IE:221 (Food Sci. Technol. Abstr. 12P1690 2(12) (1970).
117. S. Parkash, and J. G. Armstrong, Measurement of the dielectric constant of butter, Dairy Industries, 35(10):688–689 (1970).
118. J. F. O'Connor and E. C. Synnott, Seasonal variation in dielectric properties of butter at 15 MHz and 4°C. J. Food Sci. Technol. 6:49–59 (1982).
119. A. D. Green, Measurement of the dielectric properties of cheddar cheese. J. Microwave Power and Electromagn. Energy 32(1):16–27 (1997).
120. A. K. Datta, Cornell University, Ithaca, NY, unpublished data.
121. M. Kent, Complex permittivity of white fish meal in the microwave region as a function of temperature and moisture content, J. Phys. D: Appl. Phys. 3:1275–1283 (1970).
122. M. Kent, Microwave dielectric properties of fishmeal, J. Microwave Power 7(2):109–116 (1972).

123. M. Kent, Complex permittivity of fish meal: a general discussion of temperature, density and moisture dependence, J. Microwave Power 12(4)341–345 (1977).
124. M. Kent and D. Anderson, Dielectric studies of added water in poultry meat and scallops. J. Food Eng. 28:239–259 (1996).
125. M. Zheng, Y. W. Huang, S. O. Nelson, P. G. Bartley, and K. W. Gates, Dielectric properties and thermal conductivity of marinated shrimp and channel catfish. J. Food Sci. 63(4):668–672 (1998).
126. P. L. Jones, Radio frequency processing in Europe, J. Microwave Power Electromagn. Energy 22(3):143–153 (1987).
127. R. V. Decareau, Microwaves in the Food Processing Industry, Academic Press, Orlando, FL, 1985.
128. R. F. Schiffman, Microwave technology in baking, in Advances in Baking Technology (B. S. Kamel and C. E. Stauffer, eds.), Blackie Academic & Professional, Glasgow, Scotland, pp. 292–315, 1993.
129. D. Z. Ovadia and C. E. Walker, Microwave baking of bread, J. Microwave Power Electromagn. Energy 30(1):81–89 (1995).
130. J. Zuercher, L. Hoppie, R. Lade, S. Srinivasan, and D. Misra, Measurement of the complex permittivity of bread dough by an open-ended coaxial line method at ultrahigh frequencies, J. Microwave Power Electromagn. Energy 25(3):161–167 (1990).
131. L. C. Haynes and J. P. Locke, Microwave permittivities of cracker dough, starch and gluten, J. Microwave Power Electromagn. Energy 30(2):124–131 (1995).
132. Y.-R. Kim, M. T. Morgan, M. R. Okos, and R. L. Stroshine, Measurement and prediction of dielectric properties of biscuit dough at 27 MHz, J. Microwave Power Electromagn. Energy 33(4) (1998).
133. A. R. von Hippel, Dielectric Properties and Applications, The Technol. Press of M.I.T. and John Wiley, New York (1954).
134. M. A. Stuchly and S. S. Stuchly, Dielectric properties of biological substances—tabulated, J. Microwave Power 15(1):19–26 (1980).
135. W. R. Tinga and S. O. Nelson, Dielectric properties of materials for microwave processing—tabulated. J. Microwave Power 8(1):23–65 (1973).
136. M. Kent, Electrical and Dielectric Properties of Food Materials, Science and Technology Publishers, Essex, England (1987).
137. A. K. Datta, E. Sun, and A. Solis, Food dielectric property data and their composition-based prediction, chap. 9 in Engineering Properties of Foods (M. A. Rao and S. S. H. Rizvi, eds.), Marcel Dekker, New York, 1995, pp. 457–494.
138. Construction of a Database of Physical Properties of Foods. EU Project ERB FAIR CT96-1063 [WWW document], URL:http//www.nel.uk/fooddb/ (accessed Dec. 7, 1999).

4
Fundamentals of Heat and Moisture Transport for Microwaveable Food Product and Process Development

Ashim K. Datta
Cornell University
Ithaca, New York

I. INTRODUCTION

From bread baking to food sterilization, most differences between the performance of a microwave product or process and its conventional counterpart can be attributed to the time-temperature history and its spatial variation in food. A product's adequate flavor or excessive moisture can be attributed to the same cause. Thus, time-temperature history and its spatial variation are critically important to microwave product and process development. A multitude of food and oven factors determine the rate of microwave absorption and its spatial variation, as discussed in Chapter 2. Additional food properties, such as density, specific heat, and thermal conductivity, as well as conditions surrounding the food (boundary conditions), determine the time-temperature history and its spatial variation.

Fundamentally, microwave heating is volumetric and nonuniform. Its characteristics compare to those of conventional heating as follows:

1. It's quick. The rates of heating are much higher than in conventional heating.
2. It's generally more uniform than conventional heating.
3. It's selective; moist areas heat more than the dry areas. Such selectivity is absent in conventional heating.
4. Unlike conventional heating, significant internal evaporation inside the

microwave-heated material leads to additional mechanisms of moisture transport that enhance moisture loss during heating.
5. It can be turned on or off instantly, unlike conventional heating.

The characteristics of microwave heating can be beneficial or detrimental, depending on the application. For example, the selective heating of microwaves is extremely useful when the wet interior areas of a material need to be heated; they heat moist areas faster than the drier areas, driving out the moisture. On the other hand, when a food with a crispy surface is reheated, the wet interior areas are heated more by the microwaves and too much moisture is transported to the surface, making the food soggy. The volumetric and fundamentally nonuniform heating of microwaves is dependent on many food and oven characteristics. This relationship is the subject of this chapter.

General concepts involving heat and mass transfer during microwave processing are covered in extensive review papers [1, 2]. Discussion of heat and mass transfer during microwave processing is also included in comprehensive books [3–5] and articles (e.g., Refs. 6, 7). Microwaves are used in many applications besides food processing. Each has some unique aspects, and the reader can obtain important insight by reviewing literature covering these applications (e.g., ceramics [8], polymers [9], hazardous waste [10], and biomedical uses [11]).

This chapter is restricted to heat and mass transfer during food processing. Also, the chapter will explain the physics of the heat and mass transfer processes using results from experimental data and mathematical models. Details of the measurement techniques for heat and mass transfer are discussed in Chapter 8, and details of modeling can be seen in Ref. 12. A final comment is that microwave heating is inherently quite complex. What follows are simplified heat and mass transfer descriptions of commonly observed physical phenomena in food processing.

II. NATURE OF MICROWAVE HEATING

A. Volumetric Nature of Microwave Heating

As discussed in the chapter on dielectric properties, the water and the ions are the primary food components that absorb the microwaves, leading to volumetric heating. The volumetric heating rate, or the power deposition of the microwaves Q is related to the electric field strength, E by [see Eq. (4) in Chapter 1]

$$Q = 2\pi f \epsilon_0 \epsilon'' E_{\text{rms}}^2 \qquad (1)$$

where f is the frequency of microwaves, ϵ_0 is the permittivity of free space and ϵ'' is the dielectric loss of the material, and E_{rms} is the root-mean-square average value of electric field at a location. The electric field E_{rms} can vary significantly with location and is discussed later. Representative values of heat generation in microwave heating are shown in Table 1, along with those for ohmic (electroconductive) heating.

B. Nonuniformities of Microwave Heating

1. Nonuniformities Due to Electromagnetic Field Patterns

Microwave heating is fundamentally nonuniform. As the energy penetrates a lossy material, it is absorbed and less of it remains to penetrate further. Thus, energy absorption is nonuniform. The shape, size, and properties of the load, as well as the design of the microwave oven, complicate this scenario for energy absorption, but nonuniformity remains the rule. Nonuniformities in the electromagnetic patterns inside a domestic microwave oven and other resonant cavities are discussed in Chapters 1 and 2 on electromagnetics, and the reader is referred to those chapters for further detail.

2. Nonuniformities in Heating Due to Spatial and Time Variation in Food Properties

Compositional and temperature variation in the food, present initially or developed during the heating, can contribute to spatially nonuniform energy absorption. For example, in microwave thawing, the outside layer can thaw first since the outside typically absorbs more energy. Once the material thaws, its energy absorption (dielectric loss) increases tremendously, as shown in Fig. 5 in Chapter 3. This thawed outer layer essentially shields much of the microwave energy, and heating rates inside drop significantly. Thawing is discussed further in Sec. VIII.A. Nonuniformity of heating due to temperature variation during heating is also evident in salty foods where the microwave absorption (a dielectric property) in-

Table 1 Comparison of Heating Rates in Microwave and Ohmic Heating[a]

Parameter	Microwave heating	Ohmic heating
Dielectric loss	2–30	
Conductivity (S/m)	0.25–4	0.005–1.2
Heat generation for $E = 20$V/cm (W/cm^3)	1–16	0.02–5
Rate of temperature rise (°C/s)	0.25–4	0.005–1.2

[a] To contrast the rate of temperature rise during initial times near the surface of a can during conventional heating in 0.2°C/s.

creases with temperature. In a drying process, typically more moisture is lost from the outer regions. The microwave absorption in the drier outer regions will reduce, and the microwaves will be preferentially absorbed in the wetter regions, leading to more efficient evaporation of the moisture (faster drying)—a very desirable situation.

3. Quantitative Descriptions of Nonuniformity

Although the phrase "nonuniform microwave heating" is common and is used in this article, it is important to note that there is no universal quantitative definition of such nonuniformity. Even the variable of interest is not universal—sometimes "nonuniformity" refers to nonuniformity of electric fields, while at other times it refers to nonuniformity in temperature. Several indices of temperature nonuniformity are possible. The simplest index of nonuniformity, especially when using measured data, can be defined in terms of temperature range, i.e.,

$$T_{nu,range} = T_{max} - T_{min} \quad (2)$$

where $T_{nu,range}$ is the nonuniformity in temperature. Another possibility is to use the volume-weighted standard deviation of temperatures, defined by

$$T_{nu,\sigma} = \sqrt{\int_{vol} \left[T - \left(\int_{vol} T \, dv_f \right)^2 \right]^2 dv_f} \quad (3)$$

where $T_{nu,\sigma}$ is the standard deviation, v_f is the volume fraction of material at temperature T, and vol is the total volume. When the heating pattern mainly consists of a very small hot region near the surface, lower temperature values are predominant, making the standard deviation a poor indicator of nonuniformity.

Other indices of nonuniformity can be considered. One possibility is to define percentile values. For example, the difference between the average temperatures of the hottest 5% of the food and the coldest 5% is defined as

$$T_{nu,p} = \frac{\int_{hottest\ 5\%} T \, dv - \int_{coldest\ 5\%} T \, dv}{5\% \text{ of total volume}} \quad (4)$$

Such a definition would reduce errors due to the extreme values in Eq. (2), but it is more sensitive than the standard deviation [Eq. (3)] to changes in heating conditions.

Note that although nonuniformity can only increase with time in Eqs. (2) to (4), if a moist food is heated long enough, all temperatures will tend to reach the boiling point of water, and hence nonuniformity will eventually begin to drop. When significant evaporation is present, nonuniformity, as defined by Eqs. (2) to (4), suffers from this drawback.

III. LAMBERT'S LAW AS A SIMPLIFIED DESCRIPTION OF MICROWAVE POWER ABSORPTION

As discussed in Chapters 1 and 2, the spatial distribution of energy deposition in a food is quite complex for heating in a microwave oven. To describe accurately the heat and mass transfer that occurs, we would need to obtain the profile of energy deposition in the food from detailed electromagnetics and subsequently combine it with heat transfer. As the temperature rises, the dielectric properties change and the electromagnetic (energy deposition) patterns can change significantly. Thus, the electromagnetics and heat transfer are also coupled, as shown schematically in Fig. 1. As was described in Chapters 1 and 2, the electromagnetics is already quite complex and the addition of heat (and mass) transfer would further increase them (see also Sec. X). For a simplified description of heat and mass transfer in foods, accurate only for some restricted situations, it is instructive to study situations in which, far from the source and over a small area, microwaves can be approximated as plane waves.

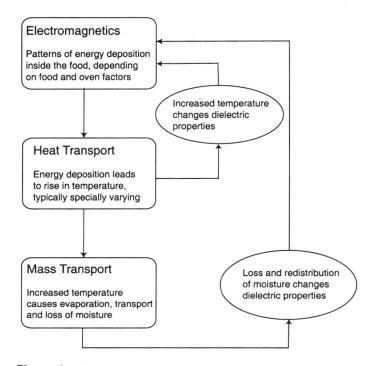

Figure 1 A conceptual schematic showing how electromagnetics, heat transfer, and mass transfer are coupled in microwave heating.

When plane waves are incident on a thick material (food), the energy level drops exponentially, as shown by

$$Q = Q_0 \exp\left(-\frac{x}{\delta_p}\right) \tag{5}$$

where Q_0 is the rate of heat generation at the surface and x is the distance into the material from the surface. The relationship given by Eq. (5) is shown graphically in Fig. 2. The quantity δ_p is called the power penetration depth and is related to the dielectric properties by the equation (see Chapter 1, Sec. II.D)

$$\delta_p = \frac{c}{2\pi f \sqrt{2\epsilon'} \left(\sqrt{1 + \left(\frac{\epsilon''}{\epsilon'}\right)^2} - 1\right)^{1/2}} \tag{6}$$

The quantity δ_p is interpreted as the distance over which 63% of the incident microwave energy is lost through absorption in the material. (Note that the power penetration depth δ_p is *not* the skin depth, which is the distance over which the electric field decays to 63% of its incident value, as explained in Chapter 1.) Representative ranges of penetration depth for food materials is provided in Fig. 3. As was noted in the previous section, even this simplified description (Fig. 2) shows that microwave heating is inherently nonuniform, depositing varying amounts of energy spatially. With the exception of a few studies (see Sec. X), most tempera-

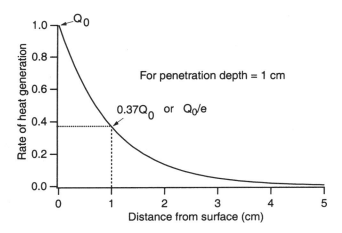

Figure 2 Graphical description of Lambert's law showing how energy of microwaves would decay inside a thick material when plane waves are incident on its surface. Note that about 63% of the energy is lost in the distance equal to the penetration depth from the surface.

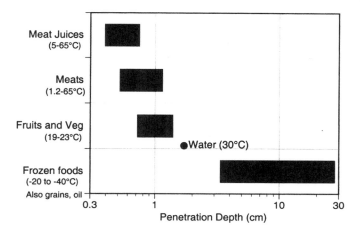

Figure 3 Typical ranges of power penetration depth in various groups of food materials. Dielectric properties data used in this figure correspond to those provided in Fig. 22 of Chapter 3.

ture modeling of microwave heating of foods has used the exponential decay description shown in Eq. (5). It is important to note that Eq. (5) serves as a qualitative description of microwave energy absorption in a microwave oven and is valid only for some restricted situations (see Chapter 1 on electromagnetics for more details). In fairly lossy foods that have a large amount of water with salt, a detailed electromagnetic description shows a rapid drop in microwave absorption from the surface into the material, somewhat similar to an exponential decay [13]. When the material is not so lossy, Eq. (5) often cannot describe the energy absorption accurately. Some researchers [14, 15] have used Eq. (5) to "predict" focusing at the center of curved geometries such as cylinders. Such a formulation is clearly qualitative and is not based on fundamental electromagnetic considerations. See also discussions in Sec. XII.

IV. HEAT TRANSPORT IN MICROWAVE HEATING: GENERAL DESCRIPTION

A. Governing Equation for Heat Transport in Microwave Heating

The governing energy conservation equation for microwave heating uses the rate of heat generation, Q from Eq. (1), as a source term, as shown in the following one-dimensional equation:

$$\underbrace{\rho c_p \frac{\partial T}{\partial t}}_{\substack{\text{rate of} \\ \text{energy} \\ \text{accumulation}}} + \underbrace{\rho c_p u \frac{\partial T}{\partial x}}_{\substack{\text{convective} \\ \text{energy flow}}} = \underbrace{k \frac{\partial^2 T}{\partial x^2}}_{\substack{\text{diffusive} \\ \text{energy}}} + \underbrace{Q}_{\substack{\text{microwave} \\ \text{heat} \\ \text{generation}}} - \underbrace{h_{fg} \dot{I}}_{\substack{\text{energy} \\ \text{used in} \\ \text{internal} \\ \text{evaporation}}} \tag{7}$$

Here T is the temperature at a position x and time t. Other symbols are explained in the List of Symbols. For the purpose of simplifying and emphasizing the microwave heat generation aspect, thermal properties such as the thermal conductivity k are treated as constant in the above equation. Note that although Eq. (7) is for a transient heating process, Q used here from Eq. (1) is based on a time average value of the electric field E_{rms}. Use of a time average value is justified since time variations in the electric field are extremely fast (of the order of 10^{-10} s) compared with the characteristic time scales for thermal diffusion and convection.

The order of magnitude of the individual transport mechanisms in Eq. (7) depends strongly on many food and oven factors, both electromagnetic and thermal. The relative magnitudes of the transport mechanisms can also change during heating as the temperature and moisture in the food changes. For example, at the onset of heating, thermal gradients within the food may be negligible, and both convection and diffusion terms are small. Diffusion becomes important as temperature gradients develop in the food. The temperature profile depends on internal diffusion, surface heat exchange, and the spatial variation of Q. Convection becomes important if the food is liquid. For example, in a stagnant fluid, temperature gradients will set up buoyant flow patterns. Convection is also important in a porous food material containing water. As the food temperature rises, internal evaporation can become quite significant. Such internal evaporation can raise the internal pressure high enough to cause pressure driven flow of liquid water and vapor in the porous food. All of these scenarios, where some transport mechanisms become more important than others, are covered in the following sections.

Complications arise because the rate of heat generation Q, which is a function of the electric field and the dielectric properties, depends strongly on food parameters, such as composition, temperature, size and shape, and oven parameters, as explained in Chapters 1 and 2. Thus, Q is generally a function of location in the food. To complicate matters further, Q also changes with time as the food temperature increases, since the dielectric properties are temperature sensitive. The moisture loss and/or redistribution that accompanies heating also can change the heating pattern significantly.

B. Boundary (Surface) Conditions During Microwave Heating

The surface of microwave-heated food typically exchanges heat by either natural or forced convection. The heat transfer coefficient at the surface will be discussed in detail later. When a wet material is heated, evaporative cooling at the surface can also have a strong effect on the temperature profile, as illustrated in Fig. 4, where surface evaporative cooling lead to significantly lower temperatures and a microbiological safety problem. Since the temperatures typically do not reach high values (more than 100°C, the boiling point of water), radiative heat loss from the food surface to the oven wall may not be significant. However, if susceptors are used, radiative heat gain can become the dominant mechanism at the surface. Surface conditions can be expressed in terms of a generalized boundary condition, as follows:

$$-k\frac{\partial T}{\partial n} = \underbrace{h(T - T_\infty)}_{\substack{\text{convective} \\ \text{heat gain or loss}}} + \underbrace{m_w h_{fg}}_{\substack{\text{evaporative} \\ \text{heat loss}}} + \underbrace{\sigma_{rad}\epsilon_{rad}T^4 - \alpha q_{rad}}_{\substack{\text{radiative} \\ \text{heat gain or loss}}} \qquad (8)$$

where T is the food surface temperature, n represents the normal direction to the surface, h is the surface convective heat transfer coefficient, m_w and h_{fg} are the rate of evaporation and the latent heat of vaporization of the evaporated liquid, respectively, σ_{rad} is the Stefan-Boltzmann constant, ϵ_{rad} is the radiative surface emissivity of the load, α_{rad} is the surface absorptivity of the load, and q_{rad} is the incident radiant heat flow rate per unit surface area.

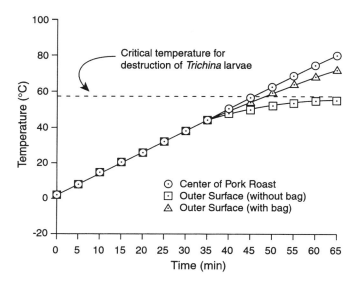

Figure 4 Surface evaporation can lead to microbial safety problems. (From Ref. 96.)

1. Heat and Mass Transfer Coefficients Inside a Domestic Microwave Oven

Detailed measurements of air flow or heat transfer coefficients in microwave ovens are hard to achieve. Forced air cooling is used to cool the magnetron that generates the microwaves. Although the magnetron sits outside the microwave cavity, some of the air passing the magnetron is drawn from the inside of the microwave oven through small holes in the oven walls (see also Chapter 7), resulting in low-velocity air flow inside the oven. This airflow is primarily to reduce the condensation of water. The heat transfer coefficient based on these very small velocities approaches that of natural convection; a heat transfer coefficient of 2.6 W/m²K was used in Ref. 6. Even in a microwave tunnel pasteurizer, where food slowly moves in a tunnel, air is considered relatively still; a heat transfer coefficient of 5–10 W/m²K was used in Ref. 17.

C. Thermal Properties

The important thermal properties of a food material are thermal conductivity, density, and specific heat. For thermal conductivity, it is useful to think of a food material as a composite of its ingredients: air, water, and ice, each having a distinct thermal conductivity value. Figure 5 shows the approximate magnitudes of thermal conductivity of different food materials in terms of the air, water, and ice contained within them. As shown in Table 2, the smallest values of thermal conductivity arise in the freeze-drying of food which leads to a very porous structure of the solid, and thus the presence of a significant amount of air. The freeze-drying process is also conducted below atmospheric pressure, which reduces the thermal conductivity of air. The highest thermal conductivities are for frozen foods, because of the high thermal conductivity of ice. The product of density and specific heat [ρc_p in Eq. (7)] is known as volumetric heat capacity. This quantity measures a material's ability to store thermal energy. Representative specific heat data for

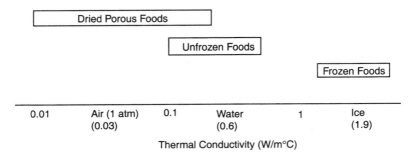

Figure 5 Ranges of thermal conductivities of various foods.

Table 2 Typical Thermal Conductivities of Food Materials (Representative Values)

Material	Thermal conductivity, W/mK
High moisture, frozen	
Apple[a]	1.669
Beef, lean	1.42
Strawberries, tightly packed	1.1
High moisture, unfrozen	
Apple	0.418
Beef, lean	0.506
Egg white	0.558
Low moisture, non-porous	
Butter	0.197
Low moisture, porous	
Apple, dried	0.219
Apple, freeze-dried	0.0405
(at pressure 0.2880×10^4 Pa)	
Beef, freeze-dried	0.0652
(at pressure of 1 atm)	
Egg albumin gel, freeze-dried	0.0393
(at pressure of 1 atm)	

[a] From Ref. 18. The rest of the data are from 1998 ASHRAE Refrigeration Handbook.

foods are shown in Table 3 and that for density are shown in Table 4. A wealth of information, including tabulated data about thermal conductivity, density, and specific heat, is provided in a number of references [18–20].

V. HEAT TRANSPORT IN MICROWAVE HEATING OF SOLIDS

This section will provide some simple descriptions of heat transfer in microwave heating valid only for restricted situations. More realistic description of heating in a cavity is provided in Sec. X. The heating of solid foods will always involve a certain amount of evaporation both in the interior of the food and at the surface. When evaporation inside the food can be disregarded, the heating process can be described relatively simply. Thus, for the rest of this section, convection and evaporation terms are dropped from Eq. (7).

Table 3 Typical Values for Specific Heat of Foods

Food	Specific heat, kJ/kgK
Air	1.01
Apple (80% water)	3.87
Fat, beef (51% water)	2.89
Beef, lean (72% water)	3.43
Bone	2.09
Bread, white (45% water)	2.84
Flour (12% water)	1.8
Rice (12% water)	1.8
Fish, fresh (80% water)	3.6
Butter (14% water)	2.05
Vegetable oil	1.67
Carrots, boiled (92% water)	3.77
Ice	2.04
Potato, boiled (80% water)	3.64
Potato, dried (8% water)	1.92
Water	4.18

Source: Adapted from Ref. 18.

Table 4 Typical Values for Densities of Foods

Food	Density, kg/m^3
Apple	576
Apple juice (13% solids)	1060
Beef, lean	1076
Beef fat	953
Carrot	640
Cauliflower	320
Cabbage	449
Fish, cod	1100
Fish, pike perch ($-18.9°C$)	910
Flour (bulk)	449
Ham	1070
Ice (0°C)	917
Milk (cow)	1010
Water (4°C)	1000

Source: Adapted from Ref. 18.

A. Temperature Profile in a Solid When Diffusion Can Be Ignored

During initial or short periods of intense heating, thermal diffusion and surface heat losses can be minimal in comparison to volumetric microwave absorption. In such situations, the heat conduction in food is generally very small compared to the rate of volumetric heating. Power absorption in the food may be uniform or might vary spatially. The energy equation given by Eq. (7) becomes the following if the convection and diffusion terms are dropped:

$$\rho c_p \frac{\partial T}{\partial t} = Q(r) \tag{9}$$

where Q varies with location r. For a given location r, if the absorbed microwave power density Q does not vary with time, the rate of temperature rise at the location is constant, giving rise to a linear temperature rise with time. Such linear rise of temperature with time, finally reaching the boiling temperature of water, has been observed in heating of moist food (e.g., Ref. 21).

Equations (9) and (1) can be combined as

$$E_{\text{rms}} = \sqrt{\frac{\rho c_p}{2\pi f \epsilon_0 \epsilon''_{\text{eff}}} \frac{\partial T}{\partial t}} \tag{10}$$

Equation (10) can be used, for example, to estimate the electric field E_{rms} from the measured temperature rise $\partial T/\partial t$. Again, it is emphasized that Eq. (10) is valid only in specific situations and is generally not a good approximation for long periods.

B. Temperature Profiles in a Thick Slab

As discussed in Sec. II.E of Chapter 1, for a thin slab, microwaves incident on one face is reflected from the other face of the slab, setting up oscillatory and complex power absorption behavior that cannot be described by Lambert's law. When the slab thickness L_{crit} is above the value given by [22]

$$L_{\text{crit}} = 5.4\delta_p - 0.0008 \tag{11}$$

the slab can be considered thick; i.e., reflections from the distant face of the slab can be ignored and the microwave power deposition is given by Lambert's law [Eq. (5)]. In Eq. (11), both L_{crit} and δ_p are in meters. Temperature profiles that can result from a Lambert's law type of heating in a thick slab are shown in Fig. 6 over a range of heating times. These profiles are calculated from the solution to the energy equation [Eq. (7) without the convection term and with Q given by Eq. (5)]. Plane waves are considered incident on both faces of the slab with a total thickness of $2L = 0.1$ m and $\delta_p = 0.0091$ m. Details of the solution process are provided in Ref. 23.

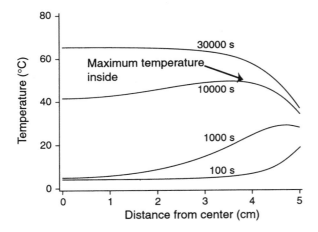

(a)

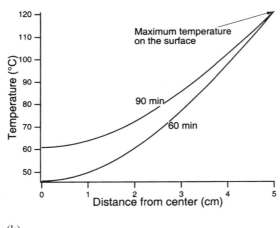

(b)

Figure 6 Developing temperature profiles in (a) a thick slab during microwave heating compared with (b) temperature profiles in a slab during conventional heating. (From Ref. 23.)

We can identify three qualitatively different temperature profiles in Fig. 6, each representing a heating time period. The concave-up profile of initial times is simply a reflection of the exponential decay of heat generation, as assumed by Eq. (5). This type of profile can exist for small durations of heating when sufficient temperature gradients have not developed for the surface cooling effect to be significant. Eventually, this profile develops into the second type, where a peak temperature develops slightly inside the food. This is the most common type of temperature profile reported from experimental measurements and theoretical calculations of microwave heating of solid foods. For extended heating, a steady state can be theoretically achieved where the peak migrates to the center of the slab. This last type of profile is unlikely to be present in foods since it is true only after a very long duration of heating, during which time the moisture and biochemical changes in the food will modify the situation. Except for the initial concave-up profile, the interior of the food being heated always has higher temperatures. The popular expression of microwaves heating "inside out" probably stems from the fact that the surface is always slightly colder due to the surrounding colder air.

VI. HEAT TRANSPORT IN MICROWAVE HEATING OF LIQUIDS

This section provides a simplified description of microwave heating of liquid using Lambert's law for energy deposition. The results are valid only for restricted situations, as discussed at appropriate places.

A. Unagitated Batch Heating of a Liquid in a Container

Microwaves deposit energy that varies with location. Such spatially nonuniform heating leads to buoyancy-driven recirculating flows (Fig. 7) during microwave heating of a liquid [24–26]. Note that the direction of circulation depends on electric field variations inside the container. For example, the focusing effect mentioned earlier would cause liquid to rise at the center of a cylindrical container, thus reversing the flow pattern shown in Fig. 7 [26]. The flow of liquid leads to mixing and generally some reduction of nonuniformity, as shown in radial profiles for heating of tap water in a cylindrical container (Fig. 8). Since the hot liquid rises while colder liquid sinks, the liquid becomes thermally stratified, as shown in the axial temperature profiles in Fig. 8. The extent of flow and mixing depends on thermal as well as dielectric properties. An increase in viscosity reduces the natural convection and associated mixing, making temperature profiles more nonuniform, and the core liquid becomes colder [26]. An increase in the dielectric loss also causes more energy to be deposited near the surface, making the core region colder [26]. Selective electrical shielding of the top, bottom, or sides of the container can be used to modify the flow and temperature profiles [27].

B. Continuous Heating of Liquids

Although continuous pasteurization of liquids using microwaves has been attempted [28–32], it is not used commercially to any significant extent. Successful

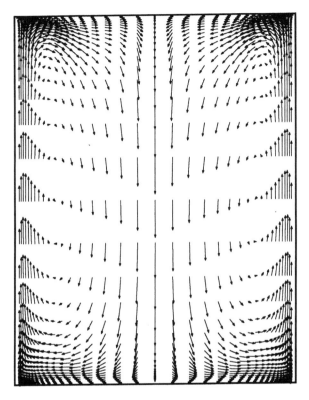

Figure 7 Flow patterns during microwave heating of water in a cylindrical container in the absence of any focusing effect, computed for power absorption following Lambert's law.

radio-frequency heating for continuous flow through a tube was reported by Ref. 33 with applications for sausage emulsions. Heat transfer during microwave heating of liquids flowing through a microwave-transparent cylindrical tube has also been studied [25]. Lambert's law approximation for the decay of microwave power in the radial direction was used to solve the governing energy equation and to produce temperature profiles. The tube surface was considered to be thermally insulated. A steady, developing laminar flow was used. It was shown that the velocity variation from the wall to the center of the tube essentially negated any uniformity advantage of the microwaves. To benefit from the more uniform heating of microwaves, the tube diameter and flow rates had to be kept very small. Otherwise, a very large temperature differential developed between the wall and the

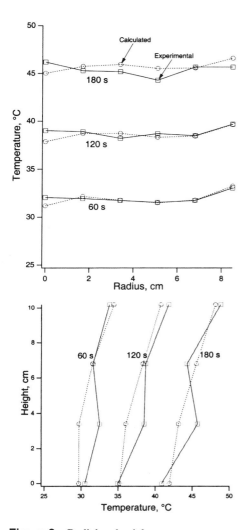

Figure 8 Radial and axial temperature patterns during microwave heating of tap water in a cylindrical polypropylene container (radius 8.5 cm) in the absence of any focussing effect. The top and bottom surfaces of the liquid were covered with aluminum foil.

center due to the combined effects near the wall of much higher heating rates and much reduced flow velocities. Clearly, microwave heating can provide significantly better uniformity of fluid temperatures than conventional heating, with the caveat that the process must be designed carefully.

C. Superheating (Inhibition of Boiling) in Microwave Heating

Nucleation during boiling is affected by pits and cavities on a surface. In surface convective heating, the walls are always hotter than the core of the liquid, and nucleation rates are optimal compared to the heating rate. In microwave heating, however, the container walls are cold compared with the core of the liquid, and the nucleation rates are never equivalent to the heating rate. Consequently, a microwave-heated fluid having the same core temperature as a conventionally heated fluid can have a much lower boiling rate. Such superheating can result in eventual violent and periodic vaporization, a known safety problem in domestic microwave heating. Superheating during microwave heating of liquids was reported in Ref. 34 and later analyzed in Ref. 35.

VII. EFFECT OF CHANGES IN TEMPERATURE AND FREQUENCY ON HEAT TRANSPORT

A. Effect of Food Temperature on Heat Transport

As was discussed using Fig. 1, the dielectric properties change with temperature and this, in turn, changes the electromagnetic field and thus couples heat transport with electromagnetics. An increased dielectric loss with an increased temperature can happen in two situations in foods—a phase change from ice to water and the presence of significant ions (salt). In both cases, the dielectric loss strongly increases with temperature. The dielectric properties during a thawing process change as shown in Fig. 5 of Chapter 3. The effect of this change in properties is discussed in Sec. VIII.A. Also, as shown in Fig. 3 of Chapter 3, ions in a salty material increase its microwave absorption as temperature increases. The coupling of heat transport with electromagnetics for this situation has been achieved and is described in Sec. X. This section also discusses under what conditions coupling is necessary to describe the temperature effect.

Thermal runaway refers to a very significant effect of food temperature change on heating. Thermal runaway is the process in which a material absorbs increasing amounts of energy as its temperature rises, leading to a very rapid increase in temperature. In non-water-containing materials, temperatures can become very high, and thermal runaway describes this situation. In water-containing food materials, temperatures reach only slightly higher than the boiling point of water; in this case thermal runaway indicates a rapid but limiting increase in temperature.

B. Effect of Microwave Frequency on the Heat Transport

The effect of change in microwave frequency on heat transfer is through the change in the electromagnetic (heat generation) pattern in the food and is appropriately discussed in Chapters 1 and 2. For simple systems where Lambert's law or exponential decay can be assumed, change in frequency from 2450 to 915 MHz can be simply an increase in penetration depth. Increased penetration is exploited in some applications to achieve a more uniform heating. Larger variations in frequency can lead to more dramatic changes in power deposition patterns, as discussed under variable frequency microwave heating in Chapter 2.

VIII. SOME UNIT OPERATIONS INVOLVING PRIMARILY HEAT TRANSPORT

Two of the unit operations involving primarily heat transport, microwave thawing and microwave sterilization, are discussed in this section. Although some moisture transport is always present in heating processes, such transport is considered small in these operations.

A. Heating Frozen Foods: Thawing and Tempering

Thawing and tempering (bringing the temperature up to a few degrees below complete thawing) are some of the most effective uses of microwave heating of food. However, thermal runaway effects from phase changes can lead to severe nonuniformities. Ideally, coupled solutions of electromagnetics and energy transport should be considered (see Fig. 27) for a food heated in a microwave cavity with changing thermal and dielectric properties. This solution has not appeared in the literature. Some coupled electromagnetic and heat transfer models have considered either one- or two-dimensional solutions of Maxwell's equations for a slab or a cylinder as opposed to those for a cavity [2, 36, 37], or assumed an exponential spatial heating profile [Eq. (5)], as in Refs. 38–40. Phase change over a range of temperatures, as is true for food materials, was included in Refs. 39, 40, 37. A power law relationship between thawing time and slab thickness, L, is reported in Refs. 36, 37:

$$t_{\text{thawing}} \propto L^n \tag{12}$$

where the exponent n is 2 for no microwave power and decreases with increasing microwave power. A value of 1.56 was reported in Ref. 37 for slabs heated by microwaves from one side. Such power law relationships have not been determined under resonant conditions (see the chapters on electromagnetics) when the inside of the slab can be heated directly at high intensity, resulting in quick thawing. For thickness values greater than a certain value (see Chapter 1 on electromagnetics), resonance can be avoided and heating rates would be higher at the surface than in the interior.

Thawing studies for heating in a cavity have been reported, but they assumed an exponential decay of energy from the surface [39, 41]. In the experimental and numerical study of Ref. 41, microwave thawing was shown to be faster, but it also could be significantly nonuniform. For the exponential decay of power, the temperatures near the surface are higher than inside, as shown in Fig. 9. At higher power levels, the outside thaws faster. Since the thawed material has a much higher dielectric loss, microwave penetration depth at the surface is significantly reduced, in effect developing a "shield." This shielding considerably slows down the reduction in thawing time at higher powers, as shown in Figs. 10 and 11. An increase in load volume increases thawing time, as expected (Fig. 12). For a given volume of load, the smallest aspect ratio (height divided by width for a rectangular load), or the flattest load, thaws the fastest (Fig. 12).

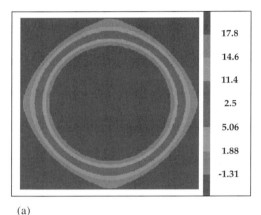

(a)

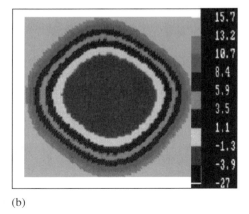

(b)

Figure 9 Computed (a) and experimental (b) temperature (°C) contours at midheight after 20 min during microwave thawing of rectangular blocks (5 × 5 × cm) of tylose with 2% salt at 10% power level. (From Ref. 97.)

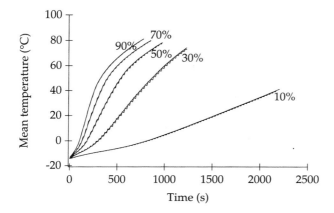

Figure 10 Computed mean temperature profiles of rectangular blocks of tylose (5 × 5 × 5 cm) with 2% salt during microwave thawing at various power levels. (From Ref. 40.)

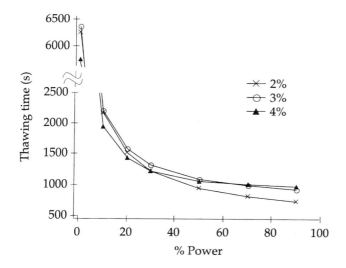

Figure 11 Computed thawing time for rectangular blocks (5 × 5 × 5 cm) of tylose (with % salt as noted), as a function of microwave oven power levels. Thawing with conventional heat (0% power) at the same surrounding temperature of 26°C and $h = 30$ W/m^2 °C is added for comparison. (From Ref. 40.)

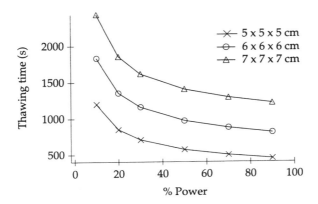

(a)

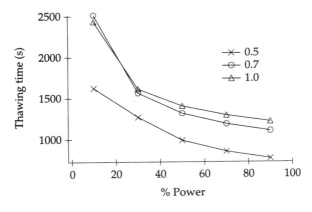

(b)

Figure 12 Computed thawing time for rectangular blocks of tylose with 2% salt for (a) three different volumes and (b) three different aspect ratios (height/width) for a volume of 7 × 7 × 7 cm. (From Ref. 98.)

B. Pasteurization and Sterilization

For a comprehensive discussion of microwave pasteurization and sterilization, see Ref. 42. Microwave heating for pasteurization and sterilization is preferred over conventional heating for the primary reason that it is *rapid* and therefore requires less time to achieve to the desired process temperature. This faster heating of microwaves is particularly true for solid and semisolid foods that depend on a slow thermal diffusion process in conventional heating. They can approach the benefits of high temperature-short time processing in conventional heating, where bacterial destruction is achieved but thermal degradation of the desired components is reduced.

To illustrate the possible advantages of microwave heating in the context of pasteurization or sterilization, first some quantities describing bacterial and nutrient destruction are defined. Both of these destruction processes are typically assumed to follow first-order reaction (see Chapter 6), given by

$$\frac{dc}{dt} = -k_T c \tag{13}$$

where c is the concentration, t is time, and k_T is the reaction rate constant at temperature T. The rate of reaction is assumed to vary with temperature according to the Arrhenius law (see Chapter 6). When temperature is varying with time during a process, Eq. (13) can be integrated to obtain

$$\ln \frac{c_i}{c} = \int_0^t k_0 e^{-E_a/RT} \, dt \tag{14}$$

where c is the final concentration after heating for time t, k_0 is frequency factor, and E_a is the activation energy for the reaction in Arrhenius law. Note that in Eq. (14), temperature T needs to be in absolute scale. To describe the extent of reaction, instead of the final concentration c, an alternate but equivalent quantity called an equivalent heating time, F_0 is used. The equivalent heating time F_0 leads to the same final concentration when temperature is held constant at a reference temperature T_0. Thus,

$$F_0 = \frac{\ln(c_i/c)}{k_{T_0}} \tag{15}$$

Using Eq. (14), Eq. (15) can be simplified as

$$F_0 = \int_0^t 10^{(T - T_0)/Z} \, dt \tag{16}$$

where $Z = 2.303 R T_R^2 / E_a$ and $T T_R \approx T_R^2$ has been used. Equation (16) is used extensively in food literature to describe the extent of bacterial destruction. This equation can also be used to describe nutrient destruction or any other reaction that follows first order. Thus, a heating process would have two F_0 values for bacterial and nutrient destructions, respectively.

Figure 13 illustrates the possible advantages of faster heating of microwaves by computing bacterial and nutrient destruction F_0 values for typical time-temperature histories of microwave and conventional heated processes. Because of the rapid rise in temperature for microwave heating, no significant destruction of nutrient or bacteria takes place during the come-up time. Once the

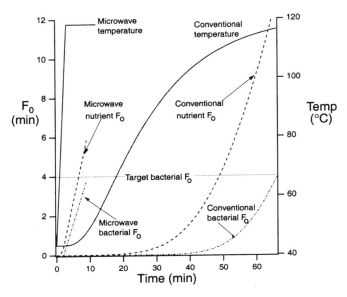

Figure 13 Typical point temperature and quality parameters (bacterial and nutrient destruction) in microwave and conventional heating, showing rapid heating in microwaves can reduce the thermal destruction of nutrients (lower nutrient F_0 value) at that location. (From Ref. 43.)

processing temperature is reached, food can be held at this temperature while the lethality rapidly accumulates. Note that the target bacterial F_0 of 4 is achieved for microwave heating in a much shorter time than conventional heating. For this target bacterial F_0 value, the nutrient F_0 value for microwave heating is considerably lower than that for conventional heating, signifying lower destruction of nutrients. This lower destruction for microwave heating is due to a combination of the reduced duration of heating and the relatively lower sensitivity of nutrient destruction to temperature.

The other significant advantage of microwave and radio-frequency processes is that they can be relatively more uniform depending on the particular heating situation [43], although this advantage is hard to predict. Under some conditions, they can have worse uniformity than conventional heating. Figure 14 explains a scenario in which microwave heating is spatially more uniform than conventional heating. The information shown in the figure is computed from mathematical models of a conventional heating process and a comparable microwave heating process for a solid for input parameters given in Ref. 43. Figure 14a shows that the temperatures reached by the two processes have an approximately similar range (as read from the horizontal axes) at the heating times shown.

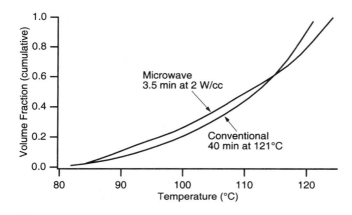

(a)

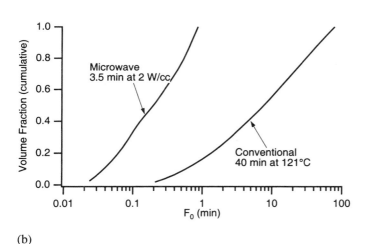

(b)

Figure 14 Volumetric distribution of (a) temperatures and (b) quality parameters (thermal times) in microwave and conventional heating. (From Ref. 29.)

The vertical axis shows the cumulative volume fractions of the food associated with various temperatures; i.e., for any temperature, the value on the curve signifies the volume fraction of food that has temperatures at or below this value. Figure 14b shows that, for the same conventional and microwave heated food, having approximately similar temperatures shown in Fig. 14a, the F_0 values (signifying time-temperature histories) have quite different ranges. The conventional heated process shows a much larger spread of F_0, which primarily signifies its tremendous nonuniformity of temperatures and long processing times, which will lead to significant overprocessing of the surface regions of the food.

Thermal effect is generally assumed to be the sole lethal mechanism in the microwave pasteurization and sterilization processes. Thus, temperature-time history at the coldest location will determine the microbiological safety of these processes.

Industrial microwave pasteurization and sterilization systems have been reported for over 30 years [4, 44–48]. Heat and mass transfer during pasteurization and sterilization have been studied by a number of authors [17, 49–51]. Although much is known about microwave pasteurization and sterilization, the present writer could locate only two commercial systems in operation worldwide as of this

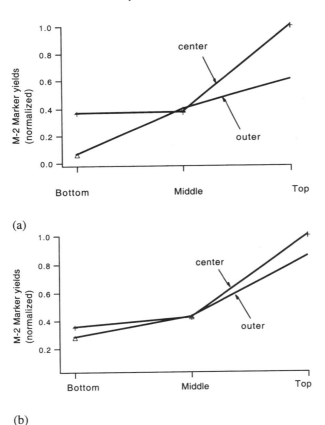

Figure 15 Time-temperature history effects (M-2 marker yields) in cylinders of ham with 0.7% salt from (a) experimental measurement and (b) numerical prediction. (From Ref. 51.)

writing [47, 48]. Implementation of microwave sterilization processes can vary significantly with various manufacturers. For example, in one implementation [47] the process design consists of heating, equilibration (in hot air), holding and cooling stages. The equilibration stage between heating and holding equilibrates the temperatures and prevents nonuniformities within the product. Another system [48] consists of microwave tunnels with several microwave launchers over the packaged product (prepared meals). Exact positioning of the package is made within the tunnel and the package receives a precalculated, spatially varying, microwave power profile optimized for the package. Sterilization requires a temperature above the boiling temperature of water, and therefore the operation must be carried out in pressurized systems. The concept of microwave sterilization has also been extended to the sterilization of food-processing waste [52].

Ensuring lethality of the cold point can be difficult in microwave heating, due to its dependence on geometry and properties of the food, although it is more uniform than conventional heating. The cold point is not as easily predicted as in conventional heating, where it is the deepest location inside a solid food. For example, for a cylindrical load heated in a pressurized vessel in a microwave oven, the spatial distribution of temperature can change qualitatively over time (Fig. 30), and for some materials the integrated time-temperature histories (F_0 values) can be higher at the center of the cylinder, compared to the surface (Fig. 15). Heating uniformity can be improved, for example, by appropriately placing multiple microwave sources [48] and holding (equilibration) in hot air [47]. Research also demonstrated improvement in uniformity due to increased penetration depth when using the 915 MHz frequency [53] or radio frequency [54].

IX. MOISTURE TRANSPORT IN MICROWAVE HEATING OF SOLID FOOD

At the commonly used frequencies for microwave heating, it is mostly the water component in food that makes heating possible. Thus, food heated in a microwave oven will have a varying amount of water evaporation (or sublimation of ice) due to increased thermal energy (absorbed from microwaves). Internal evaporation can generate significant pressure, depending on the resistance of the solid matrix of the food to the transport of liquid water or vapor. The resulting pressure-driven flow becomes an additional mechanism of internal moisture and heat transport. Pressure-driven Darcy flow is a mechanism different from diffusion; i.e., it is not concentration driven, and therefore it *cannot* be successfully described by the diffusion equation

$$\frac{\partial M}{\partial t} = \frac{\partial}{\partial x}\left(D_{\text{eff}}\frac{\partial M}{\partial x}\right) \qquad (17)$$

This equation lumps all transport mechanisms in an effective diffusivity value D_{eff}, and is generally inadequate for describing moisture transport during microwave heating, one of the main reasons being the presence of pressure-driven flow. Although difficult to justify from fundamental considerations, several literature studies [14, 55, 56] have also used Eq. (17) with a constant value of D_{eff}. Often, such studies are essentially a parameter-fitting exercise where the value of D_{eff} is obtained by using predictions that best fit the measured data. See, for example, Fig. 16 from the study of Ref. 57 which concluded that the diffusion model with constant initial and boundary conditions is unable to describe the drying kinetics.

Models of moisture transport are available that include other transport mechanisms like the pressure-driven flow and temperature effects on moisture

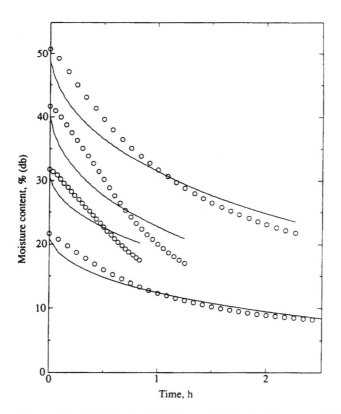

Figure 16 Inability of a diffusion model (lines) with constant initial and boundary conditions in predicting the experimental moisture changes (circles) during microwave drying of maize. (From Ref. 57.)

transport. The temperature effect on moisture transport is studied in [58–60] using a mathematical formulation known as the Luikov model [61]. The pressure-driven transport of moisture is studied by [62] using another mathematical formulation known as the mechanistic model (e.g., Ref. 63). Since experimental spatial moisture profiles are rare (see Chapter 2), computational results from a mechanistic mathematical model are used to describe moisture transport during microwave heating in the following sections. Schematics of the one-dimensional model used to make the computations are shown in Fig. 17. Further details of this model can be seen in Ref. 62.

A. Moisture Transport in Conventional Heating

In order to understand microwave moisture transport, it is instructive to note how it occurs in conventional heating (Fig. 18). The surrounding medium (typically air) in conventional heating is at a very high temperature. As the moisture evaporates from near the surface, a drier, porous region develops that has a lower thermal conductivity, which combined with evaporation, leads to significant drop in temperature from the surface to the interior. Surface moisture drops quickly at higher temperatures while moisture from inside diffuses to the surface to be convected away, leading to the typical moisture profiles shown in Fig. 18. Although high internal temperatures produce some internal evaporation, internal pressures

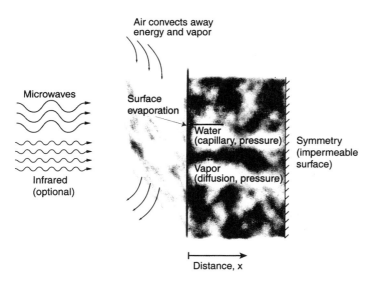

Figure 17 Schematic of the one-dimensional model for moisture transport study. (From Ref. 99.)

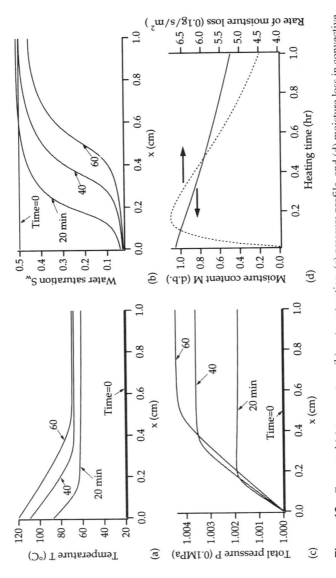

Figure 18 Computed (a) temperature, (b) water saturation, (c) pressure profile, and (d) moisture loss in convective heating of low moisture potato slab (Fig. 17) at an air temperature of 177°C.

are small (Fig. 18) and pressure-driven flow of liquid vapor is not as significant as shown in microwave heating in the next section.

B. Soggy Surface in Microwave Heating of Both High and Low Moisture

Temperature, moisture, and pressure profiles in microwave heating are shown in Figs. 19 and 20. Higher rates of internal evaporation and resulting pressure generation modify the moisture transport considerably. Also the moisture-removal capacity of the air is drastically reduced due to its cold temperature (air is not directly heated by the microwaves). During the initial heating period, when sufficient pressure has not built up because significant evaporation has not begun, pressure-driven moisture flow is insignificant and the moisture profile is similar to that in a surface convective drying situation. As the internal temperature approaches 100°C, evaporation increases and pressure starts to build. Even small amounts of pressure in a low-moisture material can cause too much moisture to reach the surface, and if the moisture-removal capacity of the air around the surface is insufficient, moisture accumulates resulting in soggy foods that were crispy prior to microwave heating. Moisture profiles after any significant pressure build-up are fundamentally different from profiles resulting from capillary (or other) diffusion mechanisms in surface convective heating, where moisture moves from a higher concentration to a lower concentration. The moisture profile after surface convective heating decreases from high values in the inside region to low values near the surface.

C. Excessive Moisture Loss in Microwave Heating of a High-Moisture Material

During microwave heating of a high-moisture food, pressure rises much faster and reaches a much higher value than during conventional heating (Fig. 20c). At higher pressures, enough moisture is pushed to the surface that the surface is saturated and cannot retain additional water. After this, liquid is pumped across the open boundary without undergoing a phase change. Such pumping causes a large drop in internal moisture in about 3 min (Fig. 20d). This pumping effect can cause excessive moisture loss in microwave heating of a high-moisture material.

D. Other Consequences of Pressure Development: Explosions and Microwave Puffing

The magnitude of the pressures developed during microwave heating is important for material integrity and safety. The explosion during microwave heating of food is well known [64]. It is often the consequence of intensive heating concentrations

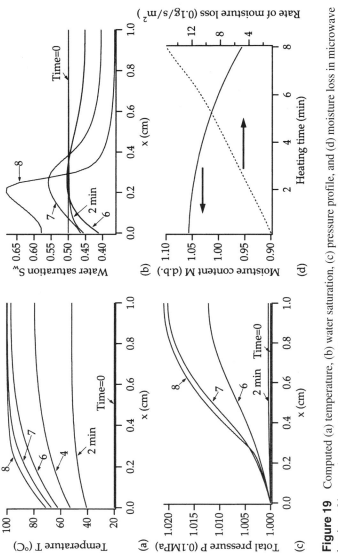

Figure 19 Computed (a) temperature, (b) water saturation, (c) pressure profile, and (d) moisture loss in microwave heating of low moisture potato slab (Fig. 17). Surrounding air temperature is 20°C and surface microwave flux is 3 W/cm^2. (From Ref. 62.)

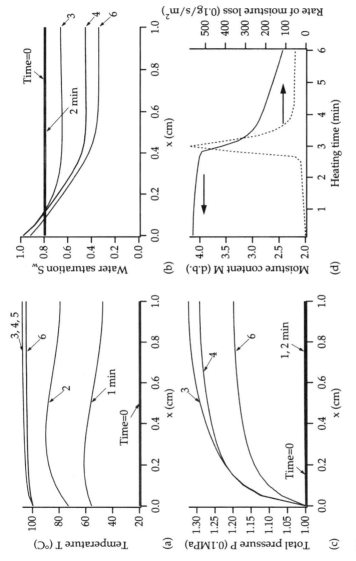

Figure 20 Computed (a) temperature, (b) water saturation, (c) pressure profile, and (d) moisture loss in rapid microwave heating of a high-moisture potato slab (Fig. 17). Surrounding air temperature is 20°C and surface microwave flux is 3 W/cm². Increased moisture loss after 3 min is due to a "pumping effect" whereby liquid water leaves the boundary without being evaporated. (From Ref. 62.)

(due to, for example, the focusing effect), which lead to very rapid evaporation and thus the generation of high pressures. Pressure generation has been used to advantage in microwave processes other than drying. Popping corn, one of the most successful uses of the microwave oven, is based on this rapid pressure developement [65]. Exploitation of pressure generation has resulted in many patents on microwave puffing [66].

E. Moisture Leveling for Higher Internal Moisture

One of the unique advantages of microwave heating over conventional heating is the selective heating of locations in a material where more moisture is available. As can be seen from Eq. (1), for a given electric field E, the rate of volumetric heating Q increases with the increased dielectric loss, ϵ'' at higher moisture content. These increases have important applications in the drying of foods. Selective heating of high-moisture locations is desirable toward the end of a convective drying process, because much of the moisture is deep inside the material and the surface is relatively dry. It is the inside that needs to be heated more for increased evaporation. Conventional surface heating will continue to heat the surface and deteriorate the food's quality. Conventional heat also has a particularly difficult time reaching the wet inner areas, because the dry surface layers typically have low thermal conductivity. Using microwaves to selectively heat the wet interior/internal areas would increase moisture transport from inside toward the surface, somewhat equilibrating the moisture in the product. This equilibration of moisture is sometimes referred to as "moisture leveling."

F. Geometry and Other Electromagnetic Effects

The electromagnetic effects discussed in Chapters 1 and 2 influence the temperature and therefore the moisture transport. For example, the focusing effect, discussed in Chapter 2, leads to lower moisture near the center of a sphere (see Fig. 12 of Chapter 8). Corner and edge heating leads to more moisture loss at those places. Also, increased nonuniformities of heating lead to increased moisture loss, as illustrated in Fig. 21, where higher moisture loss results when starting from a frozen material that heats less uniformly.

G. Microwave Drying

Drying has been a very successful application of microwave energy and is a very active research area [67, 68]. The unique characteristics of microwaves as they relate to food moisture—selective spatial absorption of energy and enhanced moisture loss due to pressure-driven flow—are often exploited in microwave drying. Combined convection-microwave drying is the more common process. In the later

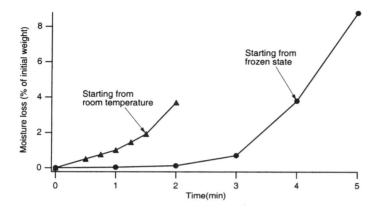

Figure 21 Experimental moisture loss during microwave reheating of food in a bowl, showing increased moisture loss for a frozen material due to higher nonuniformities of heating. (From Ref. 100.)

stages of a surface convective drying process, the remaining moisture is generally deep inside the material. In conventional heating, internal thermal penetration is slow, thus slowing down the drying process. Microwaves applied during the later stages of convective drying, on the other hand, penetrate easily through the drier surface layer and are absorbed selectively in the wet layers, where a higher rate of heating is desired. Thus, much higher drying rates can be maintained toward the later part of a drying process. Microwave drying is also energy efficient, and process control is easier.

Enhanced moisture loss is one of the primary effects of using microwave heating in the drying process. Figure 22 illustrates the typical curves for microwave drying at various power levels. As expected, higher absorbed microwave power leads to increased rates of evaporation and moisture loss. Drying rate curves, which are derivatives of moisture vs. time curves, also illustrate this effect, as shown in Fig. 23. In the specific experiment in Fig. 22, increasing absorbed power caused greater than a linear decrease in the rate of moisture loss. This may not be true for all materials, moisture contents or power levels.

Nonuniformities in the microwave field and associated heating patterns can lead to nonuniformities in drying which can be a significant problem. For example, nonuniform drying among various regions of the product can lead to high temperatures in regions dried earlier, causing product degradation [69]. Various ways of averaging the microwave field to improve uniformity have been achieved by such means as mechanical movement [70], pneumatic agitation such as in a fluidized bed dryer [71], or spouted bed dryers [72]. Industrial drying equipment and processing applications are discussed in Chapters 7 and 9.

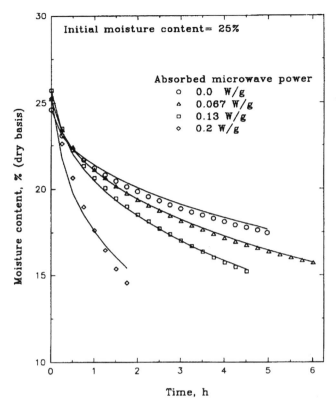

Figure 22 Effect of absorbed microwave power on transient moisture changes in soybean with air blowing at 30°C. (From Ref. 101.)

H. Microwave Freeze-Drying

Freeze-drying is used to remove moisture from temperature-sensitive materials. It is extensively used for foods, e.g., coffee. Typically, the material stays frozen at a low temperature while the surrounding pressure is reduced to a level which allows water to sublime to the vapor phase without going through the liquid phase. Microwaves can significantly accelerate drying.

Using a heat and mass transfer model and associated experimentation, Ma and Peltre [73, 74] explain the freeze-drying process, as shown in Fig. 24. They considered a frozen and a dry region with a sublimation front. The water vapor from the sublimation of ice diffuses through the dry layer across a water vapor concentration gradient to the surrounding atmosphere under vacuum. A vapor trap

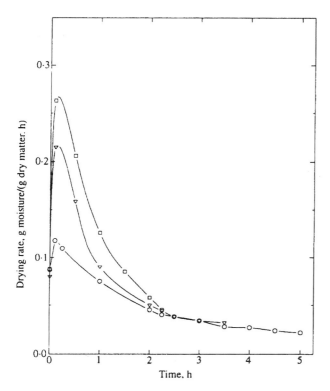

Figure 23 Rates of drying at various absorbed power for corn at 50% initial moisture content with air blowing at 30°C. (From Ref. 57.)

maintains a low partial pressure of water vapor and acts as a mass transfer sink. Figure 24 shows temperature and concentration profiles as functions of relative positions from the sublimation interface, which itself moves with time. Note that the temperature profiles in the dried layer are qualitatively similar to the description for a slab with Lambert's law. The concentration profiles are linear, representing a pseudo steady state, since the process is slow. In a separate experimental study, as shown in Fig. 25, moisture profiles showed the interior moisture leveling and qualitative agreement of the computed profiles in Fig. 24.

Ma and Peltre [73, 74] showed that the drying time can be significantly reduced by increasing the microwave power level. A similar reduction in drying time is reported in an experimental study [75] of the freeze-drying of a layer of peas, as shown in Fig. 26. Reduced drying time is the primary advantage of using microwaves in the freeze-drying process.

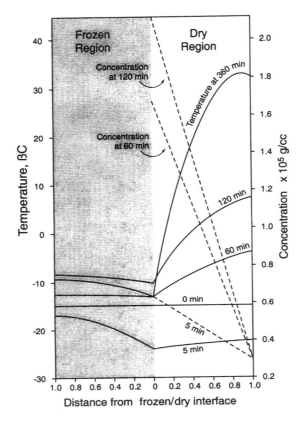

Figure 24 Computed temperature and moisture profiles in a frozen slab during microwave freeze drying, shown as a function of relative position from the frozen/dry interface. Microwaves are incident from the right edge of the figure and the left edge is the line of symmetry for the slab. (From Ref. 102.)

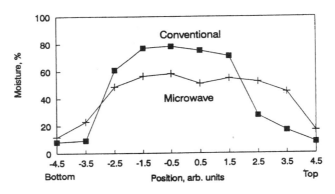

Figure 25 Measured moisture profiles during conventional and microwave freeze-drying. (From Ref. 103.)

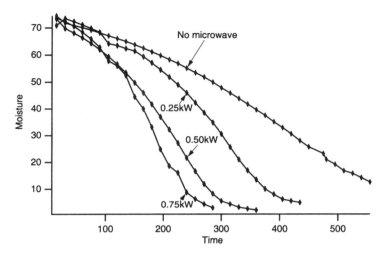

Figure 26 Measured moisture content as a function of time for freeze drying of a layer of peas at various levels of microwave power. (From Ref. 75.)

X. COUPLING OF THE TEMPERATURE AND MOISTURE VARIATION OF DIELECTRIC PROPERTIES DURING PROCESSING

Changes in dielectric properties due to changes in temperature and moisture during heating couple the electromagnetics with heat transfer and moisture transfer. The effect of temperature changes during heating has been studied in detail and is elaborated in this section. Significant changes in moisture in the food also change its dielectric properties, which can change the magnitude and uniformity of microwave absorption in a very significant way. See, for example, the drying study of Ref. 57. The spatial variation of microwave power absorption evolves with time, as shown in a one-dimensional coupled electromagnetics–moisture transport study of wood drying [76]. Such coupling of the effect of moisture changes during processing with electromagnetics has not been studied for food systems and is not elaborated here.

A. Coupled Electromagnetics and Heat Transfer Studies in the Literature

Coupled electromagnetics and heat transfer studies in simpler one- and two-dimensional systems [77–79] are not adequate for a three-dimensional microwave cavity because most of the modes vary in three directions. These studies are generally able to give only qualitative results that are insufficient for the applications of microwave processing of foods. Relatively few researchers [17, 80–82] have performed three-dimensional electromagnetic studies on cavity heating. Although some of those studies [17, 80, 81] included the energy equation, only Ma et al. [80] have discussed coupling issues in their studies. The works by Ma et al. [80] and Torres and Jecho [81] do not consider strong variations in dielectric properties that can lead to extensive variation in heating patterns. The study of coupling for a three-dimensional cavity containing a food load would require extensive computing resources and development time. It is not explicitly included in most of the commercial codes of today (see discussion later).

B. Example of a Coupling Procedure for Electromagnetics and Heat Transfer

The coupling of electromagnetics and heat transfer for microwave heating of foods in a three-dimensional cavity using a finite element code is discussed in this section. As the food heats up nonuniformly, its dielectric properties vary spatially and change with time. As shown in Chapters 1 and 2, Maxwell's equations for electromagnetics imply that electric fields inside food also change accordingly, thus changing the energy deposition and temperature profiles. This interaction is shown schematically in Fig. 4.27.

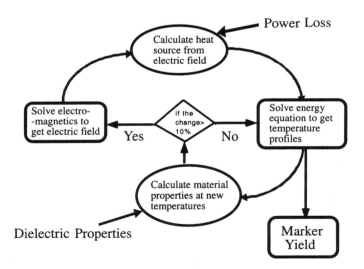

Figure 27 One implementation of coupling of electromagnetics and heat transfer in microwave oven heating using two commercial software. Biochemical changes such as yield of marker chemicals are computed from temperature values.

The coupling was implemented [83] using two software programs, EMAS and NASTRAN, that have no builtin ways to be coupled for the thermal-electromagnetic analysis; the only way to couple them was at the operating system level. To develop the coupling, as shown in Fig. 27, two modules called power loss and dielectric properties were developed. The coupled solution starts with solving the electromagnetic fields in foods using EMAS. The E field obtained this way is converted into power loss data using Eq. (1) and inserted as a module into the NASTRAN input data file for temperature calculation. Temperature distributions are calculated using NASTRAN. In order to complete the coupling, the dielectric properties of foods are modified in the dielectric properties module (shown in Fig. 27), which reads the temperature data on each node, calculates the mean temperature for each element by averaging the temperature values, and updates the dielectric properties in the EMAS input data file. For further details of the implementation, see Ref. 83.

C. When Are Coupling Effects Significant?

The importance of coupling effects is now illustrated for three sets of dielectric properties that change with temperature. Dielectric properties as a function of temperature and salt content are taken from Sun et al. [84] for one group of food materials (meats and meat juices), as shown in Fig. 28. The salt contents in this correlation were varied to represent three sets of dielectric properties. For comparison with the coupled solutions, uncoupled solutions were obtained where the electric field pattern and the heating potential calculated from solving the electromagnetics were kept constant in the thermal analysis.

A salt content of 0.5% in Fig. 28 represents low-loss foods. The predicted temperature ranges ($T_{max} - T_{min}$) are shown in Fig. 29a. The range of tempera-

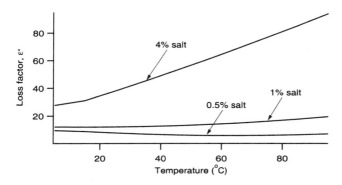

Figure 28 Changes in dielectric loss with temperature [84] used for calculations in Fig. 29.

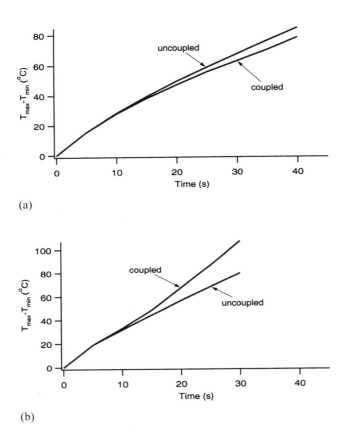

Figure 29 Ranges of temperatures ($T_{max} - T_{min}$) computed using coupled and uncoupled model for (a) a low-loss food and (b) a high-loss food showing greater inaccuracy in temperature calculations for high-loss foods.

tures calculated for the coupled solution is lower than that for the uncoupled solution. The dielectric loss factor for the low loss foods decreases slightly in the temperature range (Fig. 28), decreasing the heating rate and temperature nonuniformity for the coupled solution (Fig. 29a).

A salt content of 4% in Fig. 28 represents high loss foods. As seen in Fig. 29b, the range of temperatures is significantly higher for the coupled solution, compared to the uncoupled. In contrast with Fig. 29a, and for low loss foods, the coupling effects are more significant for high-loss foods such as ham and other salted meats.

D. Migration of Cold Points as Revealed by a Coupled Solution

The effect of coupling, which is the true description of the heating process, can also make qualitative (and dramatic) differences in the heating pattern, as can be seen in Fig. 30 in the context of microwave sterilization. In this figure, the relative locations of cold and hot areas change. At low temperature, the material is less lossy, allowing microwaves to penetrate more. Due to the curved geometry in this case, the greater penetration of microwaves leads to a focusing effect, and, the highest temperatures occur further inside. As the material heats up, the dielectric loss increases considerably and most of the microwaves are absorbed near the surface, changing the location of highest temperature close to the surface. As will be discussed later in the context of microwave pasteurization and sterilization, such changes in heating pattern have important implications for the microbial safety of processed foods. Such qualitative changes in heating pattern will be missed if electromagnetics and heat transfer are not coupled; i.e., the locations of highest (and lowest) temperatures will be predicted erroneously at their initial values.

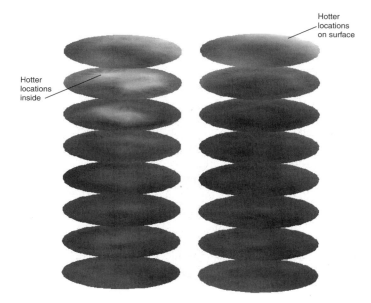

Figure 30 Computed changes in temperature patterns during microwave heating of a cylindrical ham (0.7% salt), showing a qualitative change in the spatial distribution of power absorption during heating, due to dielectric properties changing with temperatures.

XI. QUALITY IMPROVEMENT

Improvement of quality in microwave heating can mean different things, depending on the process. Broadly speaking, two areas control the quality of the microwave heating process: (1) control of heating uniformity and rate, and (2) control of moisture loss and distribution. These two are often related; for example, nonuniformity of heating is often the cause of excessive moisture loss.

A. Control of Heating Uniformity and Rate

In applications such as reheating and sterilization, the goal is to provide the most uniform and rapid heating. Most uniform heating leads to the minimum loss of moisture, which is desirable in these situation. A number of techniques have been tried to obtain more uniform temperature profiles during microwave processing with varied success, which are discussed below.

1. Turntables

Most domestic microwaves include rotating turntables, which physically transport loads through high- and low-intensity electric field regions. Depending on the size of the load (which can vary significantly in domestic use) and its dielectric and thermal properties, turntables have had limited success. Turntables with a shaking action to improve the uniformity of heating in liquids have also been proposed [85].

2. Power Cycling

Power cycling is also routinely used to improve heating uniformity. In most cases, cycling of power is obtained by turning the microwaves on for a fraction of a cycle and off for the rest of the cycle. This strategy provides time for the temperatures to equilibrate by diffusion; i.e., colder areas receive heat from warmer areas, averaging out the temperatures. However, cycling effectively slows down the rate of heating, partly negating the advantages of microwave heating. As shown in Fig. 10, the average rate of heating in cycled power is equivalent to the rate of heating at continuous power at a level equal to the time average of cycled power. The magnitude of the on/off time for a given power level can influence the temperature somewhat, as shown by the experimental data for the 50% power level in Fig. 31. This difference can be attributed to effects such as evaporation.

3. Mode Stirrers

Perturbation of the modal pattern also changes the power deposition pattern and could be used to increase heating uniformity. Mode stirrers, metallic fan blades

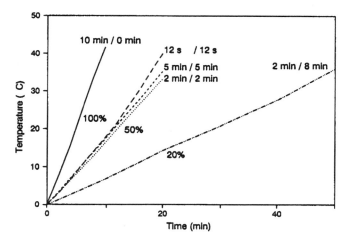

Figure 31 Heating profiles in a starch model at different power levels. The dish was covered with a lid not touching the starch surface. (From Ref. 104.)

that rotate near the feedport, produce continuously changing electric intensity patterns in the applicator as they turn. Since the time scale to produce significant changes in the electric pattern as the blade rotates is large compared to the time it takes to establish the electrical pattern, a number of electric intensity distributions are traversed over a single rotation. However, since electric intensities in resonant conditions are much greater than for nonresonant conditions, a number of resonance patterns must be traversed for the stirrer to be effective in uniformly heating the load. If these patterns are not traversed, the stirrer will act as a pulsing agent, with a time scale equivalent to that of the rotating blade [1].

4. Oven Design Changes

Changes in oven design can provide significant improvements in the uniformity of heating. For example, if the power entry port is rotated steadily, a more uniform heating pattern may be achieved [86]. Many patents exist (see, for example, Ref. 87) for improvements of uniformity within the oven.

A combination of microwave and other types of heating, such as hot air or infrared, can provide some improvements in heating uniformity, although the primary goal of hot air and infrared is to control the moisture levels. A combination of microwave oven and induction heating has also been reported. In induction heating, the metal pan on which the food is placed is magnetically excited (heated), which in turn heats the food. The primary goal of induction heating is surface browning of the food. In one model of a domestic microwave oven, microwaves were combined with hot air, infrared, and induction heating [88].

Another approach to uniformly heating a load is to physically move one of the cavity walls to predetermined locations known to produce different resonance patterns. For example, increasing the size of the cavity for a fixed frequency would increase the number of modes, spreading the number of electric intensity peaks over a larger area in the applicator. While this step may not be practical for domestic ovens, side plungers have been employed in the sintering of ceramics in single-mode cavities [89], for example. Perhaps the most effective method for obtaining uniform heating is by changing the operating frequency of the microwave source during processing, a step that has not yet been tried for foods commercially because the equipment to produce variable frequency microwaves is quite expensive.

5. Active Microwave Packaging

Active microwave packaging, such as the use of susceptors and aluminum foil reflectors, can change the heating pattern considerably. Susceptors are considered below in the discussion of control of moisture loss. An example of the use of aluminum foil is Micro Match, a specialized plastic cover that has strips of aluminum foil on it [90] that make use of an effect called *field intensification*. They act the same way as an antenna and modify the microwave field, redistributing the microwave energy to improve uniformity or to selectively heat various components of a load, such as a frozen dinner.

B. Control of Moisture Loss and Distribution

In some heating applications, such as the drying or reheating of a product with a dry crust (such as French fries), the quality is measured primarily in terms of moisture loss and redistribution. In drying applications, both the rate of total moisture loss and the uniformity of moisture can be important. For reheating products with a crust, the primary goal is to retain the crust, i.e., emulate moisture distribution that would result from conventional heating. Moisture migration can be modified by a number of ways—adding hot air, infrared heat, susceptors, pressurizing, and covering with moisture impermeable material. These are now discussed in detail.

1. Use of Hot Air

As shown in Secs. IX.B and IX.C, moisture migration to the surface due to pressure driven flow and the inability of moisture removal from surface due to cold can cause moisture to accumulate at the surface. One obvious improvement would be to use hot air that has a higher moisture removal capacity. Desirable aspects of moisture removal from surface in conventional heating can be retained this way.

This has been studied by Datta and Ni [91]. As shown in Fig. 32, hot air can reduce surface moisture and increase surface temperature. However, the improvement is limited by its lower power density compared with the use of infrared. Larger air flow is needed to convect moisture away from the surface. Otherwise, higher heat conduction into the material can produce internal pressures that can push more moisture to the surface, increasing the surface moisture. A combination of microwaves with hot air is used in many food processes such as drying and baking [92].

2. Use of Infrared Heat

The use of infrared heat is another possible way to remove moisture from the surface, keeping it crisp [91]. This, however, depends on how far from the surface the infrared penetrates the material. Figure 33 shows how the surface moisture can be kept very low compared to the microwave heat when all the infrared heat is absorbed very near the surface, as for foods with small penetration depth (<1 mm). Depending on the material, however, not all infrared can be absorbed at the surface. If the infrared heat penetrates significantly, it behaves closer to microwave heating and can increase the surface moisture instead. Thus, as shown in Fig. 33 for foods with IR penetration depths comparable to microwaves (>4 mm), IR can actually increase the surface wetness. When IR is assisted in some food with larger IR penetration, appropriately reducing MW power level is likely to lower surface moisture, creating a better end product.

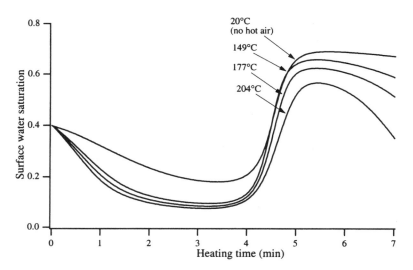

Figure 32 Adding hot air can remove surface moisture, but not as effectively as infrared (see Fig. 33) with very small penetration depth. (From Ref. 105.)

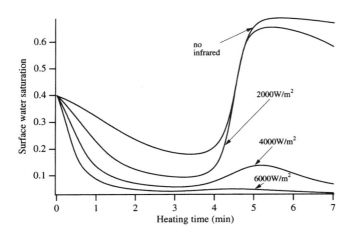

Figure 33 Adding infrared can remove surface moisture more effectively than adding hot air (see Fig. 32). (From Ref. 105.)

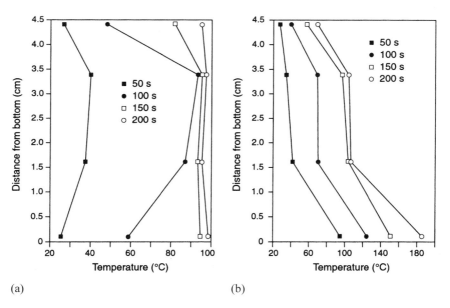

Figure 34 Experimentally measured transient temperature profiles in a dough (a) without any susceptor compared with (b) placing on a susceptor under the product. (From Ref. 93.)

3. Use of Susceptors

Susceptors are yet another attempt to add some of the desirable aspects of conventional heating [93]. These are discussed in detail in Chapter 12 on packaging techniques. Susceptors are generally thin metallized films deposited on the packaging surface. Susceptors can be as wraps or as trays which are included in some ovens. The thin metallized film absorbs microwaves and heat to temperatures as high as 200°C, producing a heating surface that emulates conventional heating.

An example of how the temperature profiles can be modified with the use of a susceptor is shown in Fig. 34 from Ref. 93. The actual temperature change of a food material will depend strongly on the type of susceptor, the food material, and other factors. In Fig. 34, the susceptor was a metallized polyethylene terepthalate (PET) film laminated to a paperboard using a polyurethane-based adhesive. An industrially prepared dough (short cut pastry) was heated that was cylindrical with a radius of 5 cm, height of 4.5 cm, and a weight of 400 g. As can be seen in this figure, the temperature at the dough surface, in the presence of a susceptor, reaches much higher value and makes crisping and browning of the dough possible.

XII. COMPUTER-AIDED ENGINEERING OF HEAT AND MASS TRANSFER PROCESSES

The use of computer modeling as an alternative way to check various scenario in product and process development will become more commonplace in the future. For a general discussion on such computer-aided engineering, see Ref. 94. Microwave food processing can particularly benefit from this trend since it is inherently quite complex, and comprehensive measurement of process parameters in real time is quite difficult. However, as mentioned in Fig. 1, heat and moisture transport under microwave heating are coupled processes where electromagnetics, energy transport, evaporation, and moisture transport are involved. This is inherently complex. Due to the dependence of dielectric property on temperature and moisture levels, the microwave pattern changes and in general cannot be assumed a constant. For an overview of modeling of microwave food processing, see Ref. 12. There are three levels of modeling possible based on the complexity of the process or the desired accuracy:

1. Coupled model where electromagnetics and heat transfer (or moisture transfer) are included, and temperature (or moisture) variation of properties require repeated solution of Maxwell's equations during heating. This provides the most accurate description and, as expected, is the most complex.
2. Electromagnetics and moisture transfer are included; however, the electromagnetic model is solved once and the resulting microwave pattern

is used as a heat source term to compute transient heat and/or moisture transfer. The complexities of electromagnetic modeling are removed in this scenario. Under some conditions, for example, short durations of heating, this can provide a reasonably good description of the process.
3. Electromagnetics is simplified as Lambert's law [Eq. (5)] and used as a heat source in a heat or moisture transfer model. This is the simplest and the least accurate description of the process.

Of the above possibilities in computer modeling of microwave heating processes, items 2 and 3 can be implemented in a software that is purely heat and mass transfer. Many such software are available today, listed in websites such as http://www.math.psu.edu/dna/CFD_codes_c.html.

Implementation of item 1 requires that the electromagnetics and the heat and mass transfer software are integrated in a simple way. Very few commercial software have this capability today. Implementation of item 2 would require that the heat and mass transfer software have the ability to read the electric field (or power deposition) pattern calculated in an electromagnetics software. Once this is completed, the rest of the process is limited by the capabilities of the heat and mass transfer software. Implementation of item 3 is the most common in modeling of food processing. Use of Lambert's law has some limitations worth mentioning.

A. Treatment of Variable Penetration Depth

For variable penetration depth, Eq. (5) can be modified as

$$Q = Q_0 \exp\left(-\int \frac{dx}{\delta_p}\right) \tag{18}$$

where δ_p (T, M) is a function of temperature and moisture and x is distance from the surface. Implementation of Eq. (18) requires that T and M data are available for calculation of δ along the path of integration from the boundary to any location. This can pose problems with some commercial software.

B. Inclusion of Focusing Effect

Prediction of focusing effect in a curved geometry (see Chapter 2) while still using Lambert's law has been reported in a number of situations [15, 14]. Their idea is that as the energy decays exponentially toward the center, the area also becomes smaller in a geometry such as a cylinder or a sphere. Thus, if the area reduces faster than the decay, there will be accumulation of energy. As discussed in Ref. 12, although this formulation seems to make sense qualitatively, the detailed formulation is fundamentally flawed. The quantitative details do not follow from the solutions of Maxwell's equations for a curved geometry and is not appropriate for predicting focusing effect. If focusing is present, the complete set of Maxwell's equations need to be solved to predict it.

C. Moisture Transport with the Pressure Driven Flow

It is important to note that most of these software generally implement diffusional and convective heat and moisture transfer processes. However as shown in Sec. IV, microwave heating of moist material involves internal evaporation. This couples the liquid and the vapor phase. The pressure development also leads to Darcy flow. In particular, the coupling of the liquid and vapor phases is not implemented in the typical commercial software of today.

Thus, the use of a standard heat and mass transfer software to generate computer models of microwave processing of food has major limitations today. They can be used without the electromagnetics when (1) penetration depths are much smaller than the geometry (so Lambert's law can be applied); (2) materials such as active packaging that change the microwave fields locally are absent; and (3) no strong rate of evaporation is present so latent heat effects are less important and pressure-driven flows are not important for energy transport. In the future, ideas generated in specialized noncommercial codes applicable to microwave heating, such as Ref. 95, would probably find its way into commercial codes.

ACKNOWLEDGMENTS

Much of the work presented here is the work of former graduate students Hanny Prosetya, Steven Lobo, Haitao Ni, Montip Chamchong, and Hua Zhang. They made possible the advances reported here. Reviewer John Roberts also provided many useful comments.

LIST OF SYMBOLS

c	Velocity of light, 3×10^8 m/s; concentration
c_p	Specific heat, J/kgK
D_{eff}	Effective diffusivity, m^2/s
E	Electric field, V/m
E_a	Activation energy, J/mol
E_{rms}	Root mean square (average) electric field, V/m
f	Frequency of waves, Hz
F_0	Equivalent processing time at reference temperature T_0, s
h	Convective heat transfer coefficient, W/m^{2}K
h_{fg}	Latent heat of vaporization, J/kg
k	Thermal conductivity, W/mK

k_0	Frequency factor, 1/s
k_T	Rate of reaction at temperature T, 1/s
L	slab half thickness
L_{crit}	Minimum half thickness of slab for using Lambert's law
M	Moisture content
Q	Volumetric rate of heating, W/m^3
R	Gas constant, 8.314 J/K mol
t	Time
T	Temperature
T_∞	Surrounding fluid temperature
u	Velocity, m/s
v_f	Volume fraction
x	Distance, m
Z	$= 2.303 R T_R^2 / E_a$

Temperature difference for the reaction rate to increase by a factor of 10, K (or °C)

Greek Letters

δ_p	Penetration depth, m
ϵ_0	Permittivity of free space, 8.854×10^{-12} F/m
ϵ''	Dielectric loss factor, dimensionless
ϵ'	Dielectric constant, dimensionless
ρ	Density, kg/m^3

REFERENCES

1. C. Saltiel and A. K. Datta, Heat and mass transfer in microwave processing, Advances in Heat Transfer, vol. 32, 1998.
2. K. G. Ayappa, Modelling transport processes during microwave heating: A review, Reviews in Chemical Engineering, vol. 13, no. 2, pp. 1–68, 1997.
3. R. V. Decareau, Microwaves and food, FNP Newsletter, vol. 2, no. 7, pp. 1–6, 1992.
4. R. V. Decareau, Microwaves in the Food Processing Industry. New York: Academic Press, Inc., 1985.
5. C. R. Buffler, Microwave cooking and processing: Engineering fundamentals for the food scientist. New York: Van Nostrand Reinhold, 1992.
6. U. Rosenberg and W. Bogl, Microwave pasteurization, sterilization, blanching, and pest control in the food industry, Food Technology, pp. 92–99, 1987.
7. U. Rosenberg and W. Bogl, Microwave thawing, drying, and baking in the food industry, Food Technology, pp. 85–91, 1987.
8. D. Clark and W. Sutton, Microwave processing of materials, Annual Reviews of Materials Science, vol. 26, pp. 299–331, 1996.
9. D. Lewis and J. Shaw, Recent developments in the microwave processing of polymers, MRS Bulletin, vol. 18, no. 11, pp. 37–40, 1993.

10. S. J. Oda, Microwave remediation of hazardous wastes: A review, Material Research Society Symposium Proceedings, vol. 347, pp. 371–382, 1994.
11. J. Thuery, Microwaves: Industrial, Scientific, and Medical Applications. Boston: Artech House, 1992.
12. A. K. Datta, Mathematical modeling of microwave processing of foods: An overview, in Food Processing Operations Modeling: Design and Analysis (J. Irudayaraj, ed.), (New York: Marcel Dekker) 2001, pp. 1–67.
13. F. Peyre, A. K. Datta, and C. E. Seyler, Influence of the dielectric property on microwave oven heating patterns: Application to food materials, Journal of Microwave Power and Electromagnetic Energy, vol. 32, no. 1, pp. 3–15, 1997.
14. L. Zhou, V. M. Puri, R. C. Anantheswaran, and G. Yeh, Finite element modeling of heat and mass transfer in food materials during microwave heating—model development and validation, Journal of Food Engineering, vol. 25, pp. 509–529, 1995.
15. D.-S. Chen, R. K. Singh, K. Haghighi, and P. E. Nelson, Finite element analysis of temperature distribution in microwaved particulate foods, J. Food Engineering, vol. 18, pp. 351–368, 1993.
16. S. R. Lobo, Characterization of spatial non-uniformity in microwave reheating of high loss foods, Master's thesis, Cornell University, 1988.
17. D. Burfoot, C. J. Railton, A. M. Foster, and R. Reavell, Modeling the pasteurisation of prepared meal with microwave at 896 MHz, Journal of Food Engineering, vol. 30, pp. 117–133, 1996.
18. D. R. Heldman and R. P. Singh, Food Process Engineering. AVI Publ. Co. Westport, Connecticut, 1981.
19. M. A. Rao and S. Rizvi, Engineering Properties of Foods. Marcel Dekker, 1995.
20. M. R. Okos, Physical and Chemical Properties of Food. ASAE, St. Joseph, Michigan, 1986.
21. A. K. Datta, Heat and mass transfer during microwave processing of food materials, Chemical Engineering Progress, vol. 86, no. 6, pp. 47–53, 1990.
22. K. G. Ayappa, H. T. Davis, E. A. Davis, and J. Gordon, Microwave heating: An evaluation of power formulations, Chemical Engineering Science, vol. 46, no. 4, p. 1005, 1991.
23. J. J. Dolande and A. K. Datta, Temperature profiles in microwave heating of solids: a systematic study, The Journal of Microwave Power and Electromagnetic Energy, vol. 28, no. 2, pp. 58–67, 1993.
24. H. Prosetya and A. K. Datta, Batch microwave heating of liquids: an experimental study, The Journal of Microwave Power and Electromagnetic Energy, vol. 26, no. 3, pp. 215–226, 1991.
25. A. K. Datta, H. Prosetya, and W. Hu, Mathematical modeling of batch heating of liquids in a microwave cavity, The Journal of Microwave Power and Electromagnetic Energy, vol. 27, no. 1, pp. 38–48, 1992.
26. R. C. Anantheswaran and L. Liu, Effect of viscosity and salt concentration on microwave heating of model non-newtonian liquid foods in a cylindrical container, Journal of Microwave Power and Electromagnetic Energy, vol. 29; no. 2, pp. 119–126, 1994.
27. R. C. Anantheswaran and L. Liu, Effect of electrical shielding on time-temperature distribution and flow profiles in water in a cylindrical container during microwave heating, Journal of Microwave Power and Electromagnetic Energy, vol. 26, no. 3, pp. 156–159, 1991.

28. Y. Emami and T. Ikeda, System and method for sterilization of food material, UK patent application GB2 098 040, 1982.
29. A. K. Datta and J. Liu, Thermal time distributions in microwave and conventional heating, Transactions of the Institution of Chemical Engineers, vol. 70, no. C, 1992.
30. H. S. Ramaswamy and S. Tajchakavit, Continuous-flow microwave heating of orange juice, ASAE Paper 93-3588, pp. 1–16, 1993.
31. T. Ohlsson, In-flow microwave heating of pumpable foods, in Presented at International Congress on Food and Engineering (Japan), pp. 1–7, 1993.
32. J. J. R. Thomas, E. M. Nelson, R. J. Kares, and R. M. Stringfield, Temperature distribution in a flowing fluid heated in a microwave resonant cavity, Material Research Society Symposium Proceedings, vol. 430, pp. 565–569, 1996.
33. J. Houben, L. Schoenmakers, E. van Putten, P. van Roon, and B. Krol, Radio-frequency pasteurization of sausage emulsions as a continuous process, Journal of Microwave Power and Electromagnetic Energy, vol. 26, no. 4, pp. 202–205, 1991.
34. D. R. Baghurst and M. P. Mingos, Superheating effects associated with microwave dielectric heating, J. Chem. Soc., Chem. Comm., pp. 674–677, 1992.
35. D. Stuerga, A. Steichen-Sanfeld, and M. Lallemant, An original way to select and control hydrodynamic instabilities: Microwave heating 3: Linear stability analysis, Journal of Microwave Power and Electromagnetic Energy, vol. 29, no. 1, pp. 3–30, 1994.
36. B. J. Pangrle, K. G. Ayappa, H. T. Davis, E. A. Davis, and J. Gordon, Microwave thawing of cylinders, AIChE Journal, vol. 37, no. 12, pp. 1789–1800, 1991.
37. T. Basak and K. G. Ayappa, Analysis of microwave thawing of slabs with the effective heat capacity method, AIChE Journal, vol. 43, no. 7, pp. 1662–1674, 1997.
38. P. Taoukis, E. A. Davis, H. T. Davis, J. Gordon, and Y. Talmon, Mathematical modeling of microwave thawing by the modified isotherm migration method, Journal of Food Science, vol. 52, no. 2, pp. 455–463, 1987.
39. X. Zeng and A. Faghri, Experimental and numerical study of microwave thawing heat transfer for food materials, Journal of Heat Transfer, vol. 116, pp. 446–455, 1994.
40. M. Chamchong, Microwave thawing of foods: Effect of power levels, dielectric properties, and sample geometry. PhD thesis, Cornell University, 1997.
41. M. Chamchong and A. K. Datta, Optimization of heating rates and non-uniformity in microwave thawing of foods, in Proceedings of the 7th International Congress on Engineering and Food (R. Jowitt, ed.) (Sheffield, UK: Sheffield Academic Press Ltd.), 1997, pp. C41–C44.
42. IFT, Microwave and radio frequency pasteurization and sterilization, in How to Quantify the Destruction Kinetics of Alternative Processing Technologies (Dallas, TX), June 2000.
43. A. K. Datta and W. Hu, Quality optimization of dielectric heating processes, Food Technology, vol. 46, no. 12, pp. 53–56, 1992.
44. M. R. Jeppson and J. C. Harper, Microwave heating substances under hydrostatic pressure, US Patent No. 3,335,253, 1967.
45. E. M. Kenyon, D. E. Westcott, P. LaCasse, and J. W. Gould, A system for continuous thermal processing of food pouches using microwave energy, Journal of Food Science, vol. 36, pp. 289–293, 1971.

46. W. Schlegel, Commercial pasteurization and sterilization of food products using microwave technology, Food Technology, no. 12, pp. 62–63, 1992.
47. L. Harlfinger, Microwave sterilization, Food Technology, no. 12, pp. 57–61, 1992.
48. R. Tops, Industrial implementation: Microwave pasteurized and sterilized products, Presented at the Symposium on Microwave Sterilization during the IFT Annual Meeting (Dallas, TX), 2000.
49. M. A. K. Hamid, R. J. Boulanger, S. C. Tong, R. A. Gallop, and R. R. Pereira, Microwave pasteurization of raw milk, Journal of Microwave Power, vol. 4, no. 4, pp. 272–275, 1969.
50. T. Kudra, R. F. V. D. Voort, G. S. V. Raghavan, and H. S. Ramaswamy, Heating characteristics of milk constituents in a microwave pasteurization system, Journal of Food Science, vol. 56, no. 4, pp. 931–934, 937, 1991.
51. H. Zhang and A. K. Datta, Experimental and numerical investigation of microwave sterilization of solid foods, American Institute of Chemical Engineers Journal, 2000.
52. J. Casasnovas and R. C. Anantheswaran, Thermal processing of food packaging waste using microwave heating, Journal of Microwave Power and Electromagnetic Energy, vol. 29, no. 3, pp. 171–179, 1994.
53. M. Lau, J. Tang, I. A. Taub, T. C. S. Yang, C. G. Edwards, and F. L. Younce, HTST processing of food in microwaveable pouch using 915 MH microwaves, Presented at the AIChE Annual Meeting (Dallas, TX), 1999, pp. T3017-66D.
54. T. Wig, J. Tang, F. Younce, L. Hallberg, C. P. Dunne, and T. Koral, Radio frequency sterilization of military group rations, Presented at the AIChE Annual Meeting (Dallas, TX), 1999, pp. T3017–66B.
55. C. H. Tong and D. B. Lund, Microwave heating of baked dough products with simultaneous heat and moisture transfer, Journal of Food Engineering, vol. 19, pp. 319–339, 1993.
56. L. Lu, J. Tang, and L. Liang, Moisture distribution in spherical foods during microwave drying, Drying Technology, vol. 16, no. 3–5, pp. 503–524, 1998.
57. U. S. Shivhare, G.S.V. Raghavan, and R. G. Bosisio, Modelling the drying kinetics of maize in a microwave environment, Journal of Agricultural Engineering Research, vol. 57, pp. 199–205, 1994.
58. L. Zhou, V. M. Puri, and R. C. Anantheswaran, Effect of temperature gradient on moisture migration during microwave heating, Drying Technology, vol. 12, no. 4, pp. 777–798, 1994.
59. L. Zhou, V. M. Puri, and R. C. Anatheswaran, Measurement of coefficients for simultaneous heat and mass transfer in food products, Drying Technology, vol. 12, no. 3, pp. 607–627, 1994.
60. D. Mukherjee, V. M. Puri, and R. C. Anantheswaran, Measurement of coupled heat and moisture transfer coefficients for selected vegetables, Drying Technology, vol. 15, no. 1, pp. 71–94, 1997.
61. A. V. Luikov, Systems of differential equations of heat and mass transfer in capillary-porous bodies (review), International Journal of Heat and Mass Transfer, vol. 18, pp. 1–14, 1975.
62. H. Ni, A. K. Datta, and K. E. Torrance, Moisture transport in intensive microwave heating of wet materials: a multiphase porous media model, International Journal of Heat and Mass Transfer, 1998.

63. S. Whitaker, Simultaneous heat, mass, and momentum transfer in porous media: A theory of drying, Advances in Heat Transfer, vol. 13, pp. 119–203, 1977.
64. Y. C. Fu, C. H. Tong, and D. B. Lund, Microwave bumping: Quantifying explosions in foods during microwave heating, Journal of Food Science, vol. 59, no. 4, pp. 899–904, 1994.
65. W. Lin and C. Sawyer, Bacterial survival and thermal responses of beef loaf after microwave processing, Journal of Microwave Power and Electromagnetic Energy, vol. 23, no. 3, pp. 183–194, 1988.
66. P. Whalen, Half products for microwave puffing of expanded food product, US Patent 5102679, 1992.
67. P. Bhartia, S. S. Stuchly, and M. A. K. Hamid, Experimental results for combinational microwave and hot air drying, Journal of Microwave Power, vol. 8, no. 3, pp. 245–252, 1973.
68. T. Kudra, U. S. Shivhare, and G.S.V. Raghavan, Dielectric drying—a bibliography, Drying Technology, vol. 8, no. 5, pp. 1147–1160, 1990.
69. L. Lu, J. Tang, and X. Ran, Temperature and moisture changes during microwave drying of sliced food, Drying Technology, vol. 17, no. 3, pp. 413–432, 1999.
70. E. M. Torringa, E. J. van Dijk, and P. S. Bartels, Microwave puffing of vegetables: Modeling and measurements, in Proceedings of 31st Microwave Power Symposium (Manassas, VA: International Microwave Power Institute), 1996.
71. T. Kudra, Dielectric drying of particulate materials in a fluidized state, Drying Technology, vol. 7, no. 1, pp. 17–34, 1989.
72. H. Feng and J. Tang, Microwave finish drying of diced apples in a spouted bed, Journal of Food Science, vol. 63, no. 4, pp. 679–683, 1998.
73. Y. H. Ma and P. R. Peltre, Freeze dehydration by microwave energy. Part I. Theoretical investigation, AIChE Journal, vol. 21, no. 2, pp. 335–344, 1975.
74. Y. H. Ma and P. R. Peltre, Freeze dehydration by microwave energy. Part II. Experimental study, AIChE Journal, vol. 21, no. 2, pp. 344–350, 1975.
75. J. S. Cohen, J. A. Ayoub, and T. C. Yang, A comparison of conventional and microwave augmented freeze-drying of peas, in Drying '92, vol. 2 (A. S. Mujumdar, ed.) (New York: Elsevier) pp. 585–594, 1992.
76. I. W. Turner and W. J. Ferguson, A study of the power density distribution generated during the combined microwave and convective drying of softwood, Drying Technology, vol. 13, no. 5–7, pp. 1411–1430, 1995.
77. N. F. Smyth, Microwave heating of bodies with temperature dependent properties, Wave Motion, vol. 12, pp. 171–186, 1990.
78. K. G. Ayappa, H. T. Davis, E. A. Davis, and J. Gordon, Two-dimensional finite element analysis of microwave heating, AIChE Journal, vol. 38, no. 10, pp. 1577–1592, 1992.
79. J. Clemens and G. Saltiel, Numerical modeling of materials processing in microwave furnaces, International Journal of Heat and Mass Transfer, vol. 39, no. 8, pp. 1665–1675, 1995.
80. L. Ma, D. Paul, N. Pothecary, C. Railton, J. Bows, L. Barratt, J. Mullin, and D. Simons, Experimental validation of a combined electromagnetic and thermal FDTD model of a microwave heating process, IEEE Trans. on Microwave Theory and Techniques, vol. 43, no. 11, pp. 2565–2571, 1995.

81. F. Torres and B. Jecko, Complete FDTD analysis of microwave heating processes in frequency-dependent and temperature-dependent media, IEEE Trans. on Microwave Theory and Techniques, vol. 45, no. 1, 1997.
82. H. Zhao and I. W. Turner, An analysis of the finite-difference time-domain method for modeling the microwave heating of dielectric materials within a three-dimensional cavity system, Journal of Microwave Power and Electromagnetic Energy, vol. 31, no. 4, pp. 199–214, 1996.
83. H. Zhang and A. K. Datta, Coupled electromagnetic and thermal modeling of microwave oven heating of foods, Submitted to the Journal of Microwave Power and Electromagnetic Energy, 1999.
84. E. Sun, A. K. Datta, and S. Lobo, Composition-based prediction of dielectric properties of foods, The Journal of Microwave Power and Electromagnetic Energy, vol. 30, no. 4, pp. 205–212, 1995.
85. N. L. Campbell, J. G. Pinto, J. A. Drewe, G. W. Ghing, and F. R. Borkat, Microwave apparatus for heating contained liquid, US Patent 4,742,202, 1988.
86. W. Fu and A. Metaxas, Numerical prediction of three-dimensional power density distributions in a multi-mode cavity, Journal of Microwave Power and Electromagnetic Energy, vol. 29, no. 2, pp. 67–75, 1994.
87. H. Sakai, H. Uehashi, K. Sakata, M. Noda, Y. Omori, K. Hayami, and M. Katayama, Microwave oven with a projection for uniform heating within the cavity. US Patent 5698128, 1997.
88. W. G. Phillips, The multitalented microwave, Popular Science, vol. 251, no. 6, p. 30, 1997.
89. Y. Tian, Practices of ultra-rapid sintering of ceramics using single mode applicators, in Ceramic Transactions, Microwaves: Theory and Applications in Materials Processing, vol. 21 (D. Clark, F. Gac, and W. Sutton, eds.), vol. 21, (Westerville, OH. The American Ceramic Society), 1991, pp. 283–300.
90. R. Lingle, Fine tuning microwave packaging, Prepared Foods, no. 11, pp. 76–82, 1987.
91. A. K. Datta and H. Ni, Infrared and hot air additions to microwave heating of foods for control of surface moisture, Submitted to the Journal of Food Engineering, 1999.
92. A. L. M. Bernussi, Y. K. Chang, and E. Martinez-Bustos, Effects of production by microwave heating after conventional baking on moisture gradient and product quality, Cereal Chemistry, vol. 75, no. 5, pp. 606–611, 1998.
93. H. Zuckerman and J. Miltz, Temperature profiles at susceptor/product interface during heating in the microwave oven, Journal of Food Processing and Preservation, vol. 19, pp. 385–398, 1995.
94. A. K. Datta, Computer-aided engineering in food process and product design, Food Technology, vol. 52, no. 10, pp. 44–52, 1998.
95. I. W. Turner and P. Perre, A comparison of the drying simulation codes transpore and wood2d which are used for the modelling of two-dimensional wood drying processes, Drying Technology, vol. 13, no. 3, pp. 695–735, 1995.
96. J. Gerling. Personal communication, May 1991. Gerling Laboratories, Modesto, CA.
97. M. Chamchong and A. K. Datta, Microwave thawing of foods: Effect of power levels and dielectric properties, Journal of Microwave Power and Electromagnetic Energy, 1998.

98. M. Chamchong and A. K. Datta, Microwave thawing of foods: Mathematical modeling and experiments, Journal of Microwave Power and Electromagnetic Energy, 1998.
99. H. Ni, Multiphase moisture transport in porous media under intensive microwave heating. PhD thesis, Cornell University, 1997.
100. H. Ni, A. K. Datta, and R. Parmeswar, Moisture loss as related to heating uniformity in microwave processing of solid foods, Journal of Food Process Engineering, vol. 22, no. 5, pp. 367–382, 1999.
101. U. S. Shivhare, V. Raghavan, R. Bosisio, and M. Giroux, Microwave drying of soybean at 2.45 GHZ, Journal of Microwave Power and Electromagnetic Energy, vol. 28, no. 1, pp. 11–17, 1993.
102. Y. H. Ma and P. Peltre, Mathematical simulation of a freeze drying process using microwave energy, AIChE Symposium Series, vol. 69, no. 132, pp. 47–53, 1973.
103. A. Prakash, A. Barrett, and I. A. Taub, Optimizing the microwave-assisted freeze-drying process to achieve a reversibly compressible product, in Proceedings of the 1993 Food Preservation 2000 Conference, vol. I (Hampton, VA: Science and Technology Corporation), 1993, pp. 241–251.
104. H. S. Ramaswamy and T. Pillet-Will, Temperature distribution in microwave-heated food models, Journal of Food Quality, vol. 15, pp. 435–448, 1992.
105. A. K. Datta, Understanding and improvements of microwave oven heating, in Proceedings of the Appliance Manufacturers Conference and Exposition (Nashville, TN), October 1998, pp. 35–45.

5
Generation and Release of Food Aromas Under Microwave Heating

Varoujan A. Yaylayan
McGill University
Ste. Anne de Bellevue, Quebec, Canada

Deborah D. Roberts
Nestlé Research Center
Lausanne, Switzerland

I. INTRODUCTION

Flavors and colors generated as a result of the Maillard reaction are of critical importance for the commercial success of microwave-processed foods. Recent interest in the microwave generation of Maillard aromas and colors was a response on the part of the food industry, based on the consumer demand for fast and convenient food products. However, food products heated under microwave irradiation lack the color and flavor associated with the Maillard reaction observed under conventional heating [1]. In addition, loss of aroma, whether added or formed during microwave processing, is even more pronounced than in conventional heating. The fundamental differences between microwave and conventional heating, the composition of the food matrix, and the design of microwave ovens all seem to play a role in the inability of microwave heating to propagate color and aroma generating Maillard reactions in food products. The increased sales of microwave ovens in the last decade, especially into the North American market, provided the food industry with the impetus for renewed interest in carrying out the Maillard reaction in microwaveable food products.

Application of microwave energy to carry out chemical reactions such as the

Maillard reaction gained considerable importance in the late 1980s after the discovery that a wide range of reactions can be completed under microwave irradiation in a much shorter period of time compared to conventional heating by reflux [2–5]. Microwave heating refers to heating by electromagnetic waves in the requency range of 300–3.0 × 10^5 MHz (see also electromagnetic spectrum in Chapter 1), while conventional heating is primarily based on diffusion of energy. The conversion of microwave energy into heat can be achieved by two mechanisms: dipole rotation and ionic conduction; consequently the ionic and dipolar molecules are the primary ones interacting with microwaves to generate heat (see Chapter 3 for details). At the frequency of domestic microwave ovens (2450 MHz), the process of dipole rotation occurs 4.9 × 10^9 times per second and results in increased rate of heat generation. The relative importance of these two mechanism of heat transfer depends to a large extent on the temperature due to its effect on ion mobility and characteristic dielectric relaxation times [6].

Microwave heating is relatively fast compared to conventional heating since it does not depend on the slower diffusion process in the latter. This property initiated the initial investigation into carrying out chemical reactions under microwave irradiation [7]. In certain cases, chemical reactions were completed in a few seconds that otherwise would have taken hours. In addition to fast rates of heating, microwaves are also more selective and components can be heated selectively in a reaction mixture compared to conventional heating. This property has been used to enhance the extraction of essential oils from plants immersed in a microwave transparent solvent [8].

Superheating of solvents is another phenomena that accompanies microwave heating and helps accelerate chemical reactions. Superheating (see also inhibition of boiling in Chapter 4) refers to the increase in temperature of liquids above their boiling points while they remain completely in the liquid phase. For example, water boils under microwave heating at 105°C and acetonitrile (b.p. 82°C) at 120°C. A chemical reaction carried out in an open vessel in acetonitrile under microwave irradiation will be accelerated by 14 times relative to conventional heating, assuming the reaction rate doubles for every 10°C rise in temperature [5]. When chemical reactions are carried out in closed containers under microwave irradiation, the maximum temperature attainable is not limited to the temperature of the heating medium, as in conventional heating, but depends only on the microwave power applied and the rate at which the sample can lose heat [5]. The extreme high temperatures attained in a closed container during microwave heating can generate extreme high pressures (especially if the reaction produces gaseous products), which can alter equilibrium product distribution according to Le Chatelier's principle [5].

The two aspects of microwave irradiation that affect the propagation of Maillard reaction in foods include the interaction of microwaves with the food matrix and its effect on the chemical reactions in general [9]. Unlike solvent-me-

Generation and Release of Food Aromas

diated Maillard reactions, the food matrix imparts special consideration, due to selective absorption of microwaves by different food components, compartmentalization of reactants, presence of components with differing loss factors, mobility of reactants, moisture content, and presence of ionic species. Solvent-mediated and "in-matrix" Maillard reactions will be discussed separately.

II. GENERATION OF MAILLARD AROMAS UNDER MICROWAVE HEATING

As the Maillard reaction is a series of chemical transformations [10], factors that influence a chemical reaction also effect the Maillard reaction. In general, the rates of chemical reactions depend primarily on temperature, pressure, time, and concentration of reactants. High temperature, pressure, and superheating of reaction solvent associated with microwave irradiation can accelerate simple or single-step chemical reactions such as esterification, hydrolysis, cyclization, Diels-Alder, and S_N2-type reactions [3, 7, 11]. If the microwave heating is performed under a closed system, then the rate of the microwave reaction accelerates up to 1000 times. However, the time factor plays a crucial role in influencing the product distribution of more complex reactions when carried out under microwave heating. The influence on competitive and consecutive reactions is an important consequence of fast rate of heating under microwave irradiation that is especially pertinent to the propagation of Maillard reaction. In the competitive (or parallel) reaction shown below, reactants A and B produce two products P_1 and P_2 by two separate mechanisms, such that P_1 is more stable than P_2 and the activation energy (E_{a2}) to produce P_2 is relatively smaller than that of P_1. The reaction mixture produced under microwave heating (fast increase in temperature) of A and B will be richer in P_1 compared to the conventionally heated (slow increase in temperature) sample for the same length of time. The reaction at the high temperature of the microwave will be under thermodynamic control, and produce the most stable product in higher yields [12].

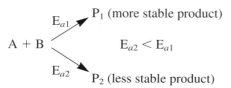

In the case of consecutive reactions of the type shown below, where the reaction produces color as the end product, the rate of formation of E from B will depend on the concentration of D and the rate of formation of D, in return, will depend on the concentration of C. Until there is considerable accumulation of C and D, very little E will be formed.

$$B \to C \to D \to E \to \text{color}$$

Heating a solution of B under microwave irradiation will produce a mixture containing relatively less of C, D, and E compared to a similar solution heated conventionally to reach the same temperature as in the microwave. If E is the precursor of color-producing species, then the microwave-heated sample will reach the same temperature as the conventionally heated mixture in a shorter period of time with a lower yield of E and hence color.

A. Solvent-Mediated Maillard Reactions: Model Systems

Given the fact that the Maillard reaction is a complex series of consecutive and competitive reactions, product distribution and intensity of browning will be most affected by microwave irradiation relative to conventional heating. Generally, the final outcome of a Maillard reaction (color, volatile aroma compounds, and nonvolatile products) depends on temperature, water content, pH, and heating time. Thus, any variation in the reaction parameters will effect the profile of the end products, and hence the perceived aroma and color. Although simple chemical reactions are fast under microwave irradiation, multistep reactions can remain incomplete or they do not proceed to the same extent as under conventional heating. They produce mixtures that contain the same products [13] but with altered distribution patterns. The flavor perception is sensitive to such variations in relative concentrations of different components, especially the character impact compounds, thus drastically changing the sensory properties.

There are few reports in the literature on the microwave-assisted generation of Maillard products using precursors or intermediates. Preparative scale microwave-assisted synthesis [14] of Amadori products from D-glucose and amino acids is feasible but has not been reported. However, Barbiroli et al. [15] observed 70–75% conversion of added glucose/leucine into Amadori compounds with a corresponding decrease in the amount of added amino acid in a bread mix when microwaved for 3 min. Steinke et al. [16] generated Strecker aldehydes from an aqueous solution of an amino acid and 2,3-butadione (diacetyl) in sealed vials microwaved for 4 min or heated in a water bath for 60 min at the same temperature. Significantly higher concentrations of aldehydes were measured in the microwave-heated samples. The effect of electrolytes and pH on the formation of Maillard products during microwave irradiation of aqueous model systems has been studied. The addition of different salts [17] such as sodium chloride, calcium chloride, and sodium sulfate increased both the intensity of browning and the concentration of flavor compounds. The total volatiles generated from a glucose/cysteine model system [18] under microwave irradiation has been found to increase with pH. It seems that increasing the pH and concentration of electrolytes enhances the rate of Maillard reactions under microwave irradiation. This observation, although consistent with the mechanism of heat generation by ionic conduction, might be influenced by the water activity changes associated with the addition of salts.

Attempts have been made to compare the chemical composition and yields of volatiles in microwaved and conventionally heated Maillard model systems. However, this type of comparison can be misleading due to the variations in the time-temperature exposure of the two systems under study. In most cases, the temperature of the microwave system is not monitored and time of irradiation is chosen arbitrarily. In order to compare the yields of two systems undergoing the same reaction at different times and temperatures, a knowledge of kinetic parameters is required to ascertain whether there are differences in the two processes. Alternatively, the intensity of brown color formation can be used as an indication that the two systems have undergone equivalent time-temperature exposure. Yaylayan et al. [13] mimicked actual cooking and surface drying of foods by subjecting the same aqueous sugar/amino acid mixtures to microwave irradiation (640 W) and to conventional heating in an open system, until all the water was evaporated and the residue was dark brown. In order to ensure that both treatments produced the same extent of Maillard reaction for comparison purposes, the conventional heating time was adjusted such that after similar dilutions, both samples had the same spectrophotometric absorption at 460 nm. On the average, 1 min of microwave heating time produced the same browning extent as 12 min of conventional heating time. With such treatment, no significant qualitative changes were observed in the composition of both samples, as identified by GC/MS. Parliment [19] studied, in sealed vials, the products of the Maillard reaction between glucose and proline formed under microwave (600 W, preheated conventionally for 3 min and irradiated for 45 s) and conventionally heated systems (150°C for 15 min). Qualitatively both systems produced similar compounds but in the microwave system N-heterocyclic compounds were present in smaller amounts. These types of experiments are inconclusive since the temperature of the microwave sample is not measured to estimate whether heating at 150°C for 15 min is equivalent to 45 s of microwave irradiation. Ji and Bernhard [20], on the other hand, calculated the pseudo zero-order rate constants for selected pyrazines produced in mixtures consisting of D-glucose and glycine in 100 ml of water that were microwaved (500 W) in an open system between 1.5 to 9.0 min. The temperature of all samples were measured after irradiation and found to be around 100°C. The calculated rate constants were found to be in agreement with those reported under conventional heating [21]. In addition, when the products of the microwave sample (9.0 min of irradiation at 100°C) were compared with those produced under conventional heating at 120°C for 4.0 h, the total yield of the products increased from 4.4 to 6.6% based on the D-glucose concentration. In general, the Maillard model studies indicate that there are no fundamental differences between microwave and conventionally heated samples in solution phase except in time/temperature exposure of the two systems and hence in the relative amounts of products formed.

B. The Maillard Reaction in Food Products: Interaction of Microwave Irradiation with the Food Matrix

At the molecular level, the mechanism of heat generation in the microwave oven relies mainly on the interaction of the microwave radiation with dipoles/induced dipoles or with ions. Proteins and lipids do not significantly interact with microwave radiation in the presence of aqueous ions that selectively absorb the radiation. However, in the absence of water, lipids and colloidal solids are known to interact strongly with microwave radiation and the observed levels of energy absorption cannot be explained by the presence of free water and by ion activity [22]. Microwave radiation can also interact with alcohols, sugars, and polysachharides. Tightly bound water monolayers do not absorb energy due to hindered molecular rotations. Overall, the extent of heat generation by microwave depends mainly on the moisture and salt content of food products (see Chapter 8).

Microwave interactions with a multicomponent system such as food can differ considerably from simple aqueous Maillard model systems, in that "matrix effects" can produce undesirable consequences. Since the core aqueous region of foods are the main sites of interaction with the microwaves, the interior vapor pressure generated as a result can actively force the vapor to the surface of the food, unlike in the conventional oven, where passive migration of water by capillary action to the surface, is diffusion controlled [23]. The water-saturated food surfaces usually remain at relatively cool temperatures of the oven during cooking (40–60°C), thus preventing browning and crisping [23]. Model studies have already indicated that there are no fundamental differences in the solution phase chemistry of Maillard reaction under microwave irradiation. However, the overall performance of food products under microwave irradiation implies the development of characteristic textural, color and aroma properties similar to that of conventional heating, which differs markedly from microwave heating due to fast rate of heating and "matrix effects."

Food products that rely heavily on Maillard flavors and colors, such as roasted and baked products, perform well in the conventional oven due to the following: (1) The high temperature of the air surrounding the product dehydrates the surface, producing a crust that protects the food from loss of moisture and important aroma volatiles. Dehydration steps are also crucial for the formation of color and aroma precursors by the Maillard reaction. (2) Long time exposure in the conventional oven ensures the completion of slow and/or multistep Maillard reactions responsible for browning and for the generation of specific aromas. (3) In the case of porous materials such as bread, the high temperature and relative low humidity of the air surrounding the product cause rapid heating of the surface of the food relative to the center, thus creating a temperature and a corresponding inward vapor pressure gradient that helps retain volatile aroma compounds inside the core. In the microwave oven, the short time exposure and the lack of hot dry air (air being transparent to microwave irradiation) surrounding the surface of the food product not only prevents crusting but also promotes sogginess due to the condensation of the moisture. On the other hand, the rapid release of moisture and its evaporation from the center of the food causes the added and formed volatiles to be "steam distilled" at temperatures below their boiling points (see Chapter 4). Hence baked and roasted food products, which rely heavily on Maillard produced flavors, usually do not perform well in the microwave oven.

Schiffmann [23] summarized the different factors related to microwave ovens that affect aroma generation during cooking of food such as variation in the type of commercial ovens (power, cavity size, etc.) and its effect on the reproducibility of performance, speed of heating, oven temperature, and vapor pressure buildup inside the food. The short time required in the microwave oven to attain the same temperature as in the conventional oven not only retards the Maillard reaction but also prevents the establishment of thermal equilibrium throughout the food and uniform temperature distribution through conductive heat transfer. These hot and cold spots in the food product aggravate further the oven hot and cold zones created as a result of standing wave patterns. In addition, different dielectric loss factors (ϵ'') associated with different components in a multicomponent food product will also contribute to the uneven heating pattern inside the microwave oven. The combined effect of these phenomena is manifested in the excessive exposure of certain parts of food to heat and diminished exposure in others, leading to undesirable textural and flavor modifications such as charring, drying, excessive evaporation, hardening, and development of burnt or raw flavor and aroma notes. The extent of these undesirable modifications is dependent on the size, geometry, thickness, and the composition of the food product [23]. Yeo and Shibamoto [24] reviewed the chemical composition of volatiles generated by microwave and conventionally heated food products. White cake batters were cooked to the same degree both in the microwave and the conventional oven. The volatiles released and sensory properties of both products were compared [25]. The number of volatiles detected and the amount of total pyrazines produced were found to be more in the conventionally baked sample. In addition, the microwave cake lacked the nutty, caramel, and browned flavors. In a similar study [26], the number of volatiles generated from boiled beef cooked by microwave for 1 h, was found to be more than the number of volatiles generated by beef boiled conventionally, for the same length of time. When both systems were compared on the basis of "doneness," the microwave sample generated only one third the amount of volatiles detected in the conventional oven. The relative success of the microwave to achieve the Maillard effect of conventional heating may depend to a large degree on the type and composition of the food product.

Most of the model studies reported in the literature compare the volatiles generated by the Maillard reaction under microwave or conventional heating to elucidate the differences between the two modes of heating. However, in an interesting study, Barbiroli et al. [15] compared the changes in the concentration of the added amino acids, amino acid/glucose mixtures, and Amadori rearrangement products (ARPs) in bread and biscuits baked by both methods. This approach provides more direct evidence than the analysis of volatiles on the differences in the chemical behavior of sugar/amino acid mixtures in food matrices when exposed to different modes of heating. Table 1 summarizes the changes observed in the content of amino acids and ARPs in bread and biscuits after baking by microwave (3 min) and conventional ovens (220°C for 30 min). In the conventional oven, most of the ARPs formed in situ from added sugar/amino acid mixtures were decomposed, whereas in the microwave oven the ARPs accumulated instead. Model studies [27] have indicated that before any significant decomposition of ARP is initiated during the Maillard reaction, the concentration of Amadori product should reach a minimum value, and it seems that this was attained after 3 min under the experimental conditions. Microwaving the samples beyond the 3 min would have initiated the decomposition process. When Amadori compounds were added to biscuits, they decomposed more or less to the same extent during both types of heating, whereas in the case of bread the decomposition of Amadori products under microwave heating was greater. The relative amounts of formation or decomposition, however, were dependent on the type of ARP and the food matrix. In the biscuit, for example, the decomposition of ARPs in the conventional oven is more pronounced than in the bread. This could be related to the differences in water content between bread (0.48 ml of water/g of solids) and biscuit (0.34 ml of water/g of solids), and the temperature of baking. Interestingly, in the microwave oven the amounts of both the accumulated and decomposed ARPs were not dependent significantly on the type of food matrix.

Table 1 Percent Loss of Added Amino Acids and Amadori Products and Percent Accumulation of Amadori Products in Bread and Biscuits Baked Conventionally (Bread: 220°C for 30 min; Biscuit: 240°C for 20 min) and Under Microwave Heating (3 min)

	Conventional oven		Microwave oven	
Added components	Amino acid (bread/biscuit)	ARP* (bread/biscuit)	Amino acid (bread/biscuit)	ARP* (bread/biscuit)
Leucine	−3%/−19%		−15%/−15%	
Leucine + glucose	−6%/−25%	+8%/+3%	−95%/−53%	+66%/+71%
Leucine ARP	Trace/+32%	−14%/−92%	+30%/+36%	−97%/−93%
Valine	−6%/−8%		−3%/−3%	
Valine + glucose	−6%/−15%	+0%/+6%	−68%/−60%	+97%/+82%
Valine ARP	+9%/+33%	−22%/−80%	+40%/+37%	−96%/−88%

* Percent weight formation relative to glucose Amadori rearrangement product (ARP) −, loss; +, formation of ARP or release of amino acid from ARP.
Source: Adapted from Ref. 15.

The result of this study clearly shows that Amadori product, the main precursor of Maillard aromas and colors, can be formed from sugar and amino acids found in food under microwave heating (3 min) but do not decompose efficiently during the short time scale of microwave exposure.

III. AROMA RELEASE DURING MICROWAVE HEATING OF FOOD PRODUCTS

Microwave heating of food products is done in a relatively quick time period as compared to conventional oven cooking. As discussed in the previous section, the aroma of the final product can result from aroma generated during microwave cooking. It can also be already contained in the food, e.g., in a precooked meal. In any case, the release phenomena is the same and it is the timing that may be different. If the aroma is only produced at the end of heating, losses due to volatilization will be diminished considerably. In microwave heating of a precooked product, the volatile aroma compounds are integral to its final aroma, and losses can imbalance the aroma. Within this chapter, "aroma" is defined as the volatile aroma compounds that contribute both to the orthonasal (sniffing) and retronasal (eating) smell of a food. This section will explain the theory and give examples of how aroma compounds present in microwave foods can be lost during cooking.

A. Methods to Measure Aroma Release

In order to measure the overall loss of aroma during heating, the food can be extracted before and after heating, for example, by different extraction techniques such as simultaneous distillation extraction for fat containing foods or solvent extraction for nonfat liquids. As analysis by these methods can be laborious, special devices, to measure the losses, by trapping have also been developed [28–30]. These methods can also be used to study loss dynamics by sampling at various time points.

B. Applications and Theory of Aroma Release

The aroma losses that occur during microwave heating are linked to the air-product partitioning as influenced by temperature, mass transfer, moisture gradient, and composition of the food. These are the same factors that influence aroma release under other situations such as product storage, release in the mouth, boiling food, or conventional oven cooking. All of these conditions have a certain temperature profile and the product-air partition coefficients can predict the release. The distinguishing factor of microwave heating is the quick heat rise followed by steam distillation.

During microwave heating, the temperature increase is not even over the entire food as seen by the presence of hot and cold spots. However after a short time, a final temperature of about 100°C is reached. The food remains at this temperature as long as it contains water; during which time steam distillation takes place. Thus, microwaving food induces a large water loss, at an especially high rate. In microwave baking where almost all of the water has vaporized, even higher product temperatures can be reached.

A strong link exists between aroma loss and the steam distillation process. During heating of microwave products, water vapor moves from the interior of the product to the surface and then vaporizes. Thus, eventually the interior of the product will become dry while the surface remains moist due to recondensation of moisture after exposure to the ambient temperature. During this process, steam serves to transport the volatile compounds from the interior of the food to the air. In work with frozen spaghetti [28], the timing of the release of the less volatile compounds correlated with the attainment of 100°C temperature, at which time water began to vaporize from the food at high rates. Highly volatile compounds, however, were released from the spaghetti before the onset of steam distillation. Likewise, high losses of volatile acids in 90/10 oil-water mixtures were observed and were related to the preferential heating of the water phase in conjunction with steam distillation [16]. Indeed, the moisture content of the system can influence the extent of aroma released. As the moisture concentration decreased below 0.1 g water/g solid during microwave heating of gelatinized flour dough, a type of encapsulation occurred that prevented limonene from being released [30]. In this system, migration rates of limonene, *tert*-butylbenzene, and pyrazine were much lower than that of water, which may be expected because of their size differences.

Shaath and Azzo [31] proposed what they termed as "delta T' theory" as a basis to design microwave stable flavor formulations. This hypothesis is based on experimentally obtained $\Delta T'$ values for individual aroma compounds used frequently in flavor formulations. These values indicate the increase in temperature of a solution of a pure aroma compound relative to an equivalent amount of water when similarly exposed to microwave irradiation. $\Delta T'$ values has been found to increase with dielectric constant (ϵ') values in a series of compounds having the same specific heat. Shaath and Azzo [31] claim that selecting aroma compounds for microwave formulations with values less than that of water ($\Delta T' = 1$) will prevent their volatilization during microwaving. "Delta T' theory" has been justifiably criticized for not taking into consideration the effect of food matrix interactions with the aroma compounds [32] and for its inability to predict the microwave behavior of a mixture of aroma compounds [33]. Experimental evidence [34] indicates that the temperature of individual aroma compounds in food cannot be significantly different from the bulk temperature of the food matrix since the heat generated by individual compounds is dissipated quickly into the surrounding medium with no significant influence on the overall temperature due to their low concentrations. In addition, Steinke et al. [16] have demonstrated the solvent (matrix) dependence of microwave volatilization of aromas. Therefore, $\Delta T'$ values relative to water may not be used to predict the microwave behavior of aroma mixtures in complex food systems.

Graf and de Roos [34] on the other hand, developed a mathematical model that takes into account the food matrix interactions and compartmentalization of aroma compounds in addition to their volatility and hydrophobicity to predict the behavior of aroma mixtures under microwave irradiation. The model considers the food matrix as an oil/water emulsion and the process of volatilization of aroma as repeated water extraction steps. The fraction f of aroma compounds remaining in an oil/water emulsion after equilibration with air in a closed system was calculated by

$$f = \frac{V_w P_{ao} + V_o P_{aw}}{V_a P_{aw} P_{ao} + V_w P_{ao} + V_o P_{aw}} \qquad (1)$$

where V_w = volume of water (ml)
P_{ao} = air to oil partition coefficient
V_o = volume of oil (ml)
V_a = volume of air
P_{aw} = air to water partition coefficient

The validity of this equation to predict the equilibrium concentrations of food volatiles was demonstrated in a model study using milk with five added aroma compounds. The experimentally determined concentrations of milk headspace volatiles were in good agreement with the calculated values using Eq. (1). This static equilibrium equation was transformed into an equation that describes dynamic nonequilibrium systems characteristic of foods during microwave heating. This transformation was based on the assumption that foods under microwave heating undergo consecutive extractions with infinitesimal volumes of moisture. The fraction f_n of aroma compounds remaining in the food after n number of extraction cycles is given by

$$f_n = [1 - f_e + f_e(f)]^n \qquad (2)$$

where f_c = fraction of food being extracted
n = number of successive extractions
f = given by Eq. (1)

In general, the model demonstrates that the rate of aroma volatilization is determined by the partition coefficients of aroma compounds, chemical and physical properties of food matrix, moisture loss, and amount of heat input. For the microwave baking of cakes, the model showed a good relationship between the product-air partition coefficient determined using Eq. (2) and the percent lost during baking. In this case, compounds with high volatility in the product such as 2,3-dimethylpyrazine were lost up to 80% during baking as opposed to compounds with low volatility such as raspberry ketone and δ-2-decenolactone that were only lost up to 20% [35].

While knowledge of product-air partition coefficients is necessary for the understanding of aroma release in complex food systems, simple flavored aqueous solutions were shown to have volatile losses that could be predicted by the application of classical physical chemistry laws such as the Henry's law [36]. Similarly, for a low-fat product such as frozen spaghetti, initial release time was found to correlate well ($R^2 = 0.98$) with the air-water partition coefficients in a model study using 6 compounds [28]. In this case, the food matrix contained little oil and the simpler partition coefficient comprising only water as the product phase gave good predictions. The release was found to be linear over time and the amount lost during heating varied between 27 and 44% [37]. Losses were also demonstrated during reheating of frozen blueberry pancakes. The losses amounted to between 10 and 56% and also were predicted from volatilities of the aroma components [38].

As demonstrated by theory [35] and by examples with actual products, the losses of aroma compounds during microwave heating can be substantial and can also be predicted using partitioning theory of the compound with the food matrix. An important point to note is that compound volatilities in the food matrix are not the same as boiling points of pure compounds. Although hydrophilic compounds such as diacetyl have low boiling points (are volatile in a pure state), they exhibit low air-water partition coefficients due to their high water solubility. It is therefore essential to use aroma compound partitioning for release predictions rather than boiling point information. This fact was also demonstrated in a study comparing the volatilites of furfural (b.p. 162°C) and methyl benzoate (b.p. 200°C) where the former was more volatile during microwave heating [39], probably due to its higher water solubility.

Aroma release due to air-matrix partitioning is compounded by the additional effect of steam distillation where steam acts as an active transport vehicle from the interior to the surface of food products and eventually to air. The next section will address some ideas of how to minimize aroma loss and maximize aroma formation.

IV. DEVELOPMENT OF NEW PRODUCTS/PROCESSES TO OPTIMIZE AROMA FORMATION AND MINIMIZE AROMA RELEASE DURING MICROWAVE PROCESSING OF FOOD PRODUCTS

In the last decade, the sales of microwave food products have experienced a larger growth rate compared to overall food sales due to the convenience offered by such products to the consumers. As a result, food products specifically formulated to perform under microwave irradiation have become priority areas of research in many food companies [40]. This trend however, seems to be changing [41], as the number of domestic microwave ovens used in functions other than heating, has fallen from almost 45% in the mid-1980s to under 30% in the early 1990s [42]. The sales projection for microwave-only food products of $4 billion for 1997 may reach only $2.7 billion from today's $2.4 billion market [43].

Different strategies have been attempted to overcome the lack of Maillard flavor and color development during microwaving. Some were aimed at packaging technologies and others at the modifications of food and flavor formulations or addition to food of specific formulations containing Maillard-active ingredients to promote flavor and/or color development. One of the earliest attempts by the food industry to promote surface browning in foods formulated for microwave was the use of susceptor packaging. Susceptors are metalized polyethylene teraphthalate (PET) films laminated on paperboard that can interact with electromagnetic radiation and generate localized heating near the surface of food when incorporated with the packaging materials of microwave food products [44]. While susceptors are effective in promoting surface browning, they do not allow the full development of flavors [32].

As has been implicated in the previous sections, the failure of microwave ovens to develop flavors and colors is due to the limited time-scale of microwave exposure of foods to elevated temperatures necessary to carry out the Maillard reaction. The main strategy used to overcome this limitation is the addition of reactive Maillard reaction precursors to the surface of food to accelerate the Maillard reaction such that it can be accomplished within the short time frame of microwave cooking (10–15 min). Most of the patent literature, however, discloses processes to generate browning rather than specific aromas and browning. The general requirement for microwave-reactive Maillard mixtures is their ability to generate, in addition to color, either no aroma or an aroma that is compatible with the intended product. Browning-only precursors could have a wider range of applications in different food products. The general composition of such mixtures reported in the patent literature include reactive sugars and amino acids/or protein hydrolysates and also a promoter (an alkaline component). These mixtures could be applied to the surface of foods either as liquids, dried powders, or as reactants encapsulated in a carrier. Bryson et al. (U.S. Pat. No 4,735,812 issued on April 5, 1988) developed a browning agent composed of hydrolyzed collagen or gelatin mixed with reducing sugars and a mixture of sodium carbonate and bicarbonate. The browning agent could be applied on the surface of foods to be microwaved either as a powder or as a liquid. Parliment et al. (U.S. Pat. No 4,857,340 issued August 15, 1989) describe aroma producing precursors encapsulated in a lipid carrier capable of releasing aroma under microwave irradiation, when placed in close proximity to a microwave susceptible material. Kang et al. (U.S. Pat. No 5,059,434 issued October 22, 1991) disclose a process for manufacturing a browning powder for muscle foods composed of separately encapsulated reactive sugars (xylose, arabinose, etc.), amino acids (lysine, arginine etc.) or hydrolyzed vegetable proteins and a Maillard reaction promoter (polyvinyl pyrolidone). A similar patent (U.S. Pat. No 5,091,200 issued February 25, 1992) for baked goods was also issued to the same inventors. Steinke et al. (U.S. Pat. No. 5,043,173 issued August 27, 1991) describe a browning water-in-oil emulsion containing only a reactive carbonyl compound and an edible hydrophilic base such that upon heating, the emulsion breaks down and brings the reactants together to initiate browning during microwave heating. Yaylayan et al. [13] reported on the development of microwave-active Maillard formulations that can generate browning and at the same time deliver specific aromas (baked, roasted, chicken, beef etc.) when coated on targeted food products. To facilitate the choice of amino acids in these mixtures, the potential of different amino acids to produce specific aroma notes in the microwave was evaluated.

Flavoring formulations are usually composed of hundreds of compounds, many of which are key to the flavor. Each compound will exhibit a different partitioning with the food matrix and the losses during microwave heating will be based on this phenomenon. There are thus two primary ways to minimize their loss: change the food matrix or change the aroma compounds. Changing the food matrix to reduce compound volatility often involves adding an oil phase or increasing the oil content. As most flavor compounds are more oil soluble than water soluble, increasing the oil phase increases the product-air partition coefficient. This was demonstrated with microwavable cakes where increasing the fat content resulted in decreased losses during microwave cooking [35]. Furthermore, the local environment could also be modified to reduce the volatility of aroma compounds, using methods such as entrapment in fat or encapsulation [45]. Food products heated under microwave irradiation may have an unbalanced flavor profile due to these losses. Re-balancing of the aroma profile could be accomplished by the addition of higher amounts of those components most easily lost before microwave heating. This option that flavoring design offers is to over-compensate for those compounds, in-order to have a better equilibrated flavor profile in the end-product. Another option is to develop flavorings that could better withstand microwave heating, by choosing compounds based on their degree of loss during microwave heating. If a flavor company develops the flavoring, they may have the ability to customize the flavor formulation for microwave applications. Finally, advances in the microwave oven technology such as the introduction of dual microwave/convection ovens should also enhance the flavor formation and retention in food products.

REFERENCES

1. J. McGorrin, T. H. Parlimant, and M. J. Morello, eds. Thermally Generated Flavors: Maillard, Microwave and Extrusion Processes. ACS Symposium Series, no. 543, American Chemical Society, Washington DC, 1994.

2. R. Gedye, F. Smith, K. Westaway, H. Ali, L. Baldisera, L. Laberge, and J. Rousell. The use of microwave ovens for rapid organic synthesis. Tetrahedron Lett. 27:279–282 (1986).
3. Richard N. Gedye, Frank E. Smith, and Kenneth Charles Westaway. The rapid synthesis of organic compounds in microwave ovens. Can. J. Chem. 66:17–26 (1988).
4. Richard N. Gedye, Werner Rank, Kenneth C. Westaway. The rapid synthesis of organic compounds in microwave ovens. II. Can. J. Chem. 69:706–711 (1991).
5. E. R. Peterson. Microwave chemistry: A conceptual review of the literature. In Quality Enhancement Using Microwaves. 28th Annual Microwave Symposium Proceedings, International Microwave Power Institute, 1993.
6. E. D. Neas and M. J. Collins. Microwave heating. In Introduction to Microwave Sample Preparation: Theory and Practice (H. M. Kingston and L. B. Jassie, eds.), American Chemical Society, Washington DC, 1988, pp. 7–32.
7. R. J. Giguere, T. L. Bray, S. M. Duncan, and G. Majetich. Application of commercial microwave ovens to organic synthesis. Tetrahedron Lett. 27:4945–4948 (1986).
8. J. R. Paré, M. Sigouin, and J. Lapointe. U.S. Pat. No 5,002,784 issued March 26, 1991.
9. V. Yaylayan. Maillard reaction under microwave irradiation. Chapter 9 in The Maillard Reaction: Consequences for the Chemical and Life Sciences (R. Ikan, ed.). John Wiley, New York, 1996, pp. 183–198.
10. V. Yaylayan. Classification of the Maillard reaction: A conceptual approach. Trends Food Sci. Technol. 8:13–18 (1997).
11. A. K. Bose, M. S. Manhas, B. K. Banik, and E. W. Robb. Microwave-induced organic reaction enhancement. (MORE) chemistry: Techniques for rapid, safe and inexpensive synthesis. Res. Chem. Interned. 20:1–11 (1994).
12. S. P. Bond, C. E. Hall, C. J. McNerlin, and W. R. McWhinnie. Coordination compounds on the surface of complexes. J. Materials Chem. 2:37 (1992).
13. V. Yaylayan, N. G. Eorage, and S. Mandeville. Microwave and thermally induced Maillard reactions. In Thermally Generated Flavors: Maillard, Microwave and Extrusion Processes. J. McGorrin, T. H. Parliment, and M. J. Morello (eds.). ACS Symposium Series, No. 543, American Chemical Society, Washington DC, 1994, pp. 449–456.
14. Shui Tein Chen, Shyh Horng Chiou, Kung Tsung Wang. Preparative scale organic synthesis using a kitchen microwave oven. J. Chem. Soc., Chem. Commun. 11:807–809 (1990).
15. G. Barbiroli, A. M. Garutti, and P. Mazzaracchio. Note on behavior of 1-amino-1-deoxy-2-ketose derivatives during cooking when added to starch based foodstuffs. Cereal Chem. 55:1056–1959 (1978).
16. J. A. Steinke, C. M. Frick, J. A. Gallagher, and K. Strassburger. J. Influence of Microwave heating on flavor. In Thermal Generation of Aromas (T. Parliment, R. McGorrin, and C. T. Ho, eds.). ACS Symposium Series, No. 409, American Chemical Society, Washington, DC, 1989, pp. 520–525.
17. H. C. H. Yeo and T. J. Shibamoto. Flavor and browning enhancement by electrolytes during microwave irradiation of the Maillard model systems. J. Agric. Food Chem. 39:948–951 (1991)
18. H. C. H. Yeo, and T. J. Shibamoto. Microwave-induced volatiles of the Maillard

model system under different pH conditions. J. Agric. Food Chem. 39:370–373 (1991).
19. T. H. Parliment. Comparison of thermal and microwave mediated Maillard reactions. In Food Flavors, Ingredients and Composition (G. Charalambous, ed.). Elsevier, Amsterdam 1993, pp. 657–662.
20. H. Ji and R. A. Bernhard. Effect of microwave heating on pyrazine formation in a model system. J. Sci. Food Agric. 59:283–289 (1992).
21. M. M. Leahy and G. A. Reineccius. Kinetics of formation of alkylpyrazines-effect of type of amino acid and type of sugar. In Flavor Chemistry: Trends and Developments (R. Teranishi, R. G. Buttery, and F. Shahidi, eds.). ACS Symposium Series, No. 388, Washington DC, 1989, pp. 76–91.
22. Y. Pomeanz and C. E. Meloan. Food Analysis: Theory and Practice, 2nd ed. Van Reinhold Nostrand, New York, 1987.
23. R. F. Schiffmann. Critical factors in Microwave-generated aromas. In Thermally Generated Flavors: Maillard, Microwave and Extrusion Processes (J. McGorrin, T. H. Parlimant, and M. J. Morello, eds.). ACS Symposium Series, No. 543, American Chemical Society, Washington DC, 1994, pp. 386–394.
24. H. C. H. Yeo, T. Shibamoto. Chemical comparison of flavours in microwaved and conventionally heated foods. Trends Food Sci. Technol. 12:329 (1991).
25. C. Whorton and G. Reineccius. Flavor development in a microwaved versus a conventionally baked cake. In Thermal Generation of Aromas (T. Parliment, R. McGorrin, and C. T. Ho, eds.). ACS Symposium Series, No. 409; American Chemical Society, Washington, DC, 1989, p. 526.
26. G. MacLeod and B. M. Coppock. Volatile flavor components of beef boiled conventionally and by microwave radiation. J. Agric. Food Chem. 24:835–843 (1976).
27. A. Huyghues-Despointe and V. Yaylayan. Kinetics of Formation and degradation of morpholino-1-deoxy-D-fructose. In Physical Chemistry of Flavors (C-H Tong, C-T Tan, and C-T Ho, eds.). ACS Symposium Series, No. 610, American Chemical Society, Washington DC, 1995, pp. 20–30.
28. D. D. Roberts, P. Pollien. Analysis of aroma release during microwave heating. J. Agric. Food Chem. 45:4388–4392 (1997).
29. S. J. Risch, K. Keikkila, and R. J. Williams. Analysis of migration of a volatile substance during heating with microwave energy. U.S. Patent 5177995, 1993.
30. Y. Fu. Microwave-assisted heat and mass transfer in food. Ph.D. dissertation, Rutgers University, 1996.
31. N. A. Shaath and N. R. Azzo. In Thermal Generation of Aromas (T. Parliment, R. McGorrin, and C. T. Ho, eds.). ACS Symposium Series, No. 409, American Chemical Society, Washington, DC, 1989, p. 512.
32. G. Reineccius and C. Whorton. Flavor problems associated with the microwave cooking of food products. In The Maillard Reaction in Food Processing, Human Nutrition and Physiology (P. Finot, H. Aeschbacher, R. F. Hurrell, and R. Liardon, eds.). Birkhäser, Basel, 1990, p. 197.
33. R. Schwarzenbach. Microwave stable flavors—development or promotion? In Flavor Science and Technology. Wiley, Chichester, 1990, pp. 281–296.
34. E. Graf and K. de Roos. Nonequilibrium partition model for prediction of microwave

flavor release. In Thermally Generated Flavors: Maillard, Microwave and Extrusion Processes (J. McGorrin, T. H. Parliment, and M. J. Morello, eds.). ACS Symposium Series, no. 543, American Chemical Society, Washington DC, 1994, pp. 437–448.
35. K. B. de Roos and E. Graf. Nonequilibrium partition model for predicting flavor retention in microwave and convection heated foods. J. Agric. Food Chem. 43:2204–2211 (1995).
36. A. Stanford and J. McGorrin. Flavor volatilization in microwave food model systems. In Thermally Generated Flavors: Maillard, Microwave and Extrusion Processes (J. McGorrin, T. H. Parliment, and M. J. Morello, eds.). ACS Symposium Series, No. 543, American Chemical Society, Washington DC, 1994, pp. 414–436.
37. D. D. Roberts and P. Pollien. Relationship between aroma compounds' partitioning constants and release during microwave heating. In Flavor Analysis: Developments in Isolation and Characterization (C. J. Mussinan and M. J. Morello, eds.). ACS Symposium Series, No. 705, American Chemical Society, Washington, DC, 1998, pp. 61–68.
38. H. C. Li, S. J. Risch, and G. A. Reineccius. Flavor formation during frying and subsequent losses during storage and microwave reheating in pancakes. In Thermally Generated Flavors: Maillard, Microwave and Extrusion Processes. (T. H. Parliment, M. J. Morello, and R. J. McGorrin, eds.). ACS Symposium Series, No. 543, American Chemical Society, Washington, DC, 1994, pp. 466–475.
39. H. Felix. 1991. Flavors for microwave food systems. Food Market and Technol., 5(5):14–19.
40. R. V. Decareau. Microwave Foods: New Product Development. Food & Nutrition Press Inc., Trumbell, CT, 1992.
41. A. E. Sloan. Top ten trends to watch and work on. Food Tech. 48:96 (1994).
42. H. Balzer. The ultimate cooking appliance. Am. Demographics, pp. 40–44, July 1993.
43. Find/SVP. The market for microwavable foods. 1993. Find/SVP, New York.
44. N. A. Shaath and N. R. Azzo. Latest developments in microwavable food and flavor formulations. In Flavors and Off-flavors (G. Charalambous, ed.). Elsevier, Amsterdam, 1990, pp. 671–686.
45. H. C. Li and G. A. Reineccius. Protection of artificial blueberry flavor in microwave frozen pancakes by spray drying and secondary fat coating processes. In Encapsulation and Controlled Release of Food Ingredients (S. J. Risch and G. A. Reineccius eds.). ACS Symposium Series, No. 590, American Chemical Society, Washington, DC, 1995, pp. 180–186.

6
Bacterial Destruction and Enzyme Inactivation During Microwave Heating

Ramaswamy C. Anantheswaran
*The Pennsylvania State University
University Park, Pennsylvania*

Hosahalli S. Ramaswamy
*McGill University
Montreal, Quebec, Canada*

I. INTRODUCTION

The microwave oven has become a common appliance in most household kitchens and competes with the conventional range as the primary cooking device. Most consumers use microwave ovens for reheating leftovers or warming cooked/chilled food systems. Over 30 million cases of food-borne illness occur annually in the United States, of which around 15% result from inadequate cooking of foods at home. Microwave heating also offers much potential in manufacturing of food products, as discussed in Chapter 9 of this handbook. Uneven heating of foods, irregular heating patterns in foods of varying heterogeneity, and the much shorter heating times encountered when using a microwave oven (as against conventional cooking methods) can result in survival of bacteria, even when the overall temperature in the product may appear to be adequate. Adequate holding/standing times need to be provided at the end of microwave heating to eliminate hot and cold spots and to allow for equilibration of temperatures within the food product.

Destruction of microorganisms by microwave heating has had considerable interest since 1940s when the first work by Fleming [1] was reported and the argument between nonthermal and thermal effects was born. Research studies have focused on different experimental procedures, approaches, experimental designs, techniques, and biological systems to distinguish thermal or nonthermal effects of microwave heating. Several theories have been proposed to explain how electromagnetic fields might kill microorganisms without heat, as summarized in a review by Knorr [2]. However, some researchers [3] refute any molecular effects of electric fields compared with thermal energy using classical axioms of physics and chemistry. Microwave heating have been shown to bring about unique structure in ceramic materials, and chemical reactions have been reported to be accelerated by using microwave heating. There is, however, a knowledge gap on the fundamental understanding of the effects of microwave and dielectric heating in food products and further studies are needed (Palaniappan and Sastry [4]).

II. KINETICS OF DESTRUCTION DURING MICROWAVE HEATING

Labuza [5] showed that the kinetics of bacterial inactivation in conventional heating processes are related to time-temperature conditions within the food and may be generally modeled as nth-order chemical reactions by the equation

$$\frac{dA}{dt} = -kA^n \tag{1}$$

where A = concentration of the species (cfu/ml) at time t (min)
k = thermal inactivation rate (rate constant) at temperature T (K)
n = generalized reaction order

The thermal inactivation constant (rate constant) can be described as a function of temperature by means of the Arrhenius equation:

$$k = k_0 \exp\left[\frac{E_a(T - T_0)}{RTT_0}\right] \tag{2}$$

where k_0 = rate constant at temperature T_0, min^{-1}
k = rate constant at temperature T, min^{-1}
T_0 = reference temperature, K
T = temperature of process, K
E_a = energy of activation, J/mol
R = universal gas constant, 8314 J/mol K

The literature indicates that bacterial destruction kinetics generally follow a first-order reaction. The rate constant k at a reference temperature and the activation energy E_a define the thermal resistance of a microorganism over the temperature range of interest for specific food products. The thermal resistance of microorganisms is also traditionally characterized in the food processing industry by means of the D and z values [6] The D value is the thermal death time required to cause a 90% reduction in bacterial numbers at a given temperature, and is analogous to the k value. The z value is the change in temperature necessary to cause a 90% change in the D value. The z value is a measure of the relative heat resistance of a microorganism to different temperatures and is analogous to the E_a value.

$$D = \frac{t}{\log(C_0/C)}$$

or (3)

$$D = \frac{2.303}{k}$$

$$z = \frac{T_2 - T_1}{\log(D_1/D_2)} \tag{4}$$

When the thermal resistance of a microorganism is known, and when the sample is subjected to a nonisothermal heating, it is possible to calculate the equivalent time necessary for thermal treatment by integration of the time-temperature history using the equation

$$F = \int_0^t 10^{[T(t)-T_R]/z} dt \tag{5}$$

This approach has been traditionally used in the thermal processes calculations. Most studies dealing with determination of kinetic parameters rely on isothermal heating conditions. A small volume of test liquid is rapidly heated to the desired temperature, samples are withdrawn at known intervals thereafter, cooled immediately and enumerated. Experiments are repeated at different temperatures and kinetic parameters are evaluated from the survivor data using Eqs. (1) to (4). A similar concept can be applied for determining kinetics parameters during microwave heating when constant temperature conditions exist. Although data for determining bacterial destruction kinetics during conventional heating are well documented, there exists a dearth of similar data for microwave heating.

Much of the data reported cannot be considered to accurately represent the kinetics of bacterial destruction entirely during microwave heating. Researchers would microwave foods containing bacteria for a given time period, switch off the microwave oven and then enumerate microbial loads from samples withdrawn at periodic intervals. The rate constant k thus cannot be considered to have been calculated entirely during microwave heating, but during the constant temperature holding period following microwave heating (in which case k may not be considered to be different from conventional heating). Some studies have recognized these limitations, and have designed sophisticated control systems [7–10] for maintaining set-point temperatures inside microwave ovens. Although these systems are very innovative and accurate, the kinetic parameters obtained may not be considered to truly represent microbial inactivation during microwave heating since microwave energy used during the testing phase (constant temperature hold conditions) is mostly to prevent heat loss rather than for sample heating.

In some test situations, nonisothermal heating conditions [as is characteristic during come-up (CUT) and come-down (CDT) times] are involved in addition to the isothermal heating periods, and the above procedure needs to be modified to correct for the CUT and CDT [11]. Simple alternatives would be to avoid taking samples during CUT and try to accomplish quick cooling so that the CDT contribution could be ignored while calculating D and z. When this is not possible, a first estimate of the kinetic parameters D and z is initially estimated using the uncorrected heating time. Then the heating time is corrected to get the effective or thermal time F using the time-temperature profile and Eq. (5). The corrected time F is then used to recalculate the D and z values. When corrections are significant, the recalculated z value will be different from the initial value. The correction procedure is then repeated several times, until the convergence of two successive z values. The parameters so obtained will be equivalent of their counterparts obtained from isothermal experiments [11].

Some recent studies [11–17] have used continuous-flow steady-state microwave heating methods to evaluate the kinetics of microbial destruction. The continuous-flow microwave treatment of liquids only involves nonisothermal heating come-up time (or residence time) inside the oven during microwave heating and a come-down time outside during cooling. The procedure for gathering kinetic parameters during continuous-flow heating is similar to what is described in the preceding paragraph. Briefly, the D value at the exit temperatures (temperature at CUT) is first calculated from the regression of log residual numbers of survivors versus uncorrected heating time (residence time), and then the z value is obtained as the negative reciprocal slope and log D versus temperature. Using the calculated z value, the heating times are corrected using Eq. (5), and D values and subsequently the z are recalculated. This step is repeated as many times as necessary until the convergence of the z value. In order to accommodate the CDT which is carried out outside the microwave oven, it is necessary to determine the cooling (thermal) contribution to microbial destruction and subtract it from total destruction. In order to do this, the effective cooling time t_c is computed using Eq. (5), with the z value obtained from thermal destruction studies. The extent of logarithmic thermal destruction (LTD) during cooling is then calculated using the relationship:

$$\text{LTD} = t_c/D \qquad (6)$$

where D is the decimal reduction value at the exit temperature obtained from thermal destruction studies. This calculated value is then subtracted from the combined destruction of microbial population due to microwave heating plus cooling. Microbial destruction data of test samples can thus be corrected for both come-up and come-down period contribution to lethality.

III. THERMAL, NONTHERMAL, AND MICROWAVE ENHANCED EFFECTS DUE TO MICROWAVE HEATING

A major area of controversy over microbial destruction centers on the distinction between thermal and nonthermal effects of microwave heating on microorganisms. This is due to the partial difficulty in determining whether experimentally detected differences in the destruction of microorganisms during conventional heating and microwave heating are due to normal experimental error or possibly due to additional destruction caused directly by the oscillating electric and magnetic fields [18–20]. Since the oscillating electric fields are responsible for the heat generation within the food product, possible nonthermal effects will always overlay the thermal effect. Conflicting reports have appeared in the literature supporting both viewpoints.

Microwaves used for domestic and industrial heating applications are part of the electromagnetic spectrum with a specific frequency of 915 or 2450 MHz. The waves have the capacity to penetrate the food and create heat by friction resulting from the oscillation of dipole molecules of water, which will try to orient and align with the field. This is the macroscopic *thermal effect* of increasing temperature within the material.

Traditionally, "nonthermal" effects under the application of electromagnetic radiation refer to lethal effects without involving a significant rise in temperature as in the case of ionizing radiation. One of the effects of such quantum energy is breakage of chemical bonds. Roughly one electron volt of energy is required to break from a molecule a covalent bond to produce one ion pair [21] and this is referred to as a direct *nonthermal effect*. Electromagnetic radiation above 2500×10^6 MHz, which possesses such a capability, is mostly referred to as ionizing radiation (example, x rays, gamma rays, etc.). As the wavelength increases and frequency decreases, not enough energy is available to break chemical bonds. Ultraviolet, visible, and possibly infrared rays have energy to break weak hydrogen bonds, but microwaves do not have sufficient energy to break any chemical bonds and therefore belong to the group of nonionizing forms of radiation.

The review of Stuerga and Gaillard [3] provides some definitions used in biological studies based on irradiation or power flux density in W/m^2 or specific absorption rate in W/kg and does not involve any assumptions relating to mechanisms of interaction. Three domains of power density in comparison to the thermal capacity of biological metabolism are defined. The threshold selected for standard definition of biological metabolism corresponds to 4 W/kg of living matter; hence *thermal effects* are defined for irradiation power density greater than 4 W/kg, when the organism cannot dissipate the energy supplied by the irradiation; *nonthermal effects* correspond to density between 0.4 and 4 W/kg. In such conditions, the thermo-regulation system is able to compensate for the effect of irradiation.

The cases where microwave heating gives a particular time-temperature profile and gradients, which cannot be achieved by other means, are only *microwave specific*. The effect of higher temperatures achieved by microwave heating was explained by Gedye [22, 23]. According to Gedye, it is possible that microwave reactions could produce different products than from reactions achieved by using conventional reflux techniques. Since microwave heating significantly increases the reaction temperature, it is possible that the microwave reaction temperature could exceed the temperature required for a new reaction that was not possible at the lower reflux temperature. According to Gedye these results are important because they confirm that microwave heating does not alter the reaction but simply provides a much faster and more efficient (higher temperature) method of carrying out organic reaction. According to Risman [24], any *nonthermal effect* must not be explicable by macroscopic temperatures, time-temperature histories, or gradients. This means that any effects which can be explained by applying verified theories to experimental data and macroscopic temperatures are *thermal*. Any effect that cannot be so explained will have to be considered nonthermal contribution. One more distinction is recognized: The effects are considered *nonthermal* if they are independent of sample temperature, and if they existed but also depended on temperature, they are considered *enhanced thermal effects* [25].

IV. FACTORS AFFECTING MICROBIAL DESTRUCTION DURING MICROWAVE HEATING

During microwave heating, heat transfer can occur without a heating medium. In addition, the initial come-up time can be shortened due the rapid rate of volumetric heating. However, limitations exist with this technology. Localized heating patterns can develop within the food product due to the shape of the food product. Some factors that will influence the heating patterns include the chemical composition of the material, the size and the geometry of the container, and the power distribution within the oven cavity. The composition will have a direct effect on the dielectric properties of the material and the penetration depth of the microwaves, as discussed in Chapter 3 in this handbook.

Several other factors must be considered when evaluating microwave heating–based food processes involving microbial destruction. A rapid temperature rise will take place and the typical come-up time will be considerably decreased during microwave heating. Therefore, a conventional system must be altered to account for this decrease in heating prior to reaching pasteurization temperatures. Due to the heterogeneous composition of food products, some of the components may provide a protective effect toward the microorganisms. It is especially important to understand the location of hot and cold spots that may be present within the product and to provide adequate heat treatment necessary to destroy all of the microorganisms. Since microbial destruction is exponentially related to the temperature, slight deviations in temperature may result in substantially higher microorganisms surviving within the food product.

There have been numerous studies investigating the effect of microwave heating on pathogenic microorganisms in foods. Bacteria shown to be inactivated by microwaves include *Bacillus cereus, Campylobacter jejuni, Clostridium perfringens, Escherichia coli, Enterococcus, Listeria monocytogenes, Staphylococcus aureus,* and *Salmonella* [26–29]. In addition, the nematode *Trichinella spiralis,* the microorganism that causes trichinosis, is also inactivated by microwave energy. Food-borne pathogens have been shown to be inactivated by microwave heating in various poultry, beef, fish, and pork products such as milk and eggs.

There does not appear to be any obvious "microwave-resistant" food-borne pathogen. Various studies have shown increased resistance of *Staphylococcus aureus, Clostridium perfringens,* or *Enterococcus faecalis* but not necessarily to the point that these could be labeled as resistant. As with heat, it is known that bacteria are more resistant to microwave heating than yeasts and molds and that bacterial spores are more resistant than vegetative cells to microwave heating.

A. Microwave Thermal Effects on Microbial Destruction

Goldblith and Wang [30] showed no differences in inactivation of *E. coli* and *B. subtilis* exposed to the same time-temperature conditions of microwave and conventional heating. They exposed spores of *E. coli* and *B. subtilis* to microwave radiation to determine the presence of nonthermal effects. It was found that bacterial inactivation closely followed first-order kinetics, with the experimentally determined microbial counts aggreeing well with predicted values for first-order destruction. They concluded that microwave heating had a purely thermal effect. Lechowich et al. [32] also studied the exposure of *S. faecalis* and *S. cerevisiae* to microwave at 2450 MHz and under conventional heating, and concluded that the inactivation by microwaves could be explained solely in terms of heat generated during the exposures. Heddlesson et al. [28] also showed that the relationship between time of heating, temperatures, and microbial destruction achieved by microwave heating followed conventional reaction kinetics and this illustrated its thermal origin.

Vela and Wu [33] exposed various bacteria, actinomycetes, fungi, and bacteriophages to microwaves of 2450 MHz both in the presence and absence of water. They found that microorganisms were inactivated only in the presence of water and that dry or lyophilized organisms were not affected even by extended exposures. This led them to conclude that microorganisms were killed only by the "thermal effect" and most likely there was no nonthermal effect. However, they added that if a nonthermal effect existed, then water would be necessary to potentiate it, as cell constituents other than water did not absorb sufficient energy to kill microbial cells.

Hamrick and Butler [34] exposed four strains of *E. coli* and one strain of *Pseudomonas aeruginosa* to 12 h of continuous, low-intensity (60 mW/cm^2 measured by a Narda probe placed at the location of the sample) microwave radiation at 2450 MHz. When they compared the growth curves of control and exposed bacteria, they found no effect on cell replication rate that could not be explained as a result of temperature variation. They concluded that at least at 2450 MHz, low-level MW radiation had no nonthermal effects on bacteria.

The destruction of *E. coli, S. aureus, P. fluorescens,* and spores of *B. cereus* by microwave irradiation at three power levels were studied by Fujikawa et al. [35, 36]. Assuming uniform temperature distributions, these authors didn't find any remarkable difference and concluded that mostly thermal effects can interpret the destruction profiles by microwave exposure. No comparison between microwave and conventional heating temperature profiles was included in the study.

Diaz-Cinco and Martinelli [37] performed experiments to see if the lethality of microorganisms is due to microwave radiation or heat generated by microwave radiation with four types of microorganisms treated, as indicated by the authors, to the same exposures of time and temperature. They concluded that killing effect was due to the heat generated by microwave energy based on the percentage of the survivors. Heddlesson et al. [28] showed that the relationship between time of heating, temperatures, and microbial destruction achieved by microwave heating followed conventional reaction kinetics and this illustrated its thermal origin.

Welt and Tong [7–9] introduced an apparatus to evaluate possible athermal effects of microwaves on biological and chemical systems and explained the limits and requirements to comparative studies between kinetics under microwave conventional heating. They compared inactivation of *C. sporogenes* and thiamin under equivalent time-temperature treatments by perfectly stirred batch treatment by conventional and microwave heating and did not observe athermal effect.

Jambunathan [10] studied the kinetics of destruction of *Listeria monocytogenes scott A* in milk during microwave and waterbath heating. A consumer microwave oven was modified to enable constant-temperature microwave oven heating using feedback temperature control. Thermal death time experiments on vegetative cells of *Listeria monocytogenes Scott A* (LMSA) suspended in sterile, whole milk were conducted at 57.8, 60.0, and 62.8°C using conventional waterbath heating and the microwave heating. Nonlinear survivor curves with distinct shoulders and tails, especially at lower temperatures, were obtained. There was no significant difference ($\alpha = 0.05$) between the survivor curves, the calculated D values for microwave and waterbath heating.

The above studies used sophisticated control systems for maintaining the test sample at target temperatures and carried out studies only after the sample has reached the designated temperature. These are excellent systems for temperature control using microwave oven; however, bulk of microwave energy input takes place during the time when the sample is heated to the target temperature rather than when when it is held. In the actual test period (after come-up time) microwave energy is only needed to cover for heat loss from the system to the environment, which is only a small fraction as compared to what is needed for heating. Hence, the heating condition will not be largely different from those under conventional heating; this could be a reason for the lack of existence of nonthermal effects in these studies.

B. Microwave Nonthermal and Enhanced Microwave Effects

In connection with additional enhancement of microbial and enzymes destruction by microwave energy, the following studies reported nonthermal and specific effects of microwaves.

Olsen [38] compared conventional and microwave heat treatment of *Aspergilus niger*, *Rhizopus nigricans*, and *Penicilium sp.* and found that for achieving a given level of kill, lower temperatures were possible in the microwave treatment than under conventional treatment. Amannus [39] also studied the relative effectiveness of conventional and microwave heating for inactivating microorganisms, and although he questioned the theory of nonthermal effects, he found that at relatively low temperatures microwave was more lethal to microorganisms.

Culkin and Fung [18] observed the destruction of *E. coli* and *Salmonella typhimurium* in microwave-cooked soups and noted that the reduction in microbial numbers did not directly correlate with the temperature achieved in the soup after irradiation. They heated tomato soup, vegetable soup, and beef broth at 915 MHz and found that for a given exposure time, the temperature was highest at the middle region, intermediate at the bottom, and lowest at the top of the soup containers. Yet, the organisms at the top of the containers were killed more rapidly than those at the bottom or at the middle. These observations led them to conclude that the killing of bacteria by microwaves was *not* entirely due to the heat produced by microwave energy. However, their methodology was not accurate, as it did not account for temperature variations in the container following microwave heating. Hence, it is possible that they could have sampled from a "cold" spot (leading to lower lethality) present in the top and from a "hot" spot present in the middle (leading to higher lethality). This could lead to an erroneous conclusion regarding presence of nonthermal effects.

Mudgett [40] calculated the lethality of *E. coli* strain in the continuous system and compared it with experimentally measured values. Experimental microbial lethality was somewhat greater than that predicted by numerical integration. The author opinioned that it could have resulted from sensitivity of the kinetic model to small differences in temperature or possibly from the second-order kinetic effects from the selective absorption of microwave energy by the test organism based on high intracellular conductivity.

Dreyfuss and Chipley [19] found that when compared to conventional heating, microwave heating resulted in significant differences in the specific enzyme activity in *Staphylococcus aureus;* e.g., malate dehydrogenase activity was about $2\frac{1}{2}$ times greater in microwave-heated cells. On the other hand, α-ketoglutarate dehydrogenase activity was about $7\frac{1}{3}$ times less in microwave-heated cells. This led them to conclude that athermal effects were at least partly responsible for differences in specific enzyme activity. However, their study did not mention the location of temperature location or the oven power. In addition, the study implicitly assumed that the measured activity levels during the period *after* microwave heating were due to microwave athermal effects, while these may well have been due to the thermal effect associated with the temperature increase.

Khalil and Villota [20] compared injury and recovery of *Staphylococcus aureus* during microwave and conventional heating. They circulated kerosene around test-tubes holding 10 ml of phosphate buffer (0.1 *M*) containing *S. aureus* cells. They found that the (sublethally) microwave-heated cells suffered greater injury as compared to conventionally heated cells (0.06% of the initial cell population recovered after heating for 30 min at 50°C for microwave treatment versus 0.17% of initial cell population recovered after conventional treatment). The lag phase for microwave-heated cells was approximately twice as long (4 h) as the corresponding lag phase for conventionally heated cells (2 h). Also, the amount of enterotoxin A produced by recovering cells following 30 min of heating showed that conventionally heated cells produced significantly more toxin (1–2 μg/ml) than microwave-heated cells at all time periods (0–72 h). They concluded that microwave-heated cells suffered a greater injury as well as greater membrane damages. They also stated that thermal effects contributed the major part of injury suffered during microwave heating.

Lund and Knox [41] microwaved poultry inoculated with between 10^6 to 10^7 cfu/g of *Listeria monocytogenes Scott A* in a 650 W consumer microwave oven for 15 min, 38 min, and 38 min plus 5 min standing time (manufacturer's recommended procedure), respectively. Temperatures were measured immediately following each heating period at eight (internal and external) locations. The 15 min cooking period caused a 2–3 log reduction while the 38 min + 5 min cooking procedure caused over a 6 log reduction of *L. monocytogenes.* Corresponding skin temperature ranged from 60 to 87°C and 80 to 99°C. Internal stuffing temperatures were 15–20°C lower. When the standing time was increased to 20 min,

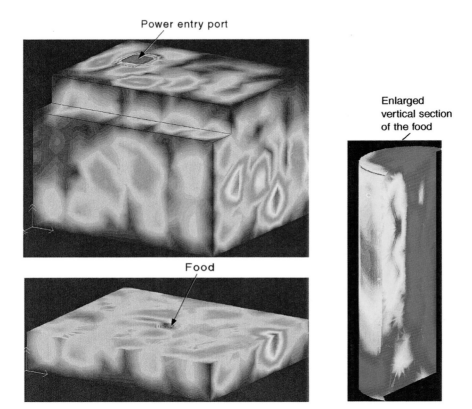

Figure 2-3 Computed electric field distributions inside the microwave cavity in Fig. 2-2b. The top figure on the left shows overall field pattern inside the cavity. The bottom figure on the left is a horizontal section exposing the inside of this cavity with a cylindrical ham placed at the center of the oven. The right figure shows the electric fields in a vertical section of the cylindrical ham. Here red signifies high values and blue signifies low magnitudes of electric fields, with other colors having intermediate magnitudes of electric fields.

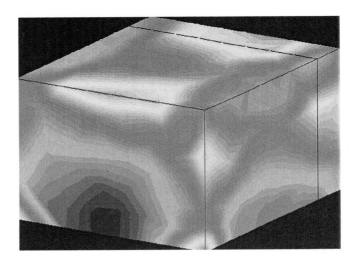

Figure 2-10 Example of overheating at edges and corners, obtained from electromagnetic simulation of a rectangular block of potato ($\varepsilon = 51.5 - j16.3$) of size $W \times D \times H = 4 \times 5.5 \times 3$ cm placed in the oven in Fig. 2-2a. Red stands for high magnitude of electric field and blue for low.

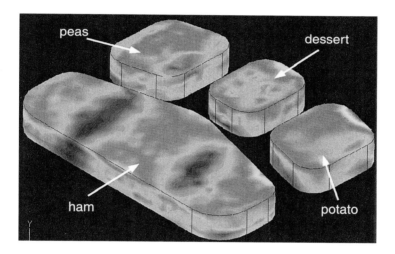

Figure 2-18 Illustration of typical nonuniformity of electromagnetic fields in a multi-compartment dinner. Computations are for the oven in Fig. 2-2a with properties of ham $(38 - j30)$, potato $(51 - j16.3)$, sauce $(75 - j24.4)$, and peas $(53.2 - j11)$. The thickness is 2 cm and the overall tray dimensions are approximately 20 × 16 cm.

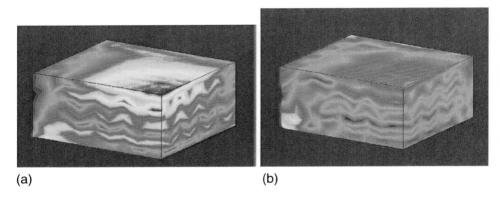

(a)   (b)

Figure 2-22 Modification of electric field pattern that would reduce edge heating using (a) aluminum foil compared to (b) no foil on the sides of a rectangular tray of whey protein gel ($\varepsilon = 46.3 - j17.3$) in a Pressurized Microwave Retort (Raytheon, Inc.). The computations are for the quarter of a tray with overall dimension 14 × 9.5 × 3.2 cm. (From Ref. 77.)

Domestic Oven (2.45 GHz)

After 10 seconds

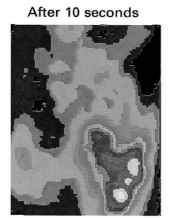

After 20 seconds

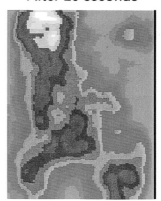

Phase Control (896 MHz)

0° for 3 seconds

210° for 3 seconds

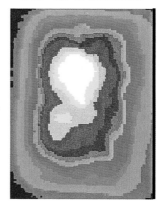

Figure 2-27 Microwave phase control heating, showing thermal images of tray of powdered foodstuff heated under phase control (lower images) and domestic microwave oven (upper images) conditions. (From Ref. 61.)

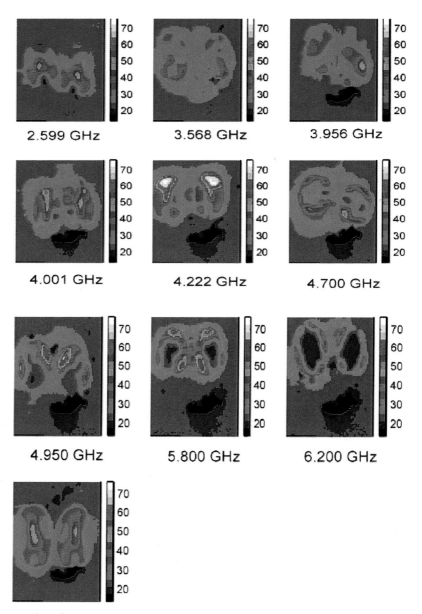

Domestic microwave oven

Figure 2-18 Illustration of typical nonuniformity of electromagnetic fields in a multi-compartment dinner. Computations are for the oven in Fig. 2-2a with properties of ham ($38 - j30$), potato ($51 - j16.3$), sauce ($75 - j24.4$), and peas ($53.2 - j11$). The thickness is 2 cm and the overall tray dimensions are approximately 20 x 16 cm. Ref. 62.)

Figure 8-8 Infrared thermography during heating of agar gel in a microwave oven.

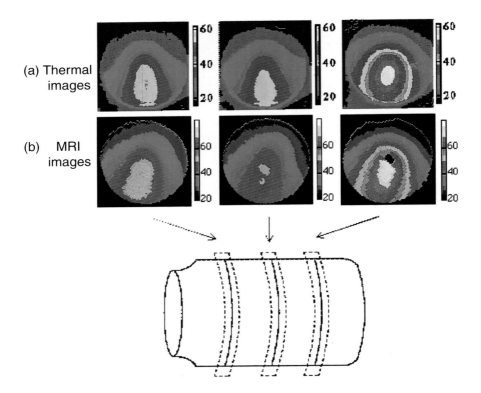

Figure 8-10 (a) Three two-dimensional-slice (dashed lines) temperature maps taken at 1/4, 1/2, and 3/4 heights of an intact jar containing TX151 gel derived from the MR phase shift parameter. Resolution 312 μm × 1.25 mm × 5 mm slice. (b) The same temperature maps obtained by infrared thermal imaging the surfaces (solid lines) exposed on separating the split jars. (From Ref. 30.)

the internal stuffing temperature reached 72–85°C, causing a 6 log or better reduction in *Listeria*. This was consistent with the study by Bunning et al. [42] for conventional heat resistance of the *L. monocytogenes* F5069 strain in milk, which reported a 6 log reduction following heating to 70°C. Lund and Knox [41] further recognized that the use of a standing time was necessary to allow temperature uniformity to cause the complete destruction of *L. monocytogenes*.

Walker et al. [43] found that when *L. monocytogenes* CRA 710, 711, and 433 were inoculated at a concentration of 10^6 cfu/ml into retail chilled foods packaged for microwave use and cooked according to the manufacturer's instructions, viable cells were recovered from three out of eight samples examined. This result was attributed to nonuniform temperature distribution in microwaved foods. Since *L. monocytogenes* is psychrotrophic, it may be unsafe to use manufacturers' recommendations to microwave foods that have been stored at low temperatures for long periods of time as these may potentially contain high levels of *L. monocytogenes* species.

Riva et al. [44] reported different z values for *Enterobacter cloacae* and *Streptococcus faecalis* in batch conventional heating (4.9 and 5.8°C) and microwave heating (3.8 and 5.2°C). The results were explained due to different heating kinetics and nonuniform local temperature distributions during microwave heating rather than to existence of specific athermal effects. The results of study by Aktas and Ozilgen [45] of the injury of *E. coli* and degradation of riboflavin during pasteurization with microwaves in a tubular flow reactor also indicated the effect of the flow behavior and other experimental conditions on the death mechanism in the microwave field. They suggested that microbial death might be caused through damage to a different subcellular part under each experimental condition.

Heddleson et al. [26, 27] examined the survival of *Salmonella* during microwave heating in the presence of various food components. The solutions were heated in a microwave oven for 47 s and the temperatures were monitored using a Luxtron fluoroptic temperature measurement system. NaCl provided the most protection, allowing a 37% survival, whereas in the buffer alone there was 0.22% survival. The final mean temperature of the solutions varied from 59 to 63°C (as measured by a mercury thermometer after mixing solutions). Surface heating was observed with the NaCl solutions (20°C greater at top). Since NaCl decreases the penetration depth, this increases the chance that microorganisms will survive below the surface due to the uneven heating patterns and a depressed temperature.

Choi et al. [46] inoculated sterilized whole milk with 10^6–10^7 cfu/ml of *L. monocytogenes* in 20 ml milk samples within glass vials. The vials were immersed in water and heated for different time periods ranging from 1 to 60 min in a 2450 MHz microwave oven set at 71.1°C (by inserting a thermoprobe supplied with the microwave oven into the water bath surrounding the glass vials). They also used 50 and 100 ml milk samples in milk dilution bottles, which were placed in the water bath. It was found that heating the 20 ml milk sample for 10 min inactivated all cells of *L. monocytogenes* and the organism was not recovered after cold enrichment incubation (4°C for up to 2 months). Inactivation was less effective for the 50 and 100 ml samples. For 50 ml samples, log reduction levels ranged from 3.3 after 5 min of heating to >6.0 (i.e., full inactivation) after 70 min of heating. The 100 ml samples had a maximum log reduction of 3.31 even after 80 min of heating. The reason for this discrepancy was attributed to the fact that in the larger samples, temperature did not reach (or reached asymptotically) 71.1°C even at the end of the heating period. This uneven heating profile in large samples, resulting in uneven microbial destruction, is a typical feature of microwave heating.

Odani et al. [47] used microwave irradiation to kill *E. coli, S. aureus,* and *B. cereus* in frozen shrimp, refrigerated pilaf, and saline. Microwave irradiation was shown to result in the release of proteins from *E. coli* as detected by gel electrophoresis of cell-free supernatants using sensitive silver staining. They suggested that the mechanisms of killing bacteria depend not only on temperature but also on other effects of microwave irradiation.

Comparing microwave and conventional methods of inactivation of *B. subtilis,* Wu and Gao [48] showed that the D_{100} of microwaves was 0.65 min but for conventional heating it was 5.5 min and demonstrated nonthermal effects of such energy on microorganisms. In a review of contributions to the study of microwave action, Joalland [49] considered the action of microwaves upon bacteria by taking into account the different cellular components such as genetic material, enzymatic activity, mitochondrias, membranes, and cytoplasma, and concluded that beside thermal affects there could exist some nonthermal effects.

In a comparative study of the inactivation of *L. plantarum* using microwave and conventional heating at 50°C for 30 min, Shin and Pyun [50] found significant nonthermal effects under pulsed microwave heating conditions. Pulsed microwave treatment was shown to achieve 2–4 log cycle greater reductions in survival counts and reported to induce irreversible damage to membrane structures with an increase of permeability in cell components. The growth of cells treated with pulsed-microwave heat treatment was delayed by about 24 h and acid production after 60 h incubation reached only about 60–80% of that achieved by the other two treatments.

Kozempel et al. [51, 52] developed a pilot-scale nonthermal flow process using microwave energy to inactivate *Pediococcus sp.* NRRLB-2354. A cooling tube within the process line to maintain the temperature below 40°C removed the heat generated by the application of microwave energy to the system. A significant reduction in microbial count was reported supporting nonthermal effects.

Several recent studies have been reported on the comparative evaluation of destruction kinetics of microorganisms and enzymes in conventional and microwave heating systems [11–15]. For comparative purposes, thermal destruction kinetics of microorganisms in apple juice were first established. Aliquots of juice inoculated with spoilage microflora (*saccharomyces cerevisiae* and *Lactobacillus plantarum*) were exposed to the various time-temperature treatments in a well-stirred water bath (50–80°C) and the survivors were evaluated. As previously described [12], the heating times were corrected taking into account the effective portion of come-up and come-down periods. As expected, the destruction behavior indicated characteristic first-order rate kinetics. Apple juice inoculated with *S. cerevisiae* and *L. plantarum* was also subjected to microwave heating under continuous-flow conditions at 700 W to selected exit temperatures 52.5–65°C [14]. Typical time-temperature profiles of test samples were gathered and corrected for both come-up and come-down time as previously described. The D and z values were corrected as detailed earlier [12]. The D values under microwave heating were significantly lower and were much more sensitive to changes in temperature as compared with thermal destruction. Semilogarithmic regression of D values vs. temperature yielded a D_{60} of 0.41 s for *S. cerevisiae* and 3.7 s for *L. plantarum* with z values of 6.1 and 4.5°C, respectively, in the microwave-heating mode, and D_{60} of 10 and 26 s, respectively, with z values of 13 and 16°C in the conventional-heating mode. This gives a relative microwave/thermal destruction ratio of 19 for *S. cerevisiae* at 60°C and 43 for *L. plantarum* at 65°C. Thus, the time required to destroy a given microbial population in the thermal mode is much longer than required in the microwave mode. These results suggest that probably there exist of some enhanced thermal effects associated with microwaves, which cannot be explained by conventional integration of thermal kinetics over the time-temperature regime.

Table 1 shows D values estimated from the experimental survivor data on *E. coli* under continuous flow conditions involving conventional and microwave heating [25]. Heating conditions are detailed in Refs. 16 and 17. Briefly, the microbial cell suspension was pumped through glass coils placed in two domestic microwave ovens (700 W, 2450 MHz) and cooled at the exit using a water-cooled condenser. Copper-constantan thermocouples were inserted into the tubes at inlet, between the ovens and at exit of the cavity and holding tube and temperature data were gathered using an HP Data Logger. The system was sanitized by circulating

Table 1 Experimental D values of *E. coli* K-12 During Microwave and Conventional Heating

	D values, s		
Temperature, °C	55	60	65
Thermal batch method	173.0	18	1.99
Continuous-flow, hot water	44.70	26.80	2.00
Continuous-flow, steam	72.71	15.61	2.98
Continuous-flow, microwave	12.98	6.31	0.78
Continuous-flow, microwave + holding	19.89	8.33	1.98

water at 70–80°C through the system before and after treatment. Prior to heating, the cell suspension was precirculated in the system to establish steady-state flow conditions and then the microwave ovens were turned on. Test samples were withdrawn during steady-state heating periods with exit temperatures in the range from 50 to 70°C and were collected into precooled sterile tubes immersed in an ice-water bath at the exit. Each exit temperature was achieved by preadjusting and changing the flow rate. For conventional heating, both coils were located inside a single steam cabinet or immersed into a water bath. The flow rates were manually adjusted to give the same exit temperatures. D values were computed from the survivor data at different temperatures based on the calculated effective times from time-temperature data (experimental or model predicted) for microwave, hot water, and steam heating conditions. A comparison shows that lowest D values were observed with the continuous-flow microwave heating section and ranged from 13 s at 55°C to 0.78 s at 65°C. The experimental D values in microwave system with hold (20, 8.3, and 2.0 s at 55, 60, and 65°C, respectively) were also lower than those obtained for steam heating (73, 16, and 3.0 at 55, 60, and 65°C, respectively) and hot water heating. Further, values in continuous systems were considerably lower than those in batch heating systems.

Possibility of nonuniform temperature distribution is generally cited as a reason for refuting claims of studies which demonstrate nonthermal effects. However, it should be recognized that, in reality, existence of such nonuniformity will only enhance the difference between the two. This is because for any given treatment time, the extent of destruction at a temperature some specified degrees higher than the mean is much more than the extent of survival at the temperature same number of degrees below the mean, giving an overall survivor greater than the mean. For example, if the experimentally evaluated destruction in microbial population during microwave heating is 5 log cycles, and, based on the integrated temperature-time profile (and spatial temperature uniformity), the computed destruction (thermal effect) is 3 log cycles, the 2 log cycle difference between the two demonstrates additive nonthermal effect associated with microwave heating. Now, if there were to be a distribution during the temperature rise rather than an assumed mean, the computed thermal distruction would be smaller than 3 log cycles, thereby increasing the difference between microwave and thermal effect!

V. IMPACT OF MICROWAVE HEATING ON INJURY OF BACTERIA

The nonuniform temperature distribution resulting from microwave heating may result in mere injury of the microorganisms at locations within the product where lethal temperatures are not encountered. A cell is considered heat injured if after an appropriate interval of time following sublethal heating it regains normal physiological activity. The duration of recovery depends on the severity of heat injury. Injured cells grow on nonselective media and are inhibited in selective media.

Detection and evaluation of heat injury has an important food-safety implication; insufficient thermal processing of foods can merely injure pathogens, which may later recover during storage of the food product. In the case of *L. monocytogenes* species, prevention of underprocessing is all the more necessary, because injured *L. monocytogenes* cells can grow at refrigeration temperatures [53]. Research has determined that microbial cells can acquire additional heat resistance upon exposure to environmental stresses such as sublethal heat treatment. When bacteria are subjected to elevated sublethal temperatures, these "heat-shocked" cells can have increased heat resistance compared to "non-heat-shocked cells." Mackey and Derrick [54] proposed that an increase in heat resistance may be due to the synthesis of heat-shock proteins in response to a heat stress.

The temperature profiles in foods that contain dissolved ions are very different from that seen in ion-free foods. Anantheswaran and Liu [55] showed that the temperature and flow profiles of NaCl solutions (up to 3.33%) were inverted as compared to a control solution of deionized water during microwave heating. This reversal in flow profiles is seen in salt solutions because of an additional heating mechanism, i.e., ionic heating which supplements the heating due to dipolar interactions. The presence of dissolved ions impedes penetration of microwave energy beyond a thin surface layer with the penetration depth being inversely proportional to the dissolved ion concentration.

When microwaves are incident in a solution containing dissolved ions, localized regions of high temperature in the immediate vicinity of the dissolved ions are created. If these localized higher temperature zones exist, then it follows that any bacterial cells in the immediate vicinity are subjected to greater heat injury than cells not in proximity to dissolved ions, even if the bulk temperatures of both systems are the same. If such a difference in injury does exist, then it could be due to this localized thermal heating or "microwave effect." The microwave effect, if it exists, could be quantified in terms of bacterial heat injury.

Jambunathan [10] studied sublethal heat injury of *Listeria monocytogenes scott A* during microwave heating using a constant-temperature microwave oven (CTMWO) heating using feedback temperature control. Heat injury experiments on vegetative cells of *Listeria monocytogenes Scott A* (LMSA) suspended in 200 ml peptone water (PW) or 1.0% NaCl solution of peptone water (PWS) were conducted at 55.0°C using the CTMWO heating apparatus. Heat injury following a period of sublethal heating was defined as

$$\% \text{ Heat injury} = \frac{N_{\text{NSM}} - N_{\text{SM}}}{N_{\text{NSM}}} \times 100 \tag{7}$$

where N_{NSM} = number of cells recovered on nonselective medium.

N_{SM} = number of cells recovered on selective medium

Percentage injury was calculated for the PW and PWS systems and also for various stages along the growth curve. The percentage injury for both systems was not significantly different ($\alpha = 0.05$) at any heating time. However, the standard deviation about the mean injury at the end of the come-up-time (CUT) and at lower heating times was large. "Injured" cells were also found in the unheated inoculum and at various stages of the growth curve.

The injury of *Salmonella* species during microwave heating was examined by Heddleson and Doores [56]. It was observed in this study that the food composition does affect the heat tolerance, with some components providing a protective effect for *Salmonella*. Regardless of the final temperature, it was determined that cells were recovered when samples were not mixed following heating.

Koutchma [57] studied bactericidal effects of microwaves and hypothermia on *E. coli* cells in batch-mode conditions. No differences were found in survival of those bacteria between the two heating modes with identical time-temperature profiles. It was shown that microwaves caused greater damages to the cell genome and resulted in different survival and interaction coefficients under combined applications with low concentrations of hydrogen peroxide.

VI. MICROWAVE INACTIVATION OF ENZYMES

Kermasha et al. [58, 59] analyzed inactivation of wheat germ lipase and soybean lipoxygenase at various temperatures using conventional and microwave batch heating, and found higher enzyme destruction rates under microwave heating conditions. In these studies, conventional heating was carried out in temperature controlled water bath, while a custom designed microwave oven was used to maintain test samples at selected temperatures either by full exposure to microwaves (microwave heating) or partial exposure by immersion of test samples in water contained in a beaker placed inside the oven (mixed mode). The temperature was maintained by an on-off feedback control [60]. Conventional and mixed mode heating was reported to produce somewhat similar results while the microwave mode resulted in a more rapid inactivation of the enzyme. Higher enzyme inactivation rates under microwave heating conditions were ascribed to possible nonthermal effects.

Thermal resistance of pectin methylesterase (PME) implicated in the loss of cloudiness of citrus beverages has been recognized to be greater than that of common bacteria and yeast in citrus juices, and PME activity has been used to determine the adequacy of pasteurization. These studies are detailed in Refs. 11–13 under continuous-flow and batch heating modes involving microwave heating, using experimental procedures detailed earlier with respect of destruction of bacteria and yeast in orange and apple juices. The results showed the time-corrected PME inactivation curves at various temperatures under conventional and microwave heating conditions characteristically demonstrated a first-order nature. At the common temperature of 60°C as the basis, the effectiveness of two systems could be compared by their D values: 154 s during thermal and 7–12 s during microwave heating experiments, the two differing by more than an order of magnitude. This difference again shows the possibility of some contributory *nonthermal* effects of microwaves for enzyme inactivation.

In order to test the observed effects PME inactivation in orange juice and destruction of *S. cerevisiae* in apple juice (discussed earlier) were nonthermal (temperature-independent additive effect) or enhanced thermal effects (temperature dependent effect not explicable under time-temperature corrected heating conditions); three different techniques as described in Ref. 15 were used to maintain temperatures below 40°C under continuous and batch mode heating (*nonthermal effects*) as well as under progressively increasing temperature conditions (*enhanced thermal effects*). In the first setup, test sample temperatures were maintained below 40°C while being subjected to full-power microwave heating (700 W) conditions by surrounding the helical coil with a jacket through which cold kerosene as a microwave-transparent liquid was circulated. Inlet (~15°C) and outlet temperatures of test sample (<35°C) and kerosene (~10°C) were monitored continuously using copper-constantan thermocouples positioned within the tubing just outside the microwave cavity.

During the 90 min treatment time, 1200 ml of test sample was circulated through the oven, but only a 110 ml portion was continuously exposed to microwaves. Some enzyme inactivation occurred during such exposure but the extent was relatively small (68%) as compared to that at higher temperatures. In terms of added microwave energy, the 90 min heating would give 3780 kJ of heat to the 110 ml orange juice, which would be sufficient to completely boil off the juice, let alone inactivate the enzyme. Hence, the nonthermal effect at the sublethal temperature was not considered to be significant.

In the second setup, microwave heating was carried out in a batch mode with a larger size of sample with continuous-flow maintained for mixing and sample removal. In order to maintain a low temperature (<40°C) for the juice, a stainless cooling coil grounded to the cavity wall, was fully submerged in the juice inside the test beaker for rapid removal of heat generated in the sample. Ice-chilled water (0–2°C) was circulated through the cooling coil at a constant flow rate. Temperature uniformity in the beaker measured using fiberoptic probes indicated good stability (±1°C). Results indicated that over a 3 h heating period, about 22% inactivation of PME occurred and it was only 0.3% up to 1.5 h. The absorbed energy during the treatment time was about 10 MJ. With respect to destruction of *S. cerevisiae* in apple juice, the effect was even smaller showing less than 1 log cycle reduction in microbial survivors. Again, it was concluded that there was no temperature-independent nonthermal effects associated with microwave heating when samples were held at temperatures below 40°C. Similar findings were reported in other studies [8, 30, 32].

To characterize the differences observed during the kinetic studies with respect to enzyme inactivation and microbial destruction between microwave and thermal heating modes, further studies were performed using the above batch system but without the cooling coil. Temperatures within the test beaker were observed to be relatively uniform ($\pm 1°C$). Test samples were varied by size and temperature of the samples linearly increased from 20 to 40, 50, 60, 65 and 70°C under nonisothermal heating condition. In order to assess the relative effects due to microwave heating, the equivalent thermal effects for similar heat treatments needed to be calculated and deducted from the total contribution. The thermal contribution was calculated from effective portion of the heating time based on thermal batch z value. To characterize the relative magnitude of enhanced effects, a microwave enhancement ratio (MER) was defined as the ratio of total inactivation/destruction under microwave heating conditions (thermal plus microwave effects) to that calculated to be due thermal effects. A value of the microwave enhancement ratio (MER) greater than 1.0 indicates the existence of enhanced thermal effects due to microwaves, while the MER value of 1.0 or below shows its nonexistence. The MER for PME was as high as 20 at higher temperature and smaller sample size, while for bacterial destruction the ratio was about 10. Thus, these studies suggest significant *enhanced effect* of microbial and enzymes destruction to be associated with microwave heating. However, *no nonthermal effect* was observed at sublethal temperatures. Such enhanced thermal effects cannot be explained temperature nonuniformity. However, it can possibly be explained by an oscillatory mean temperature, which normally results in greater destruction than a steady mean. It will further have to be assumed that a certain fraction of the sample will experience a transient higher spike in temperature, and subsequently cool off to give out its heat to the bulk. Then a different fraction of the sample will undergo a preferential absorption of the microwave energy and reach a higher temperature and so on. This way, each fraction will be subject to a temperature higher than the mean bulk temperature. A few degrees temperature spike thus achieved would give a substantial net increase in the destruction of microbial population. Even when explained by such a hypothesis, the effect could be considered "*enhanced*" effect due to microwave heating.

Boon and Kok [61] described in a patent application with alternate application of microwave pulses and cooling to achieve biochemical and chemical reactions. They indicated the procedure enabled such reactions including inactivation of microorganisms and enzymes, at a much higher reaction rate while limiting the increase in temperature of the treated material. The proposed mechanism was that high-power pulsed microwave irradiation was directly absorbed by molecules, in contrast with conventional heating where molecules are brought to a higher energy level through collision with other molecules. Thus, some molecules would be at a higher energy level or temperature than others during the short microwave pulse. Consequently such molecules would have higher reaction rates than expected based on average product temperatures and thus there would be selective heating at a molecular level [62]. Kermasha et al. [58, 59], while reporting nonthermal effects due to microwave heating, gave the following explanation. Proteins, as complex macromolecules, generally have numerous polar and/or charged moieties (i.e., COO^-, and NH_4^+) which can be affected by the electrical component of the microwave field [63]. Although the microwave energy may be insufficient to disrupt covalent bonds, the noncovalent bonds such as hydrophobic, electrostatic, and hydrogen bonds may well be disrupted. Thus the direct microwave effect could be more pronounced, immediate, and specific than the random kinetic energy mechanism associated with conventional heating.

VII. CONCLUSION

Thermal and nonthermal effects of microwave heating in foods generally appear to be inconclusive. It is difficult to precisely evaluate the effectiveness of microwave heating vs. conventional heating from the literature because of the techniques employed or a lack of detail in the methods or materials, especially in relation to temperature monitoring [56]. Lack of online temperature measurement in a microwave field, uneven heating due to microwave field distributions, inability to control temperatures of microwave-heated samples, and uncontrolled concentration of solute from the sample due to evaporative losses from the sample during heating have been cited as reasons for the difficulty in resolving the thermal vs. nonthermal effects controversy [8]. As pointed out in earlier sections, these differences may not fully explain the reasons for the existence of nonthermal effects. A oscillatory temperature rise could help explain a possible theory, but certainly additional research is needed.

The techniques employed for carrying out the kinetic studies, both conventional as well as microwave based, need good control of temperature, and literature shows considerable lack of such controls in both situations. One generality that is recurring in the literature is that nonuniform heating by microwaves may lead to survival of food-borne pathogens, including *Salmonella* and *Listeria monocytogenes*, in foods heated to internal endpoint temperatures that would otherwise be lethal to a population. For example, some studies have demonstrated that internal temperature measurement of poultry does not predict inactivation of surface inoculated *Salmonella* on poultry due to lower temperatures at the product surface during microwave heating. These observations are not limited only to microwave heating; they are equally applicable to situations of conventional heating if temperature nonuniformity exists.

Overall, the destruction of microorganisms and enzymes during microwave heating appears to be predominantly due to the heat generated by the microwaves, although additive contributions might exist. Hence it is recommended that process establishment procedures still be based on well-documented thermal effects, with the additive enhanced effects taken as safety. An understanding of microwave absorption and the resulting temperature distribution in the product can be used to predict microbial destruction during a microwave process such as pasteurization and sterilization. However, it is still a challenge to design appropriate food processes using microwaves than with conventional heating systems. This is mainly due to the additional factors that affect the heat transfer and the resulting temperature distribution within the food product during microwave heating. The shorter heating times during microwave heating necessitates better control of the heating process for microbial destruction. Any method that promotes repeatability of the process and equilibriation of temperatures should be pursued when microwave heating is utilized. Repeatability can be achieved, for example, by exact placement of the food in the oven and careful control of the composition of the food. Equilibration of temperatures can be achieved by using standing and hold times at the end of the heating process.

REFERENCES

1. H. Fleming. Effect of high frequency on microorganisms. Electrical Engineering 63(18), 1944.
2. D. Knorr, M. Geulen, T. Grahl, and W. Stitzman. Food application of high electric fields pulses. Trends in Food Science and Technology 5(3):71–75, 1994.
3. D.A.C. Stuerga and P. Gaillard. Microwave thermal effects in chemistry: A myth's autopsy. J. Microwave Power and EME 31(2):87–113, 1996.
4. S. Sastry, and S. Palaniappan. The temperature difference between a microorganism and a liquid medium during microwave heating. J. Food Processing and Preservation 15:225–230, 1991.
5. T. P. Labuza. Enthalpy/entropy compensation in food reactions. Food Technology 34(2):67–77, 1980.
6. C. R. Stumbo. Thermobacteriology in Food Processing. Academic Press, New York 2nd ed., 1992, p. 89.
7. C. H. Tong. Effect of Microwaves on biological and chemical systems. Microwave World 17(4):14–23, 1996.
8. B. Welt and C. Tong. Effect of microwave radiation on thiamin degradation kinetics. J. Microwave Power and Electromagnetic Energy 28(4):187–195, 1993.
9. B. A. Welt, C. H. Tong, J. L. Rossen, and D. B. Lund, Effect of microwave radiation on inactivation of *Clostridium sporogenes* (PA 3670) spores. Appl. Env. Microbiology 60:482–488, 1994.
10. S. Jambunathan. Kinetics of destruction of Listeria monocytogenes scott A in milk during microwave heating. M.S. thesis, Pennsylvania State University, University Park, PA. 1998, 117 pp.
11. S. Tajchakavit and H. S. Ramaswamy. Continuous-flow microwave inactivation ki-

netics of pectin methylesterase in orange juice. J. Food Processing and Preservation 21:365–378, 1997.
12. S. Tajchakavit and H. Ramaswamy. Continuous-flow microwave heating of orange juice: Evidence of non-thermal effects. J. Microwave Power and Electromagnetic Energy 30(3):141–148, 1995.
13. S. Tajchakavit and H. S. Ramaswamy. Thermal vs. microwave inactivation kinetics of pectin methylesterase in orange juice under batch mode heating conditions. Lebensm. Wiss. u. Technol 2:85–93, 1996.
14. S. Tajchakavit, H. S. Ramaswamy, and P. Fustier. Enhanced destruction of spoilage microorganisms in apple juice during continuous flow microwave heating. Food Research International 31(10):713–722, 1998.
15. S. Tajchakavit, Continuous-flow microwave heating of orange juice: Evidence of non-thermal effects. McGill University, Ph.D. thesis, 1997.
16. A. LeBail, T. Koutchma, H. Ramaswamy. Modeling of temperature profiles under continuous tube-flow microwave and steam heating conditions. Food Process Engineering. 23(1):1–24, 2000.
17. T. Koutchma, A. LeBail, and H. S. Ramaswamy. Modeling of process lethality in continuous-flow microwave heating-cooling system, Proceedings of the International Microwave Power Institute, Chicago, July 1998, pp. 74–77.
18. K. A. Culkin, Y. C. Daniel, and D. Fung. Destruction of *E. coli* and *S. typhimurium* in microwave-cooked soups. J. Milk Food Technology 38(1):8–15, 1975.
19. M. Dreyfuss and J. Chipley. Comparison of effects of sublethal microwave radiation and conventional heating on the metabolic activity of *Staphylococcus aureus*. Appl. and Env. Microbiology 39(1):13–16, 1980.
20. H. Khalil and R. Vilotta. Comparative study on injury and recovery of Staphylococcus aureus using microwaves and conventional heating. J. Food Protection 51(3):181–186, 1988.
21. W. V. Loock. Electromagnetic energy for pasteurization and sterilization. Another viewpoint. Microwave World 17(1):23–27, 1996.
22. R. N. Gedye, F. E. Smith, and K. C. Westaway. The rapid synthesis of organic compounds in microwave ovens. Can. J. Chem. 66:17–26, 1988.
23. R. N. Gedye, F. E. Smith, and K. C. Westaway, Microwaves in organic and organometallic synthesis. J. Microwave Power and Electromagnetic Energy 22:199–207, 1991b.
24. P. Risman, J. Microwave Power and Electromagnetic Energy 31(2):69–70, guest editorial, 1996.
25. H. S. Ramaswamy. Enhanced thermal effects under microwave heating conditions. Paper presented at 8th International Congress on Engineering and Food, Puebla, Mexico, April 9–13, 2000.
26. R. A. Heddleson, S. Doores, R. C. Anantheswaran, G. D. Kuhn, and M. G. Mast. Survival of *Salmonella* species heated by microwave energy in a liquid menstrum containing food components. J. Food Protection 56:637–642, 1991.
27. R. A. Heddleson, S. Doores, R. C. Anantheswaran, and G. D. Kuhn. Destruction of *Salmonella* species heated in aqueous salt solutions by microwave energy. J. Food Protection 56:763–768, 1993.
28. R. A. Heddleson, S. Doores, R. C. Anantheswaran, and G. D. Kuhn. Viability loss of

Salmonella species, *Staphylococcus aureus* and *Listeria monocytogenes* heated by microwave energy in food systems of various complexity. J. Food Protection, 1994.
29. J. R. Chipley. Effects of microwave irradiation on microorganisms. Advances in Applied Microbiology 26:129–145, 1980.
30. S. A. Goldblith, and D.I.C. Wang. Effect of microwaves on *Escherichia coli* and *Bacillus subtilis* Appl. Microbiol. 15:1371–1375, 1967.
31. S. A. Goldblith, S. R. Tannenbaum, and D.I.C. Wang. Thermal and 2450 MHz microwave energy effect on the destruction of thiamine. Food Technology 22:1267–1268, 1968.
32. R. V. Lechowich, L. R. Beuchat, K. I. Fox, and F. H. Webster. Procedure for evaluating the effects of 2,450-MHz microwaves upon *Streptococcus faecalis* and *Saccharomyces cerevisiae*. Appl. Microbiology 17:106–110, 1969.
33. G. Vela and J. Wu. Mechanism of lethal action of 2450-MHz radiation on microorganisms. Appl. Env. Microbiology 37(3):550–553, 1979.
34. P. E. Hamrick and B. T. Butler. Exposure of bacteria to 2450 MHz microwave radiation. J. Microwave Power 8:227–233, 1973.
35. H. Fujikawa, H. Ushioda, and Y. Kudo. Kinetics of Escherichia coli destruction by microwave irradiation. Appl. and Env. Microbiology 58(3):920–924, 1992.
36. H. Fujikawa. Patterns of bacterial destruction in solutions by microwave irradiation. Applied Bacteriology 76:389–394, 1994.
37. M. Diaz and S. Martinelli. The use of microwaves in sterilization. Dairy, Food and Environmental Sanitation 11(12):722–724, 1991.
38. C. M. Olsen. Microwave inhibit bread molds. Food Eng. 37:51–53.
39. I. Amannus. The effect of microwave and conventional cooking upon foodborn bacteria inoculated in foods. M.A. thesis, Mankato State University, Mankato, MN, 1979.
40. R. E. Mudgett. Microwave properties and heating characteristics of foods. Food Technology, June:84–93, 1986.
41. B. M. Lund and M. R. Knox. Destruction of Listeria monocytogenes during microwave cooking. Lancet 1989: pp. 218.
42. V. K. Bunning, C. W. Donnelly, J. T. Peeler, E. H. Briggs, J. G. Bradshaw, R. G. Crawford, C. M. Beliveau, and J. T. Tierney. Thermal inactivation of Listeria monocytogenes within bovine milk phagocytes. Applied Environmental Microbiology 54:346–370, 1988.
43. S. J. Walker, J. Bows, P. Richardson, and J. G. Banks. Effect of recommended microwave cooking on the survival of *Listeria monocytogenes* in chilled retail products. Technical memorandum, Campden Food and Drink research Association No. 548 Chipping Campden GL 55 OLD, UK, 1989.
44. M. Riva, M. Lucisano, M. Galli, and A. Armatori. Comparative microbial lethality and thermal damage during microwave and conventional heating in mussels (*Mytilus edulis*). Ann. Microbiol. 41(2):147–160, 1991.
45. N. Aktas and M. Ozligen. Injury of *E. coli* and degradation of riboflavin during pasteurization with microwaves in a tubular flow reactor. Lebensmittel Wissenschaft unt Technologie 25(5):422–425, 1992.
46. K. Choi, E. H. Marth, and P. C. Vasavada. Use of microwave energy to inactivate *Listeria monocytogenes* in milk. Milchwissenschaft 8:200–203, 1993.

47. S. Odani, T. Abe, and T. Mitsuma. Pasteurization of food by microwave irradiation. Journal-of-the-Food Hygienic of Japan 36(4):477–481, 1995.
48. H. Wu and K. Gao. Mechanisms of microwave sterilization. Science and Technology of Food Industry 3:31–34, 1996.
49. G. Joalland. Contribution a l'etude de l'effect des micro-ondes: Etude bibliographique. Viandes-et-Produits-Carnes 17(2):63–72, 1996.
50. J. K. Shin and Y. R. Pyun. Inactivation of *Lactobacillus plantarum* by pulsed-microwave irradiation. J. Food Science, 62(1):163–166, 1997.
51. M. Kozempel, O. J. Scullen, R. Cook, and R. Whiting. Preliminary investigation using a batch flow process to determine bacteria destruction by microwave energy at low temperature. Lebenm. Wiss. u. Technol. 30:691–696, 1997.
52. M. Kozempel, B. A. Annous, R. Cook, O. J. Scullen, and R. Whiting. Inactivation of microorganisms with microwaves at reduced temperature. J. Food Protection 61(5):582–585, 1998.
53. R. G. Crawford, C. M. Beliveau, J. T. Peeler, C. W. Donnelly, and V. K. Gunning. Comparative recovery of uninjured and heat-injured *Listeria monocytogenes* cells from bovine milk. Applied Environmental Microbiology 55:1490–1494, 1989.
54. B. M. Mackey and C. M. Derrick. Elevation of the heat resistance of Salmonella typhimurium by sublethal heat shock. J. Applied Bacteriology 61:389–393, 1986.
55. R. C. Anantheswaran and L. Liu. Effect of viscosity and salt concentration on microwave heating of model non-Newtonian liquid foods in a cylindrical container. J. Microwave Power and Electromagnetic Energy 29:220–230, 1994.
56. R. A. Heddleson and S. Doores. Factors affecting microwave heating of foods and microwave induced destruction of foodborne pathogens—A review. J. Food Protection 57:1025–1037, 1994.
57. T. Koutchma. Modification of bactericidal effects of microwave heating and hyperthermia by hydrogen peroxide. J. Microwave Power and Electromagnetic Energy 32(4):205–214, 1997.
58. S. Kermasha, B. Bisakowski, H. S. Ramaswamy, and F. R. Van de Voort. Comparison of microwave, conventional and combination treatments inactivation on wheat germ lipase activity. Int. J. Food Sci. Technol. 28:617–623, 1993a.
59. S. Kermasha, B. Bisakowski, H. S. Ramaswamy, and F. R. Van de Voort. Thermal and microwave inactivation of soybean lipoxygenase. Leberism Wiss. u. Technol. 26:215–219, 1993b.
60. H. S. Ramaswamy, F. R. Van de Voort, G. S. V. Raghavan, D. Lightfoot, and G. Timbers. Feed-back temperature control system for microwave ovens using a shielded thermocouple J. Food Sci. 56:550–552, 555, 1991.
61. M. E. Boon and L. P. Kok. Terinzagelegging 8702338, Octrooiraad, neverland. Cited in Shin and Pyun, 1997, 1987.
62. B. Martens and D. Knoor. Developments of nonthermal processed for food preservation. Food Technol. 46(5):124–133, 1992.
63. R. V. Decarcau. Microwaves in the food processing industry. Academic, New York, 1985.
64. H. Khalil and R. Vilotta. A comparative study on the thermal inactivation of *Bacillus stearothermophilus* spores in microwave and conventional heating. Food Engineering and Process Application 1985:583–594.

7
Consumer, Commercial, and Industrial Microwave Ovens and Heating Systems

Richard H. Edgar
Wastech International, Inc.
Portsmouth, New Hampshire

John M. Osepchuk
Full Spectrum Consulting
Concord, Massachusetts

Microwave heating equipment now plays an important and ever-expanding role in the preparation, processing, and consumption of foods. In this chapter we try to describe the available equipment which ranges from the common consumer microwave oven, the more rugged and designated commercial variety, and finally the much-higher-power equipment used for the industrial processing of foodstuffs. In this chapter we first review the historical development of these devices and systems and then the basic elements of a microwave heating system: the magnetron power source, followed by cavities and applicators. Then we scan the type of systems available today from various manufacturers with information on their features from size, weight, and power to accessories, controls, and price. Then we discuss the serious objectives of power, efficiency, and uniformity and the state of the art in seeking these goals. We survey the available controls and sensors and finish with our opinions on where this technology is headed.

I. HISTORICAL INTRODUCTION

A. Early History

The early history of microwave heating and microwave ovens has been presented by one of the authors [1]. Although there was considerable theoretical study of this subject and related work at lower frequencies, it was not until 1947 that two microwave frequencies, 915 MHz and 2450 MHz, were assigned by the FCC for microwave heating equipment [2]. These were designated ISM bands (industrial, scientific, and medical) following the first three ISM bands which were at 13.66, 27.32, and 40.98 MHz. (Note that these are harmonically related—a good practice which unfortunately was not followed at microwave frequencies.)

The early work by Percy Spencer and Raytheon in developing the microwave oven is well documented [1]. Raytheon initiated the commercial microwave oven in the 1950s; these were large, expensive wall-mounted devices licensed to Westinghouse and Tappan for the manufacturing of consumer ovens. It was not until the mid 1960s that Amana Refrigeration, Inc., a newly acquired subsidiary of Raytheon Company, first marketed a portable countertop oven at a reasonable price, below $500. The subsequent growth of the microwave oven market and its development is well chronicled [1]. Although GE had marketed a microwave oven operating at 915 MHz, a combination range model, in the late 60s and early 70s, its manufacture was halted in the mid-70s. Since then the frequency of 2450 MHz has become the worldwide frequency for all consumer and commercial microwave ovens.

In the industrial area, however, the use of the 915 MHz frequency has become dominant, although some equipment operates at 2450 MHz. The beginning of the industrial technology for microwave heating was reviewed in Ref. 1, and a recent survey (3) of the broad field of microwave ISM technology and applications has been published. The full history of the industrial business as it pertains to the food industry, however, has not been reviewed in detail. Therefore, it is useful to conduct such a review here in view of the importance of such equipment for the food industry.

B. Early Raytheon and Other Industrial Microwave Technology

In the early 1960s Raytheon's Super Power (Spencer) Laboratory, under the leadership of William C. Brown, successfully developed and demonstrated a 400 kW S-Band CW crossed-field amplifier (CFA) with over 65% efficiency. This achievement spawned the development of Raytheon's entry into the industrial microwave oven market.

W. C. Brown and his engineering team had created a technology ahead of its time. This led to attempts to develop new applications in the fields of microwave power beaming and industrial microwave heating. Power-beaming concepts under consideration included the microwave-powered helicopters, microwave-powered land vehicles, power beaming to remote locations, and later the Solar Power Satellite (SPS) concept advanced by Peter Glaser of Arthur D. Little. Early industrial applications included medical, textile, agricultural, and food products.

At Raytheon, microwave tube engineers designed the first industrial microwave systems. There was, therefore, an initial tendency to design a new tube for each new application. Many of the early applicator designs were based on the use of single-mode and periodic structures. These structures, when unencumbered by tube design constraints, can be made large enough to handle significant amounts of power. Their outstanding virtue is that the RF fields are predictable and they can be shaped to match material geometry. Their disadvantage is that they require a significant design effort resulting in a product-specific applicator, which in most cases is not suitable for general-purpose use.

Table 1 lists some of the early industrial applications. Potato chip drying [1, 4–6] is one of the best examples of an early high-power industrial application. It was developed in conjunction with major manufacturers like Frito Lay for finish drying of potato chips. The use of microwave energy for this application was based on finish drying of chips to improve the color of the finished product. The Raytheon system included a Raytheon-designed 50 kW, 915 MHz magnetron (QKH1536) feeding one end of a 5 ft wide, 22 ft long WR975 serpentine waveguide applicator. The other end of the applicator was terminated with a water-cooled matched load. The serpentine waveguide was split in half along the broad wall. This allowed it to be opened like a clamshell for cleaning and conveyor belt repair. To prevent leakage a linear choke was incorporated into the entire length of the serpentine parting line. Considering the number of new concepts involved in this system it performed remarkably well. The product opening was about 3 in high. Warm air was used to remove moisture liberated by the microwave energy. Throughput was estimated to be 1500 lb/h (wet weight) with moisture reduction from 6 to 2%. Litton sold many more potato chip systems than Raytheon did during this period. The failure of this concept falls into the realm of competing technology—the development of improved methods of color control that did not require microwave energy. Figure 1 is an illustration of the original Raytheon prototype potato chip system. Figure 2 compares the relative size of serpentine waveguides designed for operation at 2.45 GHz and 915 MHz.

Table 1 Early Industrial Microwave Food Applications

Application	F (MHz)	RF Power	Applicator	Manufacturer
Potato chips	915	50 kW	Serpentine	Raytheon/Litton
Pasta drying	915	50 kW	Multimode	Microdry
Bacon cooking	915	50 kW	Single/multimode	Microdry/Raytheon
Chicken cooking	2450	2.5 kW modules	Multimode	Litton Industries
Tempering	915	25 kW	Multimode	Raytheon
Doughnut proofing	2450	30 kW	Serpentine	DCA Industries

Figure 1 Photograph of an early Raytheon industrial microwave oven operating at 915 MHz, utilizing a folded (serpentine) applicator for potato chip processing.

Microwave Ovens and Heating Systems 219

Figure 2 Photograph of two folded (serpentine) waveguide applicators; one operating at 915 MHz and one operating at 2450 MHz.

Microdry Corporation led the development of microwave Pasta drying [7] systems. This early microwave/hot air process survived the test of time. The use of microwave energy in this process helped reduce checking or cracking of the surface of the pasta. Other benefits attributed to this process include reduced bacteria counts and shorter drying times. Most, if not all, systems were sold by Microdry Corporation, which at that time was located in San Ramon, California. These systems operate at 915 MHz. By 1973 [8] it was reported that there were six dryers operating in the United States and several more in the planning stage.

Microdry was also a leader in developing a warm air/microwave bacon cooking [7] application with Westland Foods in Concord, California. Benefits attributed to the process included improved product quality and reduced nitrosamine formation. Early Microdry systems used a single-mode waveguide applicator followed by a multimode applicator. Some users objected to the extreme length of this type of system required by high production demands. Subsequently Raytheon introduced the use of modular multimode cavities and a 3 ft-wide belt. Eventually, multimode, 3- to 6-ft-wide belt systems became the industry standard for this application.

Litton's Atherton division led the development of microwave poultry cooking systems in the 1960s.

By the early 1970s there was a proliferation of many "one-of-a-kind" systems sold into markets with a limited repeat sales potential. What many of these applications had in common was what we all know as the "shotgun" approach to marketing. Every system and application was different in concept and detail. The result was, in the case of Raytheon at least, that over a period of years a large number of one-of-a-kind systems were sold. Little or no marketing was done to determine whether or not market size would result in positive payback for Raytheon's investment. Also, because all systems were different, they were extremely difficult to maintain and support. As a result users experienced frustration and elevated maintenance costs. It wasn't until the early 1970s that the wisdom of building standard, modular systems for reasonably sized markets was implemented. It was marketing and decoupling from the components business that saved the industrial microwave business at Raytheon.

C. Later Raytheon and Amana Industrial Microwave Technology

Around 1965 Raytheon closed down its super power laboratory. At this same time Raytheon recognized the potential for this industrial technology and formed an industrial systems group in the Microwave & Power Tube Division. In the mid-1970s the group was merged under the leadership of George Freedman who nurtured it until it emerged as a separate self-contained systems business in the early 1980s. After that it ran under the leadership of R. H. Edgar at Raytheon until the early 1990s and at Amana from 1993 to 1996.

This group initiated activity in the food, foundry, and rubber markets. The foundry and the rubber-processing markets provided substantial business income for a number of years. Eventually the foundry market for core wash drying was abandoned because of an inability to compete economically with conventional gas and electric fired drying equipment. Active marketing of rubber-processing equipment (continuous vulcanization and preheating) was eventually abandoned also in spite of the fact that many rubber compounds, because of their poor thermal conductivity, temperature sensitivity, and thick cross section, are good candidates for microwave processing. However, the market consisted of many small manufacturers captive to automotive and other major industries. Plus, in many cases, special "microwave" compounding was required to make specific applications feasible. These factors dominated the Raytheon decision to back out of the market.

The first successful tempering system was shipped to the H. J. Heinz company in mid-1972 [9]. This system was initially powered by two 25 kW transmitters using RCA 8684 magnetrons. The application involved tempering frozen blocks of poultry from 0 to 27°F for a canning application that required precise product temperature control for dicing chicken. The keys to the success of this application were performance, reliability, and payback. A measure of the success of this application is found in the fact that 4 years later there were about 25 tempering systems installed throughout the United States.

Table 2 lists some of the Raytheon systems sold during the 1970s and 1980s into the food, rubber, and foundry markets. The number of systems sold is somewhat misleading in terms of establishing market size. In most cases the selling price of the food-processing systems is several times that of the rubber and foundry systems. If we assume average selling prices of $200K and $500K for tempering and cooking systems, then the cumulative sales are on the order of $85 million. Other sales are estimated at one-fourth to one-third of the food industry sales. To these numbers must be added spares and service sales which tend to run at a constant 15% of equipment sales.

Compared to consumer or commercial microwave oven sales these numbers appear small. However the impact and importance of this technology must be measured in terms of the benefits it brings to hundreds, if not thousands, of products delivered to millions of consumers.

D. Raytheon/Amana Microwave Ovens

Raytheon under Percy Spencer was the early leader in the development of the microwave oven. It was Percy who chose 2450 MHz as the suitable frequency for microwave ovens. Although the initial focus was on commercial applications, e.g., for restaurant use, after Raytheon acquired Amana Refrigeration, Inc. in the mid-1960s it triggered the birth of the modern countertop microwave oven. This was feasible at 2450 MHz while at 915 MHz, the frequency preferred by GE, the only feasible embodiment was in the larger range appliance as a combination of both conventional oven and microwave heating.

After Amana introduced the countertop oven it also introduced a variety of innovations that persist in today's ovens, e.g. the use of temperature probes, the use of microprocessors for timing and power controls of the oven, rotating feed antennas, and the slotted-choke door seal [10] which permits reliable leakage suppression in a field-tolerant door design. In addition Amana pioneered the "cook-by-weight" oven in the late 70s, a model no longer manufactured because of cost considerations. Amana has marketed a variety of combination ovens, including a

Table 2 Raytheon's/Amana Later Industrial Systems

Application	F (MHz)	Tube	Applicator	No. systems	Market
Tempering	915	RCA Magnetron	Multimode	300	Food
Cooking	915	RCA Magnetron	Multimode	50	Food
Tempering	2450	Cooker Magnetron	Multimode	25	Rubber
Vulcanizing	2450	Cooker Magnetron	Multimode	50	Rubber
Preheating	2450	Cooker Magnetron	Multimode	100	Rubber
Drying	2450	Cooker Magnetron	Multimode	20	Foundry

(a)

(b)

Microwave Ovens and Heating Systems 223

(c)

(d)

Figure 3 Photographs of some microwave ovens. (a) Old 1.6 kW commercial oven from the 1950s; (b) modern 800 W consumer microwave oven; (c) modern 1.8 kW commercial oven; (d) modern combination—convection/microwave—commercial oven. (Courtesy Amana Appliances.)

microwave/convection model for the countertop and wall-mounted market. At one time Amana and its Caloric subsidiary manufactured a combination electric or gas range with incorporated microwave heating. This model is no longer being manufactured. In the 1960s and 1970s Amana was a leader in the marketing of consumer ovens. After the price of microwave ovens was greatly reduced with offshore production [11] in the 80s, Amana's market share diminished for consumer ovens but its commercial market grew. Later, with the acquisition of the commercial microwave-oven business of Litton in the early 90's, Amana became the leading producer of commercial microwave ovens. These ovens are designed for more rugged field use and with more attention to life and reliability. They therefore command a premium price.

To add to the historical perspective, we show in Fig. 3 photographs of an early Radarange oven for restaurant use, a cost-reduced modern consumer oven with plastic/metal door, and lastly two examples of commercial ovens, one a combination oven, made by Amana Appliances. (In 1997 the Amana firm was sold to Goodman, Inc. by Raytheon Company.)

E. Other Industrial Microwave Companies

In the earliest days of industrial microwave heating major companies including Varian, Litton through its Atherton Division and Raytheon attempted to market and sell industrial microwave systems for food applications and applications in other market areas. However, over the long haul Raytheon is the only major company, through the Microwave & Power Tube Division, that remained committed to the business. During this same period there were a number of smaller companies led by visionaries who believed in the future potential for microwave heating.

Microdry Corporation led by Frank J. Smith made significant contributions, as mentioned above, to developing new applications for the food industry. This company also entered other diverse markets with innovative microwave technology. Although Microdry never became a major capital goods supplier to the food market, they have survived, with some reorganization and mergers along the way, up to the present.

John E. Gerling who led Litton's Atherton Division went on to form his own company which, under a variety of names employing his surname, survived up until the mid-1990s when it was acquired by Astex, a Massachusetts-based company focused on microwave plasma technology.

New Japan Radio Co., Ltd., founded in 1959, pioneered industrial microwave processing in Japan. In addition to its semiconductor operations, New Japan Radio had a strong microwave tube capability. As was the case with Raytheon's Microwave and Power Tube Division, they chose to enter industrial microwave equipment market with 2.45 GHz systems for food applications unique to Japanese culture. These have included rice swelling, green tealeaf drying/roasting, and vacuum drying. Although New Japan Radio has not exported its technology, they remain a viable, innovative force in Japans industrial oven market. Figure 4 is an illustration of their tealeaf dryer and roaster.

DCA industries led by Robert Schiffmannn developed and successfully marketed a significant number of microwave doughnut proofing systems to the food industry.

Figure 4 New Japan Radio 2450 MHz green tea leaf drying-roasting system. Advantages: 1) Effectively produces excellent Japanese tea still remaining innate green color with natural flavor and moisture. 2) Dries tea leaves satisfactorily and uniformly without scorching or discoloring their surface. 3) Large residual rate of contents (without spoiling their inherent composition). See table below. 4) Retards the fading or discoloring of tea during storage. 5) Good controllability to ensure stable product quality.

Vitamin C Contents of Green Tea

Hot air drying + gas drum direct roasting: (conventional method)	approx. 250 mg/100 g
Microwave drying + far-infrared roasting:	approx. 425 mg/100 g

F. Other Microwave Oven Companies

The history [1] of microwave heating shows that originally the U.S. microwave oven market was dominated by U.S. firms such as Amana, Litton, Tappan, Roper (Sears), Magic Chef, GE, Whirlpool, and a few others. As time went on, mostly economic factors shifted most consumer oven production to non-U.S. firms, although firms like GE still command a considerable share of the consumer market (GE has, for years [11], had its ovens manufactured by Samsung in South Korea).

Mergers have had their impact on the list of manufacturers. Today, after several mergers, Tappan and Electrolux brands are subsumed under the Whirlpool name. Various manufacturers market microwave ovens, e.g., Maytag, even though they are sourced from other firms, primarily in Japan or Korea. Japanese manufacturers still include Toshiba, Sanyo, Panasonic (Matsushita), and Sharp. The largest appears to be the Sharp Electronics, which maintains manufacturing facilities in the United States as well. Panasonic manufactures commercial ovens in the United States and is the largest supplier after Amana of commercial ovens. Korean manufacturers play a large role in supplying the U.S. market. Samsung Electronics is the leading supplier followed by Goldstar and Daewoo.

Microwave oven manufacturers in Europe include native firms in the larger countries, e.g., Moulinex in France. Microwave oven manufacturing also exists in eastern Europe, Russia, and China although little is known about their products.

The total sales in the United States is now estimated around 8–9 million ovens per year. Total sales worldwide is estimated around 20 million per year. The saturation figure in the United States and Japan is well over 80%, while western Europe and Korea lag behind with 50% and developing countries lag behind even further.

II. POWER SOURCES FOR MICROWAVE HEATING

A. Microwave Tubes

The first chapter of A. S. Gilmour Jr.'s [12] book describes the choices available to all microwave designers including ISM. They include CFAs (magnetrons), TWTs, Klystrons, Gyrotrons, and solid state. Although solid state power capability continues to improve it will probably be a long time before practical low-power ISM heating applications using solid state become common. So, for practical purposes, microwave tube choices are limited to magnetrons, Klystrons, and Gyrotrons. All of these devices have been used to a significant extent in communications, radar, and particle physics applications, but the magnetron has dominated all but a few ISM applications up till now.

Table 3 compares these devices in terms that are important, though somewhat oversimplified, to ISM equipment manufacturers. Simply put, magnetrons

Table 3 Microwave Power Sources

Attribute	Magnetron	Klystron	Gyrotron
Acquisition cost	$0.01–$0.1/W	Higher	Much higher
Efficiency	70–85%	30–60%	40%
Operating voltage	4–20 kV	20–50 kV	
Cooling	Air/water	Water	Cryogenic
X-rays	No	Maybe	Maybe
CW power			
2.45 GHz	1–30 kW	30–100 kW	
915 MHz	30–100 kW		
10–30 GH	Not practical		100 kW and up

are low cost, rugged, very efficient, and long lived. They can be operated from the relatively simple inexpensive unfiltered voltage doubler power supplies used in consumer and commercial ovens as well as the linear dc power supplies used for industrial systems. Low dc operating voltages ranging from 4000 to 20,000 V and low internal RF voltages almost always results in freedom from X rays and the need for X-ray shielding.

Their relatively low cost, even in the small manufacturing quantities involved in industrial applications, means that transmitters for high-power industrial applications can be sold at prices on the order of $1/W. Another important characteristic of magnetrons is high operating efficiency, which is of increasing societal importance and a strong selling point in commercial and industrial applications where energy conservation and operating cost are important issues. The combination of low initial cost, potential for long life, and high efficiency can result in a very low net cost per kilowatthour of operation.

Magnetron noise figures generally do not meet the demanding requirements of today's communications and radar systems. As a result the magnetron is used primarily in industrial, commercial, and consumer application where, to date at least, noise performance has not been an issue.

B. Cooker Magnetrons (2.45 GHz)

The early development of the magnetron for microwave ovens, the "cooker" magnetron, is reviewed in the basic history reference [1]. The early models were bulky, water cooled, and expensive. The explosive growth of the consumer microwave oven market, however, triggered the economic investments required for the ultimate evolution of this remarkable device. Today, for oven manufacturers, the price of cooker magnetrons is around $10 per tube in large quantities. Tubes for specialized applications may cost somewhat more. In Table 4 we present a list of some recent and current models of magnetrons used in consumer and commer-

Table 4 Magnetrons for Microwave Ovens

Tube manufacturer and type		Nominal power, (kW)	Nominal voltage, (kV)
Toshiba	2M229	0.850	4.0
	2M240	0.850	4.0
	2M248	1.02	4.35
	2M255	1.10	4.35
Panasonic	2M167B	0.900	4.1
	2M137	1.260	4.5
	2M210		
Samsung	2M204	0.900	4.1
	OM75S	0.870	4.1
Goldstar	2M613	1.420	4.5
	2M226	0.900	4.3

cial ovens. We note that they have nominal output power (into a matched load) of the order of 1000 W and operate at a voltage around 4.0 kV. Magnetrons with power of 2 kW and higher are available but are more expensive and used more for industrial applications.

Magnetrons found in practically any microwave oven, consumer or commercial, will generally have very similar properties in terms of efficiency, spectrum, reliability, and life. Actual values of useful life, before intermittent moding signals end of life, can be expected to be between 1000 and 2000 h, on-time. It has been recognized [13] for some time that the design of cooker magnetrons has essentially become standardized so that they all exhibit similar performance.

The tubes listed in Table 4 are only part of the complete list of tubes available for microwave ovens, but they are representative. Their size and weight are now fairly comparable. Size, outside of 1 in projections for antenna, leads, and mounting lugs, is within a $4 \times 4 \times 3$ in volume. Weights range from 1.5 to 2.0 lb, compared to over 3 lb for comparable tubes used in the 1970s and 10 times that figure in the original Radarange microwave ovens. The modern tubes are all air cooled, whereas the original Radarange tubes were water cooled. All the modern tubes are specified for a matched-load frequency at around 2455–2460 MHz. Of course, with arbitrary loads, line voltage, and temperature variations one can expect a few MHz drift, a few MHz pushing (variation with anode current), and as much as ± 30 MHz pulling with load variations, e.g., as in an empty oven or with a light load such as popcorn.

The special attention to this one component of the microwave oven is in recognition that the magnetron is the heart of the microwave oven and only its evolution as a low-cost component has made the existence of a widespread microwave oven market possible. It is the only microwave power source that is inexpensive and efficient, e.g., over 70% efficiency into a matched load. Unfortunately, such a device exists only at one frequency, 2.45 GHz. Thus, most research on microwave heating is done at 2.45 GHz, even though in principle other frequencies might yield superior results for a particular food product.

We note that the dominant manufacturers of cooker magnetrons are in Japan and Korea, e.g., Toshiba, Panasonic (Matsushita), Sanyo, Samsung, Goldstar, and Daewoo, as well as in Russia and China. Some of the Japanese and Korean firms are transferring production to locations such as Singapore and Thailand. The total production of cooker magnetrons for new ovens as well as replacement tubes is estimated to be about 25 million tubes per year. This makes the cooker magnetron the most mass-produced microwave component in the modern world.

C. 915 MHz Industrial Magnetrons

Raytheon first started using this tube around 1970 in conveyor and batch food-tempering systems. At that time the 8684 magnetron was manufactured by RCA and it was rated at 25 kW. All the transmitters used in systems manufactured by Raytheon during the period 1970–1980 were rated at this power level. It was not until 1980 or 1982 that Raytheon took a calculated risk and upped the tube rating to 40 kW. This decision was based on customer demand for increased throughput to improve payback. The first Raytheon system with 40 kW (conditionally rated) transmitters was shipped to the customer around 1981 or 1982. The conditional rating allowed Raytheon to temporarily derate the system if serious life or operating reliability problems were encountered.

Over the next 10 years the tube rating was increased to 50 kW. In 1993 Raytheon's business was transferred to Amana and a new competitor, Ferrite Components, Inc. (FCI) in Hudson, New Hampshire, entered the marketplace. As a result of competition and customer demand for greater productivity, transmitter (tube) ratings were increased first to 60 kW and then to 75 kW. During this period all tubes were supplied by Burle and by California Tube Labs (CTL). Only minor design changes, other than some minor cathode (filament) changes, were made to the tube during this entire period.

During the same period the tube manufacturers warranty has increased from about 2000 to 3000 h, the price has increased from about $3000 to about $7000 and the power rating has increased from 30 to 75 kW. Therefore, the operating cost to the user in dollars per kilowatt, assuming worst case (warranted) life, has dropped from 5 cents/kWh to its present value of about 3 cents/kWh. Thus the operating tube cost per hour in typical 500 kW systems has dropped from $25 to a little over $16/h. For a cooking system with a throughput of 2000 lb/h this equates to a reduction from 0.012 to about 0.008 cents per pound processed. A tempering system, which is less energy intensive, can have 10 times the throughput of a cooking system and incur per pound processing costs that are only a fraction of a penny per pound.

The high-power 915 MHz magnetron has dominated industrial microwave food applications for the past several decades. Without the high efficiency, reliability, and long life offered by this device, applications such as food tempering and bacon cooking would not be practical. Today there are four manufacturers for the tube: Burle Industries, California Tube Labs (CTL), EEV, and Svetlana. Each of these manufacturers produces an identical tube in terms of physical configuration and required operating electrical parameters. Competition between these manufacturers benefits both the user and the system manufacturer. This tube is also available with an 896 MHz operating frequency for applications in the United Kingdom. The current standard 75 kW tube is essentially identical to the original 25 kW RCA magnetron. Higher power performance has been achieved by operating at higher voltages and by making minor changes in interaction space and cathode design. 75 kW is considered to be the upper performance limit for this tube by all manufacturers. It operates at a typical efficiency of 85%, which is an important consideration in high-power applications. The 915 MHz magnetron is also sensitive to high VSWR caused by operation into the multimode cavities that dominate most industrial applications. Therefore three port ferrite circulators are always used to protect the tube.

The costs of these devices, expressed in terms of dollars per pound of product, are quite low when compared with other operating costs such as electricity. However, their costs receive much attention because of the dollar magnitude of annual replacement. For example, an industrial user with five lines, each line with 8–75 kW transmitters, is faced with an average annual tube replacement cost, assuming warranted life (3000 h) and one or two shift operations, ranging from $250,000 to $450,000. Fortunately, today's systems enjoy tube lifetimes that exceed warranted life by at least a factor of 2. Nevertheless, annual maintenance purchases of this magnitude receive close scrutiny by product and plant managers. Therefore, the worst case assumption that the tubes will only last through the warranty period is used only for new system, worst case, payback calculations.

At the present time, both CTL and Burle Industries are offering 100 kW versions of the 915 MHz magnetron. A small number have been incorporated into microwave plasma systems for diamond deposition. These systems generally employ switch mode power supplies (15), which offer precise operating point control and freedom from magnetron overload faults. These features are essential for deposition applications, which involve uninterrupted runs lasting many days, and in some cases, weeks.

D. Higher-Power Magnetrons at 2.45 GHz

In the past few years both Burle Industries and CTL have introduced new 30 kW industrial microwave magnetrons. These tubes can be operated over a wide power range using an external, self-excited solenoid. Efficiency is generally better than

70%. These tubes will make the design of future high-power systems easier. The availability of these tubes is particularly significant for high-power European applications where the 915 MHz band is allocated for other use.

E. Reliability and Life: 915 MHz Magnetrons

Dominant failure modes for new and used magnetrons are listed in Table 5. The first seven and possibly eight of the failure modes listed are related to improper use of the magnetron. For this reason magnetron power supplies are usually designed to automatically control filament voltage and current (failure modes 1–3). Three port ferrite circulators are used to protect the tube from a high VSWR caused by load mismatches. Arc detectors are used to prevent waveguide arcs from reaching the circulator and tube output dome (failure modes 4–6). These failure modes dominate the failure statistics.

F. The 75 kW 915 MHz Industrial Magnetron Transmitter

Most transmitters manufactured today use linear power supplies that incorporate PLC control and solid state filament and solenoid power supplies. These design features offer tremendous benefits to the user. They include user-friendly (prompted) operation, powerful local and remote modem diagnostics, data and fault logging, and more precise control of filament and solenoid operating values. They are usually operate in a "self-excited mode" in which the magnetron anode current is passed through the solenoid in order to create the magnetic field. This circuit configuration results in some regulation against ac source voltage fluctuations. Aside from these control features, the basic design of the self-excited magnetron power supply has not changed in 35 years.

Table 5 Magnetron Failure Modes

	Failure mode	Possible cause
1	π-1 Mode	Excessive filament cutback
2	Emission loss/RF noise	Filament sag/insufficient cutback
3	Open filament	Filament evaporation
4	Cracked output dome	High VSWR/waveguide arcing
5	Exterior arcing	Electrode spacing/overvoltage
6	Magnetron overload	Gassy tube/high VSWR
7	Exterior damage	User abuse
8	Thermal damage	Insufficient cooling
9	Internal arcing	Insufficient aging
10	Manufacturing defect	Improper assembly

Source: Adapted from Solomon [Ref. 14].

G. Higher-Power Industrial Magnetrons

There are arguments both for and against the use of higher-power magnetrons. These are generally based on basic magnetron device limitations, system design economics, and performance.

The basic size of a magnetron is dictated by the operating frequency. This places a fundamental restriction on the surface area available for heat dissipation and ultimately operating power. This limitation is partially offset by very good dc to RF conversion efficiency: typically 85% at 915 MHz and 70% at 2.45 GHz. Also, forced liquid cooling can be brought close to the vane tip surface to minimize vane tip temperature rise. For a given design and fixed frequency an effective method for increasing up the power of a magnetron is voltage scaling. Unfortunately, this is a losing game since the anode area only increases as the square root of the voltage. Existing 75 kW tubes are warranted for 3000–4000 h and generally experience average lifetimes on the order of 5000–6000 h of service in linear power supplies. Operation at higher-power levels involves uncertainty about life and reliability. For these reasons there are no strong imperatives at present, not technical and certainly not market size, for developing higher-power magnetrons.

High-power tubes can result in a system with fewer components, less complexity, and often lower cost. The use of circulators is mandatory with high-power microwave systems. New circulators must be designed to match the rating of higher-power tubes. The power rating and reliability of other components such as waveguides and antennas become a major concern at higher powers. They are usually significantly derated for industrial applications where ambient conditions are poor and difficult to control and maintain. Multiple feed antennas and magnetrons are often used to ensure uniform heating.

H. Trends in Future Cooker Magnetrons and Other Power Sources

The cooker magnetron is the preferred device, by far, for microwave power generation for microwave heating. It is efficient, compact, reliable, and inexpensive. Its disadvantages are the excessive noise [13] that it produces and the requirement of a high-voltage supply. This, in its low-cost picture, requires bulky and heavy transformers working at 50/60 Hz. Over the years there have been attempts to productize a low-voltage magnetron (e.g., 400–600 V) that could be driven by a simple quadrupler supply off-line voltage. These all have failed to succeed in practice. The other solution for power supply weight is the inverter supply, which is discussed in Sec. III.B.4.

Any of these approaches which still use the magnetron will still suffer from the excess noise that promises to be a regulatory problem in the future (see Secs. III.C.2, III.C.3, and VIII.E). Over the years there has been the hope that a low noise device could replace the magnetron, e.g., the multiple-beam Klystron or solid state devices. Both of these devices have been reported to yield useful efficiency of over 60%. A review [16] in 1996, however, casts doubt on the ability to reduce the costs of these devices to that of the cooker magnetron. Therefore, it is likely that for many years, at least a decade and probably more, the cooker magnetron will be the only practical choice for microwave ovens, either commercial or consumer.

There is ongoing work to further improve the efficiency of cooker magnetrons and to reduce the noise. A report [17] in 1996 indicates that an efficiency increase of up to 5% is possible with a small cost penalty. The potential for noise reduction is far more uncertain, however, because of the lack of understanding of this subject.

Ideally there would be efficient power sources at many power levels and at several frequencies. For example, if microwave ovens for the automobile are to be practical, then one would like an efficient magnetron at the 100–200 W level. This device doesn't exist. For certain food applications and possibly for microwave dryers, it would be desirable to have a low-cost 1 kW tube available at 915 MHz. This does not exist even though there is a Russian supplier of such tubes at a premium price. At other ISM frequencies, e.g., 5.8 or 24.125 GHz, the feasibility of low-cost efficient magnetrons is even more distant. When and if events conspire to a breakthrough, availability of the sought-after tubes could trigger new advances in microwave heating of foods and many other products.

III. MICROWAVE APPLICATORS AND CAVITIES

Here we describe the basic elements of a microwave heating system for both industrial equipment (Sec. A) and microwave ovens (Sec. B). The latter are about the same for either consumer or commercial ovens, the difference being in the choice of components for higher reliability and life in the commercial application.

The basic design principles of microwave ovens have been enunciated in many books, e.g., those by Decareau [18] or Buffler [19]. In addition, design considerations are covered in many of the publications of IMPI (International Microwave Power Institute, Manassas, Virginia).

Design considerations for high-power industrial systems and equipment are more involved, but there is considerable literature available from IMPI on this equipment. There are two recent books [20, 21] by Metaxas and one by Meredith [22] that describe such equipment and their design principles.

In this section we will not review these design procedures in detail but we will outline the basic concepts and embodiments to allow the reader a better understanding of some of the factors limiting performance of microwave heating systems and also where some potential for improvement exists.

A. Industrial Equipment

In Fig. 5 we show an exploded view of a conveyorized industrial microwave oven for food processing. These systems have a number of important design requirements that differentiate them from consumer and commercial ovens. They require special attention. They include the following:

   Product openings much larger than a wavelength
   Operating power on the order of 10 to several hundred kilowatts
   Severe ambient conditions including wide temperature extremes, high humidity, and water spray
   Modular construction often involving separate power generators, cavity applicator, and control system
   High reliability often demanded by three shifts, 6 to 7 day production schedules
   Operation by unskilled labor

In view of the extremely high operating power and the size of the product opening in many of these systems it is no surprise that a safety-first design approach is mandated.

1. Industrial Microwave Oven Cavities

Major U.S. industrial microwave oven manufacturers for the food industry generally use batch and continuous multimode cavities. In general, cavity design has been based on an empirical, experimental rather than an analytical microwave design process. Practical considerations such as product size, throughput requirements, sanitation requirements, code requirements, and ambient conditions play an important role in the design process. Heavy, 11 gauge, stainless is used for construction. Seams are continuously welded and finished to meet various sanitary codes such as those imposed by the U.S. Department of Agriculture (USDA).

As an example consider the Amana QMP1879 batch-tempering oven. It employs a roll in–roll out plastic pallet for holding cartons of frozen food that is to be tempered. The size of this pallet (which defines the width of the oven opening) was chosen to precisely match the size of standard pallets used in food plants for transporting boxed meat, fish, and poultry.

Tempering experiments with antenna feeds above and below the product resulted in good heating uniformity provided boxes were not stacked more than two high. Therefore, the product opening was limited to approximately 12 in. Another important consideration was cavity loading. Good efficiency ensures high-process efficiency and freedom from arcing, over the range of temperatures and dielectric loss properties encountered during the batch exposure time. Fortunately, the 9 in opening height resulted in a good tradeoff against these competing requirements.

The resulting cavity consists of a 4 ft cube. It is, based on its size, measured in wavelengths at 915 MHz, a multimode cavity. However, tempering is a difficult ap-

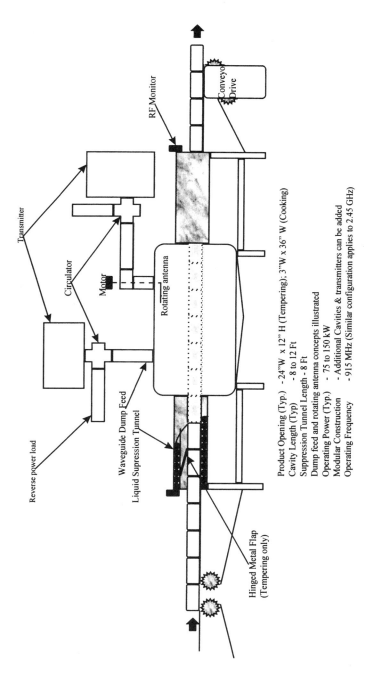

Figure 5 Exploded view of a typical conveyorized industrial microwave oven.

plication subject to thermal runaway. Excitation of many modes was not a major design consideration. Instead, uniform heating was the result of product scanning by three port antennas placed above and below the product, vertical product motion, and limiting product thickness. A typical batch tempering system is illustrated in Fig. 6.

Similar considerations dominated the design of conveyorized tempering systems, discussed in Sec. IV.C.I. The basic transverse cross section of the cavity is a 4 ft square. The cavity length was selected, as discussed above, to ensure good cavity loading resulting in efficient operation and freedom from arcing. Cavity

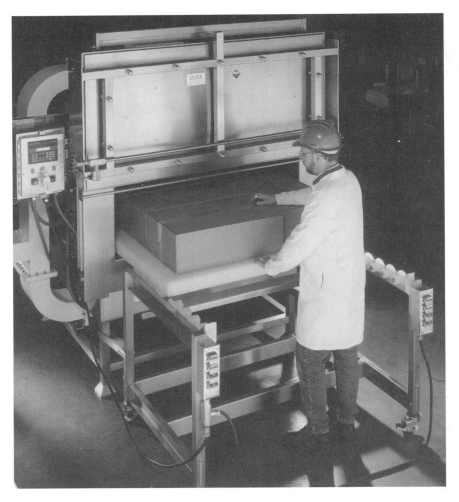

Figure 6 Amana batch tempering system operating at 915 MHz.

length has been increased in recent years, as transmitter power has increased, to maintain constant power density in the cavity.

2. Feeds

Microwaves feed systems are the critical interface between the microwave energy and the product. They must introduce energy into the oven efficiently and distribute it in a manner that results in uniform heating. Simple waveguide "dump" feeds have been used successfully in many high-power applications. The Amana tempering system illustrated in Fig. 7 is a good example. The small aperture of the

Figure 7 Amana conveyorized industrial tempering oven.

feed results in a wide beam of energy that interacts with the product before it "sees" the walls of the cavity. Cross-polarization, along with product motion, helps ensure heating uniformity. For several years Microdry has successfully used a waveguide dump feed in conjunction with a mode stirrer. This configuration depends on multiple wall reflections to excite cavity modes and, in so doing, achieve uniform heating. Although near field horn and slow-wave "fringing field" antennas have been used with some success in other industrial applications, simple, rugged, and reliable dump feed dominate high-power food applications.

3. Seals

In view of the extremely high powers and the large size of some of the product openings employed in industrial microwave ovens, seal design is a major design criterion with higher priority than either system reliability or performance. Because of the importance of this safety issue most responsible industrial microwave oven manufactures voluntarily meet the same emission standard imposed on consumer and commercial ovens. This standard is the emission standard [29] adopted by the Center for Devices and Radiological Health (CDRH; see also the chapter on safety in this handbook). It mandates limiting leakage, measured at a distance of 5 cm from the surface of an oven at the time of manufacture, to a maximum of 1 mW/cm^2 and, in service, to a maximum of 5 mW/cm^2. In practical terms, this means that a 750 kW oven, with door and tunnel open surface area about 2 m, is allowed to radiate only a few watts of energy at the time of manufacture. The importance of this emission standard is that field compliance, based on good user maintenance, virtually ensures compliance with the OSHA personnel exposure standard [30].

Industrial ovens are generally classified as batch and continuous. Batch ovens include single-door (see Sec. 1 above), double-door or indexing batch ovens, and four-door or "vestibule" ovens. Indexing batch and vestibule ovens are used when a flow-through process is required but economic limitations or large product size prohibit the use of a continuous conveyance system. Leakage suppression in these systems is limited to doors and penetrations for drive shafts, antennas, instrumentation, etc. Product and user access door closure during operation is ensured through the use of tamper-proof, redundant electromechanical interlocks that are continuously monitored and hard-wired to devices that shut off microwave power in the event of a fault. In view of the high operating power involved and the extremely demanding requirements of these applications it is common practice to employ double-door $\lambda/4$ chokes of the slotted variety [10, 27]. The chokes are often filled or covered with a dielectric to prevent damage and comply with food equipment sanitation requirements.

Leakage suppression in continuous-flow microwave ovens poses a much more significant technical problem. Reactive chokes may be used when the height of the opening is on the order of a quarter of a wavelength or less (approximately 3 in at 915 MHz). Good examples are the bacon cookers described in Sec. IV.C.2. Even though these ovens operate at power levels of hundreds of kilowatts, is possible to suppress leakage to meet the 1 mW/cm^2 standard with 5-ft-long reactive pin choke [23] suppression tunnels. It is also common practice in these systems to add several additional feet of suppression with solid "lossy" wall absorbers. The additional length reduces leakage to negligible levels, suppresses second harmonic emissions, and makes it virtually impossible (based on tunnel length and opening height) for an operator to insert his or her hand into a high field region.

Leakage suppression becomes significantly more difficult when the height of the opening is greater than a quarter of a wavelength. Reactive microwave chokes do not work. Linear tunnels with absorbing walls are impractical for two reasons. First, they will tend to absorb a significant amount of the energy impinging on them at the tunnel cavity interface. Although modest loss can help provide a parasitic load and thus reduce the potential for arcing under light load conditions, high losses raise significant heat transfer issues usually requiring the use of liquid absorbers and continuous heat exchange systems. Second, when tunnel openings approach a wavelength the length of a simple absorbing tunnel, assuming perfectly absorbing walls, the tunnels become impractical because of free space propagation.

To get a first-order approximation of the length required for very large openings, assume an isotropic point source at the cavity/tunnel interface and calculate the tunnel length (spherical radius) required to reduce the power density at the output end to 1 mW/cm^2 or less. If we assume (worst case) a 200 kW point source (200 kW installed power), the tunnel length (spherical radius) required for 1 mW/cm^2 leakage is 7000 cm or 70 m. This is clearly impractical. Consequently, a number of schemes involving absorbing walls and fixed or moving gates and other tricks for reducing leakage have been developed. Continuous monitoring of these systems is considered mandatory because these systems are not fail-safe. The tempering system described in Sec. IV.C.1 has a tunnel opening of approximately 60 × 30 cm employs an 8-ft-long tunnel with liquid absorbing walls. The liquid, which is a good microwave absorber, is continuously pumped through an external heat exchanger. Several metal gates, pushed out of the way during normal operation, block the tunnel when no product is present. Leakage under all operating conditions is well below 1 mW/cm^2. RF leakage detectors constantly monitor leakage from the input and output tunnels.

4. Power Supplies

In recent years more attention has been given to the use of switch mode power supplies [15]. Their major advantages are that they are constant current power supplies that prevent the magnetron tube from drawing excess current when an arc occurs. The drawback, up to the present, for consumer, commercial, and industrial use has been increased cost, which is difficult to pass on to the customer in a competitive environment. A few 75 and 100 kW switch mode transmitters have been built for nonfood (diamond deposition) applications. It may be that magnetrons operating in this type of power supply will achieve greater life than is possible with operation in linear power supplies. This advantage could provide additional incentive for their use. Other advantages to the use of switch mode power supplies include a modular stackable concept that allows graceful degradation and simple first echelon maintenance based on replacing the plug in modules.

5. Controls

Industrial controls have become increasing complex in order to meet the demands of more sophisticated, demanding users. These control systems are almost with out exception based on the use of programmable logic controllers (PLC). Most systems on the market today are Windows based and very user friendly. Plant electrical technicians and electricians easily understand the format of the ladder line program.

6. Safety

Protecting human life is the first priority of system safety features. To ensure personnel safety, manufacturers generally comply with safety codes mandated by local, state, and federal agencies. Also, in order to meet the requirements for entry into the European Community, many manufacturers are seeking third-party CE (Conformity European) certification and configuration control. U.S. manufacturers anticipate OSHA requirements and design systems that will, assuming proper maintenance, meet these requirements. Personnel safety hazards fall into three categories: RF exposure, high voltage exposure, and mechanical hazards.

Most internationally recognized safe electrical practices and guidelines consider voltages above 50 V as dangerous and potentially lethal. This is why most control systems today are of the 24 volt ac or dc variety. Adoption of this type of control system significantly reduces maintenance personnel hazards. Unfortunately there is no way to eliminate the high ac (typically 480 Vac) and dc voltages (10,000–20,000 Vdc) required for high-power magnetron operation. Shock from high voltages does not require contact with the electrodes. Under the wrong conditions high-voltage terminals can arc to a grounded person with life-threatening and crippling consequences. Personnel safety in these environments depends almost exclusively on mandated safe work practices that are constantly reviewed and reinforced in plant safety training classes. Manufacturers recommend annual and semiannual validation of all personnel and equipment safety interlocks.

Although no formal emission standard exists for industrial equipment, most manufacturers have adopted the CDRH emission standard [29]. To ensure compliance in the workplace, the user is encouraged to implement and record the results of periodic leakage measurements made with approved and calibrated instrumentation. Measurements are typically made at transmitter doors, applicator doors, windows, suppression tunnel openings, drain ports, and all weld joints and bolted interfaces.

Industrial microwave systems operate at very high power levels. Under light load conditions cavity, antenna, or waveguide arcs can occur. These can result in high-voltage standing wave ratios (VSWR) in the feed system. These arcs will travel back to the transmitter where they can result in catastrophic circulator and microwave tube damage. To guard against these events, arc detectors are typically installed in cavities, waveguides, and circulators. These devices will shut down or modulate the microwave power in the presence of arcing.

Another aspect of industrial processing in general is the potential for fire. Microwave systems are not exempt from this problem which can involve both equipment, plant, and personnel safety issues. Cavity fires can result from overheating of "in transit" and stationary "tramp" material captured on belt supports or on the bottom of the cavity. Although arc detectors will usually detect the light from such phenomenon, most manufacturers include or provide for exhaust manifold temperature sensors and CO_2 injection. They also warn against fire hazards and recommend reputable fire protection companies.

B. Microwave Ovens

In Fig. 8, we show an exploded view of a microwave oven. There is a cavity or metal box into which power from a magnetron is fed through a radiating aperture. With such a feed a mode stirrer is used to randomize the field pattern of the microwaves in the cavity. The power supply is shown as the most common one being a half-wave doubler circuit which uses only rectifier diode and a series capacitor between the high-voltage transformer and the magnetron tube. In the case shown there is a separate filament transformer, but in most cases this is integrated in the high-voltage transformer. We note in Fig. 8 the depiction of airflow, an important process in the operation of a microwave oven. The airflow not only cools the magnetron and transformer but the warm air also is either exhausted or drawn into the oven cavity where it helps remove builtup moisture on the walls, including transparent windows which may be made of glass or plastic. We will now note some of the other salient points about the main elements in the microwave oven. We also note that there is no circulator between the magnetron and the cavity to protect the magnetron. This is a testimonial to the ruggedness of the modern cooker tube, which can withstand operating into an empty cavity—although not recommended. The consequence of this is, of course, that the magnetron frequency and power will be randomly pulled or varied as the mode stirrer or other moding element changes in the oven cavity. In addition to the general references

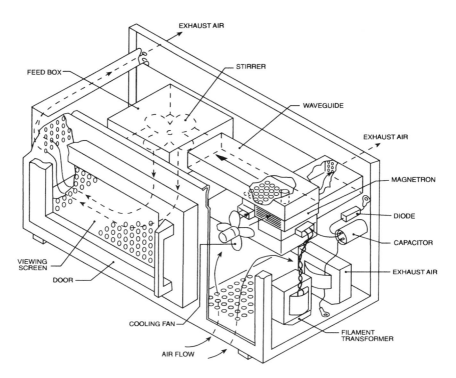

Figure 8 An exploded view of a microwave oven showing power supply, magnetron, waveguide feed, oven cavity with stirrer, and airflow pattern. (Courtesy Amana Appliances.)

[18–22], the reader is referred to a recent treatment of the basic principles of microwave ovens [24].

1. Cavities

As shown in Fig. 8, most microwave ovens employ a rectangular parallelepiped metal cavity or box to contain the microwave energy and into which the food load is placed. This is not an absolute requirement. Cavities in other shapes, e.g., cylinders or half-spheres, have also been employed. For moderate to heavy loads, it is presumed that the shape is not critical. For lightly loaded cavities, however, the nature of the cavity could be important. It is at the limit of zero or light loading that the modal structure of the cavity becomes relevant.

As pointed out in Ref. 24, it is important to avoid feeding from one end of a cavity into an effective waveguide, the cross section of which is close to the cutoff condition for a TM mode. It is shown that just below the cutoff frequency for many types of feed it will be difficult, if not impossible, to achieve a good RF match, i.e., low VSWR. This has also been pointed out by Quine [25]. This phenomenon is analogous to the blind angle phenomenon found in the operation of phased-array radar.

As pointed out in Chapter 2 of this book, the heating pattern in the food is a function of many parameters, including the size and shape of the food, nature of feed, mode stirrer, etc. For light loads it seems reasonable to assume that the mode structure of the cavity and the selection of the modes excited will be important. The literature shows that this is a controversial subject. It has been suggested that trapezoidal cavities or cavities with many walled boundaries, $n > 4$ (where $n =$ number of sidewalls), will have an advantage over the classic rectangular cross section, $n = 4$. It remains to be seen how valid this assertion is.

The metal walls of most ovens are made of cold-rolled steel or stainless steel. The nonmagnetic variety of stainless steel is preferred because of lower skin loss. Some small gain in efficiency would be obtained if an aluminum wall were used for cavity and feed waveguide. In practice this is considered a negligible factor, but if one is trying to efficiently couple to a small load it could become significant.

Dielectric parts in Fig. 8 are not shown. These could include the usual grease shield, which prevents splatter of grease on to the elements at the top of the cavity. The material of this part is usually polypropylene for ordinary microwave ovens. For microwave/convection or other designs which used heating elements, the material would be some type of high-temperature composite material. The polypropylene has low dielectric loss while the high-temperature materials will exhibit significantly higher loss.

2. Feeds

The feed shown in Fig. 8 is a rectangular waveguide from the magnetron terminating into a radiating aperture on some wall of the oven cavity. The example shown is effectively a feed at the top of the oven cavity, albeit there is a quasi (on intermediate) cavity around a mode stirrer. This type of feed is called a "dump" feed. Such a feed could also be placed at a side wall or even in the bottom wall. We note that almost all consumer ovens employ only one magnetron tube. Commercial ovens may employ two or even three tubes. In that case usually one tube feeds from the bottom, and one or two from the top.

There are a wide variety of means employed to randomize the field pattern. Besides the classic mode stirrer shown in Fig. 8, there are rotating antennas fed by a waveguide to coax transition, the radiating element being a monopole, dipole, or patch type of antenna. In most cases these rotating elements are being driven by the airflow. Thus it is important that the airflow not be obstructed by foreign objects or the bearings for the rotating element not be altered by wear, dirt, or grease.

In practice, the efficacy of a feed system is determined by the microwave measurement of the match, i.e., the reflection as seen at the magnetron. For a fixed frequency and with the mode stirrer or antenna rotating one might record input impedance contours as shown in Fig. 9. This shows the locus or contour of points on the Smith Chart, or Rieke diagram as the stirrer rotates, for the two cases of a large water load (2000 ml) and an empty oven. The general objective is to have this contour located roughly in the center of the chart or slightly in the "sink" region of the Rieke diagram. This results in good efficiency or coupling to the load. Of course, it is ideal to have this result for all stirrer positions, all frequencies, and all loads. This is, of course, impossible. Therefore the art in designing this part of the microwave oven is in finding a design that is reasonably optimum in this regard.

Besides a rotating element in the feed, there could be instead or in addition a rotating turntable sometimes called a carrousel on which the food is placed. In some very specialized applications there may be a conveyor belt which moves the food through the oven.

3. Seals

It is important to minimize leakage or radiation of the microwave energy both with respect to the in-band as well as the out-of-band part of the spectrum emitted by the oven and generated by the magnetron. Thus some thought should be given to the size of the holes in the viewing screen and in ventilating holes. The principles and formulas related to this exercise have been recorded in the literature [26]. The key problem in reducing leakage is the door seal, i.e., how to reduce the leakage between the door and the oven. Most modern ovens use a choke-type seal, which has been to be the most effective [10] especially if the slotted-choke design [27] is employed. With care this choke can be tuned so as to achieve a leakage value, relative to regulations, far below the usual limit of 1 mW/cm^2 at 5 cm from any point on the external surface of the oven. In some older ovens an absorbing material on the outside of the choke region supplemented performance.

The choke structure is usually covered by a plastic cover to prevent food and foreign objects entering the choke cavity as well as to prevent damage or alteration of the geometry, etc. The most severe test of the entire door seal is its ability to withstand no-load operation, a condition that should never be encountered except by accident. Of course, this condition might be approached when heating a small load of popcorn. For commercial use, especially, the oven must be shown to withstand such abuse.

Figure 9 Loci on the Smith chart (impedance diagram) of the impedance of a microwave oven at 2.45 GHz as the oven stirrer goes through its cycle (a) with a 2000 ml water load and (b) with the oven empty.

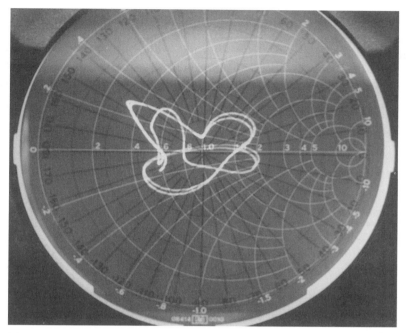

(a)

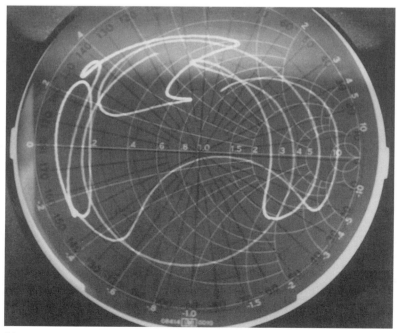

(b)

4. Power Supplies

Almost all microwave ovens operate with either a half-wave doubler supply or a full-wave doubler supply driven by a transformer at the line frequency of either 50 or 60 Hz. The half-wave case is depicted in the schematic diagram of Fig. 10. In this case the filament supply is integrated with the high-voltage transformer. Thus

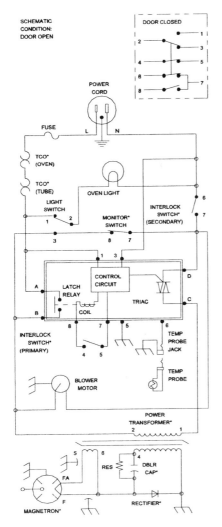

Figure 10 Schematic diagram showing a half-wave doubler power supply as used in many microwave ovens, especially consumer ovens. (Courtesy, Amana Appliances.)

such supplies are operated "cold-start"; i.e. when the line voltage is applied to the power transformer high voltage appears immediately while the filament temperature and microwave power lag in starting by about 1–2 s. This delay may be smaller at the beginning of life but as the end of life approaches this delay may be considerably greater, e.g., 2–4 s. In this case of cold start, the most common case, the initial voltage across the tube is as high as 8–10 kV until the cathode emission builds up and the voltage is clamped at the operating value of about 4 kV. This means that there is a possibility that the tube may suffer a breakdown at high voltage, causing a momentary arc followed by a voltage transient that may destroy the rectifier diode. This is a rare event but it undoubtedly occurs and is the more common cause for rectifier diode failure. Note that it normally does not reoccur and there is no damage to the tube.

In some commercial ovens where cost is less important the more expensive supply shown in Fig. 11 is used. In this case there is a separate filament supply. Thus the filament can be prewarmed before the high voltage is snapped on. In this case the tube is subject only to the normal high voltage of 4.0 kV.

It has been known for some time that a great reduction in weight can be achieved if an inverter-type of power supply is used. In this case a high-frequency transformer is employed at frequencies of 20–100 kHz. Since there is little stored energy the expected damage due to any arcs or transient is expected to be minimal. Such supplies have been studied for years and generally have been though to be too expensive for consumer use. There was marketed for a short time in the 1980s such an oven by Sharp. It operated at 80 kHz. Its power was continuously varied between 50 and 100% by varying the 80 kHz waveform. Below 50% power the power was varied by the classical on-off cycling with a period of many seconds, e.g., 10–20 s. This supply permitted the oven to be 15–20 lb lighter, a noticeable advantage. It appeared, however, to present a more nonlinear load to the

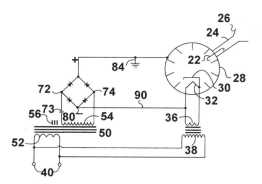

Figure 11 Schematic diagram of a full-wave power supply used in commercial ovens with separate filament transformer. (Courtesy Amana Appliances.)

line with subsequent greater harmonic generation into the line. It also appeared to exhibit somewhat higher noise because the dwell time at low anode currents was, in fact, a greater percentage of the on-time. Very recently there has appeared on the market a Panasonic oven which utilizes an inverter supply.

5. Controls

Referring again to Fig. 10, we see depicted a control circuit which is microprocessor controlled using a touchpad as seen in Fig. 3c and d. This control applies line power through a triac to the high-voltage transformer which drives the magnetron. It also controls the excitation of fans and lights and other accessories, if present. Shown in Fig. 10 is the presence of a temperature probe which in consumer and commercial ovens is almost always a metal prong with a thermistor sensor (see also the chapter on instrumentation). Thus the temperature probe is perturbing the microwave field but it is generally considered tolerable for practical purposes. The use of much more sophisticated temperature probes—e.g., the fiber-optic type—is too expensive for incorporation into products.

Also shown in Fig. 10 are the many elements of the system that monitors and controls leakage. Finally there are temperature monitors located on the magnetron and the oven, which will shut off the oven if overheating is detected. So far there has not been employed any practical detector of no-load conditions. The oven manufacturer has to rely on training and the good will of the user for the avoidance of this undesirable condition.

Other controls not shown include:

1. *Weight sensors and cook-by-weight.* Such a system was developed for Amana ovens in a top-of-the line model. It is no longer marketed.
2. *Moisture sensors.* These detect the degree of moisture in the air exhausted from the oven cavity. Since this moisture content will perceptibly increase as water begins to be driven from the food product, this signal can be calibrated relative to doneness and used to control heating time via the microprocessor.
3. *Infrared sensors.* These are mounted so as to have a view of the food surface and detect surface temperature in the usual manner of commonly available infrared sensor devices. This can again be empirically calibrated against food doneness and the recipe information fed into the microprocessor for control of heating the food, etc.

6. Safety

For a detailed discussion of chemical and microbiological safety of heated food and other safety aspects, the reader is referred to the chapter on safety.

The most commonly perceived hazard of microwave ovens is that of microwave leakage even though it is perhaps the most innocuous hazard and the most controlled hazard of the microwave oven. We have already mentioned that effective door seals and appropriated design of all gasketed joints, holes, etc. will ensure the suppression of leakage to acceptable limits. In addition, by federal mandate there are two interlocks and a "fail-safe" monitor to prevent operation of the oven when the door is open. These devices are shown in the schematic diagram of Fig. 10. The monitor switch is arranged to "blow," e.g., a fusible resistor if a selected interlock switch fails in the closed position, thus requiring a service call. Of course, nothing is perfect and even this fail-safe scheme could fail but the probability of failure is exceedingly low and is effectively zero for practical purposes.

These safety interlocks can be sensitive in their mechanical adjustment and sensitive to mechanical wear. Under some circumstances this will lead to failure to start because of a failure on interlock switch to close upon door closing, etc. Although rare it can happen and can lead to a service call.

The most serious hazard, of course, is that of electrocution which presents itself when someone attempts repair of the inner components of the oven. Such repairs should be left to accredited repair personnel. In rare circumstances electrocution hazard becomes present when someone tries to operate the oven with a two-hole wall socket that might be present in older houses. In this case excessive leakage current to the "otherwise grounded" chassis could present the hazard.

Another serious hazard is that of a potential fire. Such a fire could be ignited in many ways ranging from severe overheating of a potato to inadvertently placing a metal object near flammable materials like paper in an oven. The commonsense prevention is, of course, user caution and awareness. At the sight of any fire in the oven, the oven should be shut off and the wall plug removed. Without microwave power and with no airflow the fire will soon snuff itself out. With the passage of time the Underwriters Laboratories have mandated a test that should effectively prevent the spread of oven fires to the outside. This is usually applied to both consumer and commercial type ovens today.

Other hazards unique to the microwave oven include the phenomenon of "super-heating." This is described in Ref. [24] and elsewhere in the literature. This occurs in small objects of the order of an inch or two in diameter where an internal focusing of a hot spot occurs. In liquids this can lead to stored thermal energy in such a hot spot of above normal boiling temperature which an lead to an explosion of the liquid contents and a burn to any nearby hand. This hazard and many overheating hazards. e.g., the much-to-be-avoided overheating of a baby bottle can be mitigated by commonsense restriction of heating times, since all overheating reflects excessive heating time. In the absence of knowledge of correct heating time the user should be conservative in estimating this time.

A classic myth in the United States is that there is a hazard to patients with implanted pacemakers near microwave ovens—implying that the microwave leakage can penetrate the body and cause interference. Because of this myth warning signs around microwave ovens persist in many places around the country. A review of this myth, its origin, and its impact is thoroughly reviewed in the literature [28]. The facts about such interference are also explained.

Even though the pacemaker hazard is mythical, there is a growing realization that microwave leakage, both in and out of the ISM bands, can cause interference with other electronic devices. The FDA is exploring the full dimensions of this potential problem as it affects the growing class of medical devices and their susceptibility to radiation from wireless phones and other sources including ISM sources. The full dimensions of the interference questions remain to be assessed. Recent developments in this area are reviewed in Sec. VIII.E.

C. Agency Approvals

1. FDA

In any country there will be an agency responsible for enforcing a regulation on leakage from a microwave oven. In the United States this is the Center for Devices and Radiological Health (CDRH) of the Food and Drug Administration (FDA). Its regulation on microwave ovens [29] includes the leakage limit of 1 mW/cm^2 at 5 cm from any point of the external surface of the oven when manufactured, using a standard centered load of 275 ml water, etc. This limit is allowed to rise to a value of 5 mW/cm^2 in the field thereafter. Note that this measurement is with a standard load. It allows for the worst case of no-load operation, which yields higher leakage by as much as a factor of 5. Because of the extreme conservatism [26] of the emission limit and because of the likely short duration of no-load operation, this excess leakage is of no practical concern.

This emission limits is essentially adopted worldwide and in particular is adopted by the IEC (International Electrotechnical Commission). In Eastern Europe the measurement may be made at 50-cm distance from the door with a limit of 1/100 of the values in the FDA regulation. But these are equivalent by virtue of the inverse-square law of spreading radiation.

The U.S. regulations include many other requirements [29]. These include a check against the possibility of inserting metal objects into the oven while in operation and the use of warning signs to warn the user against empty oven operation and operation when the door is obstructed or damaged.

These regulations apply to consumer and commercial microwave ovens. Although they do not strictly apply to industrial ovens or equipments, industry generally voluntarily adheres to the same leakage limit.

There are various parts of the FDA regulations that apply to all type of microwave heating equipment. These include the requirement of an "initial report" to CDRH and the filing of reports on "accidental radiation occurrences." The reader is referred to the regulations for details.

Although the FDA regulation applies to the equipment for which the manufacturer is responsible, this does not absolve the user under occupational use conditions of its responsibilities under OSHA regulations in the United States and probably other safety agencies in other countries. In the United States although OSHA has not promulgated a specific regulation, and even though obsolete guidelines remain on the books of OSHA, this agency generally expects users of any microwave equipment to assure that their employees are not *exposed* beyond the limits of the most up-to-date exposure standards. In the United States that would be the recently published C95.1 Standard [30].

2. FCC

In the United States the Federal Communications Commission is responsible for regulating the out-of-band emissions from ISM equipment including microwave-heating equipment. These are described in the Part 18 of the Commission's rules [31]. We will not review requirements on conducted emissions, but we point out that the limits on radiation are 25 μV/m at 1000 ft from the oven using an average-signal responding instrument with a 5 MHz bandwidth. This is a fairly liberal limit, it turns out, but is still about 60 dB lower than the equivalent limit imposed on in-band leakage by the FDA.

There are very recent trends that could change this status quo. Recent trends of using ISM bands for communications purposes has led to proposals to drastically limit in-band emissions from microwave lighting devices. This trend could lead to serious restrictions on microwave heating equipment in the future if adopted. This subject is expanded later on in this chapter.

3. EU/CISPR

It is well known that regulations on ISM emissions outside the United States are more strict than in the United States by the FCC. In Europe regulations are being harmonized through the European Union (EU) and both in the EU and elsewhere in the world, there is the trend toward adopting the specifications promulgated by the CISPR (Comité International Spécial des Perturbations Radioélectriques) organization that is part of the IEC and reporting to the ITU (International Telecommunications Union). The CISPR organization is presently adopting new limits for microwave ovens operating at 2.45 GHz [32]. These are based upon measurement of "peak" values in a 1 MHz bandwidth coupled with a stringent limit when tested with a 10 Hz video bandwidth. These new limits might impact future consumer and commercial ovens. CISPR has yet to address the class of industrial equipment. The CISPR organization is proactive for protection of communications services. This posture coupled with trends in the FCC could lead to more stringent future regulations and possible cost burdens on such equipment.

4. Underwriters Laboratories

In the United States, the UL promulgates standards on performance of a variety of electronic equipment and appliances to ensure safe use. The latest version [33] of the UL Standard 923 on microwave ovens includes many requirements for safety especially with regard to protection against electrical shock and electrically induced fires. In addition there are abuse tests as well as a fire-containment test. In this test a fire is deliberately triggered by overheating a potato. It becomes dehydrated, then chars, then overheats, and finally exhibits arcing and ignition of a fire which usually can quickly spread to the plastic grease shield above, etc. In the test a piece of cheesecloth is placed just outside all exhaust vents. Compliance with the test is signified by the failure to ignite the cheesecloth while operating the oven indefinitely until such time as it shuts itself off. In a well-designed modern oven (see Fig. 10) there is usually a temperature sensor which will monitor the exhaust air and shut the oven off if a present limit is exceeded.

5. National Sanitation Foundation (NSF)

When microwave ovens are used in restaurants, they must comply with the requirements of the NSF, which are intended to minimize unsanitary conditions in the handling of food. Thus ovens should be designed to avoid mechanical features that lead to the trapping of food particles. If these areas are not easily subject to simple cleaning, then undesirable decayed matter and odors can result, etc.

IV. REVIEW OF AVAILABLE OVEN SYSTEMS AND PROPERTIES

A. Consumer Ovens

There are literally hundreds and maybe even thousands of models of microwave ovens in the world today. Roughly about 200 million ovens exist with about half in the United States. The annual sales in the United States vary between 8 and 9 million ovens compared to a peak of close to 11 million in the late 1980s. The variety of models available to the consumer today is so immense that we do not attempt a detailed encyclopedic survey. Instead, we will cite a few typical models that illustrate the diversity of ovens available. Then we will make some general remarks about the range of values available in the different performance specifications and the different design approaches followed by manufacturers.

Despite its limitations, a microwave oven is a staple for the modern home in one of its fundamental variations—countertop, built-in, over-the-range, or combination. In the 1970s combination electric ranges (under the counter) with microwaves at either 2450 or 915 MHz (before 1975) were available. Today, they are not generally available. Instead, combination ovens for the home are usually microwave/convection in any of the mounting options, and in some cases radiant elements are present. These could simply be a calrod unit, or in a few cases modern quartz halogen lamps with reflectors.

An example of the simplest oven available is the Amana CW65T It yields 600 W microwave power per the accepted IEC 705 test procedure, and heating time is controlled by one simple mechanical knob. It is small with 0.6 ft^3 cavity space and weighs 26 lb. It employs a dump feed at the top of the right wall of the cavity and a classical mode stirrer for uniformity of heating. Such ovens at the "bottom of the line" often are retailed as "loss leaders." Thus it is not uncommon that such ovens would sell for around $100 or less.

An example of a moderate-size modern countertop microwave oven is the Sharp R-320BK. (Sharp is by far the leading brand in the United States consumer oven market.) This oven weighs 33 lb and provides a 1.0 ft^3 cavity space. It employs a $13\frac{1}{4}$ in diameter glass turntable and 1100 W microwave power (IEC). Like all ovens in the Carousel™ line of Sharp, this oven, though small, includes sophisticated controls and sensors. Besides the standard set of touch pads for setting heating time, there are 6 one-touch settings that relate to the use of the sensor, a humidity sensor, in controlling heating time, etc. In this model the display is a seven-digit type.

An example of a top-of-the-line model is the Sharp Multiple Choice Line Model R590BK. This oven weighs 41 lb and has a cavity space of 1.8 ft^3. It has a 16 in diameter glass turntable and an ample amount of controls and computer-based interactive displays. The microwave power is 1100 W with at least three power levels for cook, reheat, and defrost. A humidity sensor is used to automatically control heating time for the desired result. The display is LCD with 11 lines of 16 characters for text and illustrations. Besides the standard touch panels there are autotouch tactile panels programmable to four sequences. There are 76 sensor-controlled cooking settings with an automatic popcorn key.

Most countertop ovens have dimensions varying around an average of 22 × 13 × 19 in. On the other hand, over-the-range models are larger since they are designed to the width of a range and must perform and exhaust function. Typical dimensions are in the range of 30 in width × 16 in height × 14 in depth. One can see that the different form factor here shows smaller depth but larger width and height than the typical countertop model. An example of an over-the-range model is the Maytag MMV5000. This operates with 850 W into a 1.1 ft^3 cavity. The internal cavity height of about 8 in is shorter than the usual height of a countertop oven. It has a recessed turntable with on/off control and also a humidity sensor and 10 power levels. A host of control options are available with standard time touchpads and six one-touch cook pads, e.g., for popcorn, dinner plate, drink, frozen pizza, baked potato, and a favorite choice. Other features include a digital clock and kitchen timer, oven rack, and a two-speed exhaust system with charcoal filter.

An example of a microwave-convection oven is the Sharp R820BK model. It employs a 0.9 ft^3 cavity with 900 W microwave power and convection temperature adjustable in 25° increments up to 450°F. A porcelain enamel turntable of $12\frac{3}{4}$ in diameter is used in a stainless steel cavity. (Most countertop ovens utilize painted cold-rolled steel.) The oven weight is 44 lb. This oven utilizes two radiant grills, one above and one below the food. This oven does not employ a humidity sensor, but there are ample other modes of programming and controlling heating. A key tool in the system is a 2-line, 12-digit interactive display which provides step-by-step programming instructions and cooking hints. An autotouch control is available programmable to four sequences.

Options for this and many other ovens include a kit for built-in installation, control of audible signals, a child-lock feature, and even language options. As one can see, much of the variation and diversity in microwave ovens centers not so much on the microwave properties as much as the sensors and controls where there is a fast-changing evolution following on the heels of the personal-computer revolution. Indeed, there are proposed systems to computerize all microwave oven functions and merge them into a master computer system. Discussion of these ideas is beyond the scope of this chapter but are to be found in the proceedings of recent meetings of IMPI and other societies.

All ovens described and probably almost all consumer ovens operate using the standard three-prong plug for 120 V 15 amps at 60 Hz—in the United States. An example of an oven used in Europe is the Whirlpool MT265 model, which is a combination oven. This operates with the standard European electrical supply of 230/240 V at 50 Hz. In this case the oven yields 900 W microwave power and 1300 W convection heat as well as 900 W radiant heat. This is provided by a quartz grill reflector system using two quartz halogen bulbs. The microwaves are fed in at the right wall by two apertures, one at the top and one at the bottom of the wall. The left wall is designed to perform a scattering as well as reflecting function. Finally there is a stoppable turntable in this oven. This oven does use a humidity sensor with six available power levels and digital timer and dial. Touch pads are not used but the display provides feedback on time and weight settings. A specific special accessory for this oven is a crisping or browning plate in which the active layer is on the bottom of the plate. The oven weighs a bit over 52 lb.

Because consumer ovens are little used, e.g., less than $\frac{1}{2}$ h per day and usually considerably less, considerations about life, reliability, and warranty are different from commercial ovens. Since even the most cost-reduced magnetron is expected to have a life of more than 1000 h, one can see that most ovens should not experience tube failure (wear out) for at least 10 years. Thus it is common for consumer ovens to offer a 5 year and even a 7 year warranty on the magnetron tube. On the rest of the oven, however, the warranty is limited to 1 or 2 years for parts and labor (carry-in service). (In special cases some manufacturers may offer warranted service on site for certain built-in models.)

Consumer ovens must be certified to meet FDA/CDRH and FCC labels in the United States and most conform to UL requirements. As pointed out above, an important part of the UL requirements is the fire-containment test.

B. Commercial Ovens

Commercial ovens are similar to the consumer oven except for special design and features to accommodate more frequent use and also some abuse. They are used in many thousand of fast-food establishments or wherever items require fast warming, e.g., near vending machines. Thus the ovens are generally more rugged, often have dedicated features, e.g., for repeated set heating programs. They vary in power from 800 W to as much as 2700 W. Of course, they are much more expensive than consumer ovens and special warranty aspects are associated with the product. The time periods for warrantees are generally less than for consumer ovens in view of the more frequent use. Provision is usually made through a service contract for on-site repair. The ovens require more testing and certification than their consumer counterparts. Thus, in addition to compliance with FDA/CDRH emission limits and the FCC or CISPR limits on out-of-band emissions, the ovens must also meet stringent requirements of UL, NSF (National Sanitation Foundation), and CSA. And because of their use in businesses, there must be available fast consultation services by phone in order to maintain an efficient field experience.

In Table 6 we tabulate the principal characteristics for ovens made by two manufacturers in the United States (the only two who responded to our request for such data). We present these data in full so that the reader can appreciate the wide range of options available. (It must be remembered there are other manufacturers, e.g., Panasonic, in the United States.) Most of the information in this table is self-explanatory. It must be remembered that there are more details on performance beyond what is presented in the table. For example, most ovens have a defrosting capability even though not noted specifically. Most ovens have rotating antenna feed systems. The rotation is a key factor in helping to even out heating patterns. Some ovens are noted, however, to employ rotating stirrers or a rotating turntable. There are generally many fine details about programming abilities, which are too diverse to be tabulated here. We note that operation at 208/230 V is required for the higher powers. Some ovens include an "auto voltage sensor" that detects whether the line voltage is 208 or 230 V and adjusts the power supply appropriately. This then allows for automatic compensation for severe brownout conditions when the normal voltage is 230 V. Often there are multifeeds with two or more tubes. Very often these two feeds are at the bottom as well as the top of the cavity. In some cases, however, the two feeds are at the top. The ovens tend to be much heavier than the consumer counterparts because of more rugged construc-

Table 6 Characteristics of Commercial Microwave Ovens

Mfg. Model	Power (W)[a]	Power levels	Line power[b]	Size (ft³)	Weight (lb)[c]	Control oven	Special feature	Warranty
Sharp								
R-21HC	1000	1	120V/15A	1.0	44	D.T.	S, bf	W13
R-21HV	1000	1	120V/15A	1.0	44	T.P.(10)	S, bf, r	W13
R-21HT	1000	11	120V/15A	1.0	44	T.P.(10)	S, bf, r	W13
R-22GV	1000	1	120V/20A	0.7	66	T.P.(10)	S, tbf	W3
R-22GT	1000	11	120V/20A	0.7	66	T.P.(10)	S, tbf	W3
R-23GT	1000	11	120V/20A	0.7	69	T.P.(10)	S, tbf	W3
R-24GT	1000	11	120V/20A	0.7	72	T.P.(10)	S, tbf	W3
Amana								
LD100	1000	1	120V/15A	1	40	D.T.	S, tf,tt	W1
LD10	1000	1	120V/15A	1	40	T.P.(10)	S, tf,tt	W1
RCS100	1000	1	120V/15A	1.2	58	D.T.	tf	W3
RCS10	1000	1	120V/15A	1.2	58	T.P.(10)	tf	W3
RCS10MP	1000	5	120V/15A	1.2	58	T.P.(10)	tf	W3
RFS9B	900	1	120V/20A	1.2	68	T.P.(10)	s, tf,teo	W3F
RFS11B	1100	1	120V/20A	1.2	68	T.P.(10)	s, tf,teo	W3F
RFS11MP2	1100	5	120V/20A	1.2	68	T.P.(20)	tf	W3F
CRC10T2	1100	11	120V/20A	0.6	64	T.P.(20)	s, tbf	W3F

Microwave Ovens and Heating Systems

Model	Power (W)	Pads	Voltage	Cu ft	Shipping wt (lb)	Timer	Features	Warranty
CRC12T2	1200	11	120V/20A	0.6	72	T.P.(20)	s, tbf	W3F
CEC18T2SD	1800	11	208*V/20A	0.6	74	T.P.(20)	s, tbf	W3
CRC18T2	1800	11	208*V/20A	0.6	74	T.P.(20)	s, tbf	W3
CRC21T2	2100	11	208*V/20A	0.6	74	T.P.(20)	s, tbf	W3F
RC17SD	1700	11	208*V/20A	1.0	105	T.P.(20)	2tf, avs	W3
RC17	1700	11	208*V/20A	1.0	105	T.P.(20)	s, 2tf, avs	W3F
RC22	2200	11	208*V/20A	1.0	118	T.P.(20)	s, 2tf, avs	W3F
RC27	2700	11	208*V/20A	1.0	118	T.P.(20)	s, 2tf, avs	W3F
Menumaster (Amana)								
GSAND8LW	800	1	120V/15A	0.5	37	D.T.	rs	W3
SAND8LW	800	1	120V/15A	0.8	43	D.T.	rs	W3
SNAC9LW	800	1	120V/15A	0.8	43	T.P.(7)	rs	W3
FS11	1200	5	120V/20A	0.75	88	T.P.(10)	rs	W3
FS17	1700	5	208*V/20A	0.75	88	T.P.(10)	s, rs	W3F
FSC10	1000	1	120V/20A	0.6	64	T.P.(10)	s, rs	W3
FSC12VP	1200	11	120V/20A	0.6	72	T.P.(20)	s, tbf	W3
FSC18VP	1800	11	208*V/20A	0.6	74	T.P.(20)	s, tbf	W3

[a] Power as measured by IEC 705-1988 test procedure.
[b] Electrical plug requirements at 60 Hz; export models at 50 Hz available generally.
[c] Shipping weight; net weight is 4–7 lb lighter.

D.T., dial timer; T.P., touch pads with number of pads available, some adjustable; s, stainless-steel cavity walls; br, with Braille; bf, bottom feed; tf, top feed; rs, rotating stirrer; teo, time entry option; tt, turntable; tbf, top and bottom feeds, two magnetrons; 2tf, two top feeds; two magnetrons; *208/230 V; avs, auto voltage sensor; W1, limited one-year warranty; W13, one-year warranty parts and labor, 3 years limited warranty on magnetron; W3, limited 3-year warranty on parts and labor.

tion, the presence of more than one power supply, and special features. Because of dedicated touch pads, the ovens have a distinctive commercial look although a few ovens also have a time entry option. Many of the ovens have door screens but many do not.

In most cases the ovens do have a digital display of time (countdown) and possibly power level. Many other options are available for such ovens including security kits, mounting kits, and various other things, especially for the more expensive models. There are a variety of computing chores offered such as monitoring usage, adjusting heating time for more than one food unit, adjusting for appropriate audible signals, and even a demonstration mode (Sharp) to practice programming without microwave power being energized.

We note that warranties on commercial ovens vary from 1 to 3 years with some limitations. This reflects the realities of the use conditions in businesses. After a few years of heavy usage there is a trend to discard and buy new rather than repair and prolong. Of course, there are some exceptions and such ovens are found in operation even after 10–15 years of use. The selling prices of commercial ovens varies from about $200 for the most elemental oven to close to $3000 for the highest power (close to 3000 W) unit with a maximum of advanced controls, sensors, etc. Commercial ovens present unique field problems for the manufacturers because of the intense use of the product. Various new reliability and life problems become apparent as a wide variety of foods and heating programs are introduced. As a result, frequent consultation with tube manufacturers is the norm for manufacturers of commercial ovens.

Almost all commercial oven applications have been ground based. Years ago there were reports that a Litton oven was installed in Air Force One. Its present fate is unknown. In the 1980s, however, Toshiba produced an oven for use by Japan Airlines, which was approved by both the U.S. FAA and the Ministry of Transport in Japan. This oven operated off the airplane 400 Hz 3-phase 115/200 V supply and provided 2700 W of microwave power using three magnetrons. Its development [34] required special attention to resistance to shock and vibration and more stringent limits on leakage and out-of-band emissions.

All the ovens in Table 6 are microwaves only. There do exist, however, commercial models of microwave/convection and other combination ovens. For example, Amana offers an oven (CAM 2000 or 2300) with 100 W microwave power and convection temperatures up to 475°F. This oven operates with a 208/230 V and 20 amps wall plug. (Note that in Table 6 we cited merely the wall plug specifications. The ovens draw less than the maximum allowed but these values were not tabulated.)

Amana also offers several models (e.g., the HSL2030 or 3050) of the Amana Wave Oven™ which utilize the radiant heat (infrared and visible) from quartz halogen lamps for cooking. The radiant powers are either 5500 or 10,000 W. They operate at 208/2340 V and require 30 or 50 amp provision at 60 Hz. They are of similar construction to a microwave oven but they weigh more, e.g., over 100 lb. These ovens excel at rapid preparation of steaks and other meats.

C. Industrial Ovens

Although industrial microwave systems are expensive, the cost of systems operating at 915 MHz has actually come down in the past decade (Table 7). A major reason for this is that the microwave magnetron had had its power rating increased from 25 to 75 kW with few design changes and little cost penalty. To accommodate this increased power only a few key components in the transmitter, high-voltage transformer, for example, have had to be upgraded. Also, the size of the ovens has also been increased to keep RF power density relatively constant. Another very important factor has been the introduction of strong competition.

1. Tempering Systems

Microwave tempering is one of the early success stories of the industrial microwave industry. The first system was sold to the H. J. Heinz Company in Pittsburgh, Pennsylvania, in the late 1960s. The application involved dicing frozen blocks of deboned chicken for a canning line. It was a 50 kW system used to soften 60 lb blocks of chicken in the carton after which they were band-sawed into logs and then diced.

Since then over 300 systems have been sold worldwide into a wide variety of applications. These systems are used routinely in applications that involve tempering to permit further mechanical processing such as slicing, dicing, grinding (patty forming), pressing, and molding. Hamburger, sausage, canned meat, fish portions, and portion control dinners are some of the products that rely heavily on microwave tempering.

Assuming an average installed power of 100 kW and a tempering throughput of 7000 lb/h (90% lean beef raised from 0 to 27°F), the combined tempering capacity of all these systems is on the order of 2.1 million lb/h. This represents 4.2 billion lb annually (single shift operation). A typical conveyorized tempering system is illustrated in Fig. 7.

Capital costs (assumes installation of one 150 kW system):

Microwave system cost:	$250K
Site preparation:	$ 20K
Installation and shipping:	$ 10K
Total capital cost:	$280K

Amortization 5 years at 2000 h per year (one-shift operation)	$28.00 per hour
Electric utility cost	
Ac power at 125 KVA Electricity cost at 10 cents per kWhr Total electric cost:	$12.50 per hour
Maintenance costs	
Tube replacement– Tube warranty 3000 h Worse case = 1.3 tubes per year 1.3 × $6000 = $8000 annually	$ 4.00 per hour
Total operating cost	$44.50 per hour
Total cost per pound tempered	$ 0.445 cents/pound (assumes 10,000 lb/h)

A yield improvement (elimination of drip loss) from 95 to 98% adds 300 lb/h to the production capacity or, at 1$/lb, a savings of $300 per hour. At this rate simple payback is achieved in slightly under 1000 h or 6 months for single-shift operation.

2. Bacon-Cooking Systems

Although microwave bacon cooking began in the 1970s it wasn't until the 1980s that major food companies took an interest in the process. There are between 45 and 50 bacon cookers installed worldwide. A typical bacon cooking system is illustrated in Fig. 12. Each of these systems, with an installed microwave power of 400–500 kW, has a production capacity on the order of 50,000–60,000 slices per hour. At a nominal 30 slices to the pound, this represents a throughput of about 1900 lb of raw product per hour. These systems are producing a finished product for the HRI market that meets strict length, yield, organoleptic, and water activity level criteria. Early systems were very labor intensive because of manual pack-off. Over the years improvements in material handling have reduced labor require-

Table 7 Selling Price for Microwave Bacon Cooking Systems

Year	Installed power (kW)	Price ($)	$/W
1984	240	500K	2.00
1995	490	640K	1.30
1999	450	588K	1.30

Figure 12 Amana conveyorized bacon cooking system.

ments. These improvements plus *lower capital cost cited above* has helped maintain the economic viability of this process.

Assuming, conservatively, 16 h of operation each day the combined output of all the systems is on the order of 400 million lb (input) or 12.1×10^9 slices (output) annually. Bacon is cooked to about a 30% yield. This means that 60% by weight of the incoming product is rendered as fat that is collected for sale and reuse. In other words, central processing of bacon removes a burden of 240 million lb of fat annually. Almost all of the bacon produced today is for the HRI market. It has been the forecast of experts that sales of this product will eventually spread to the retail market.

Capital costs (assumes installation of one 450 kW system)

Microwave system cost:	$ 600K
Ancillary pack-off/slice:	$ 350K
Site preparation:	$ 100K
Installation and shipping:	$ 10K
Total capital cost	$1060K

Amortization

7 years at 4000 h per year (two-shift operation)	$55.00 per hour

Electric utility costs

AC power at 600 KVA
Electricity cost @ 10 cents per kWhr

Total electric cost:	$ 60.00 per hour
Maintenance costs	
Tube replacement– Tube warranty 3000 hours Worse case = 8 tubes per year 8 × $6000 = $48,000 annually	$ 12.00 per hour
Total operating cost	$127.00 per hour
Total cost per slice	$ 0.254 cents per slice (assume 50,000 slices per hour)

Not included in the above costs is the recovery of the rendered fat or the scrap (ends and pieces) that can be processed for bacon bits or pizza topping.

3. Potato Chip Processing

There are at least four companies that have tried to use microwave energy for processing potato chips. Frito-Lay was mentioned in the introduction. As you probably know, TGTBT (Too Good to Be True) [35] was active in the 1990s producing Louise's potato chips. According to reports, this company has gone out of business because of poor market acceptance of their "no fat" potato chip. In spite of these failures, two other companies that must remain unnamed continue to experiment with microwave energy to develop a proprietary process for producing low fat chips. One of them spent close to 3 years quietly developing and market-testing the product. Today they have installed close to 2 MW microwave equipment and are producing a low-fat product for "high-end" markets.

As is the case with bacon cooking, potato chip processing is not, at first glance, an attractive application for microwave energy. Table 8 illustrates the point. As can be seen in the table, both bacon and potato chip processing are energy-intensive applications. This problem is exacerbated by low efficiency, which, in turn, is the result of very light belt loading typical in both applications. Low belt loading also results in high RF fields in the cavity and an increased potential for arcing and

Table 8 Energy Levels in Two Industrial Microwave Applications

Attribute	Bacon cooking	Potato chip processing
RF to heat conversion efficiency	Low	Low
Product yield	30–35%	20–25%
Energy requirement	300–400 Btu/lb	300–400 Btu/lb

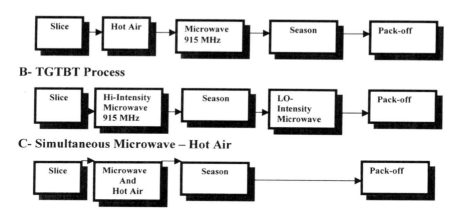

Figure 13 Schematic diagrams of potato chip processing systems.

fire. To compensate for this problem, by reducing field intensity, newer systems have employ larger cavities, dual belts (over/under), and wide belt formats for product conveyance. All processors employ warm air to remove moisture liberated by the microwave energy. Some consider the use of warm air to be a proprietary aspect of the process. It is interesting to note that the abstract for the TGTBT patent issued in 1993 describes a serpentine waveguide similar to that employed by the Raytheon system in the 1960s. These processes are illustrated in Fig. 13.

Capital costs (assumes installation of one of a 300 kW system, which can process 1000 lb raw potatoes per hour to produce 250 lb of chips per hour):

Microwave system cost:	$450K
Ancillary pack-off/slice:	$250K
Site preparation:	$100K
Installation and shipping:	$ 10K
Total capital cost:	$810K

Amortization

5 years at 400 h per year (two-shift operation) $40.50 per hour

Electric utility cost

AC power at 400 KVA
Electricity cost at 10 cents per kWhr

Total electric cost:	$40.00 per hour
Maintenance costs:	
Tube replacement– Tube warranty 3000 h Worst case = 4 tubes per year (4 × $6,000 = 24,000 annually)	$ 6.00 per hour
Total operating cost	**$86.5 per hour**
Total cost per slice	$ 2.2 cents per ounce (assumes 4000 oz/per h)

Assuming $1.50 for a 4 oz bag of chips, this cost is only 1.5% of sell or 3% of cost assuming a 100% markup.

4. Microwave Fluid* Heat Exchangers

Microwave fluid heat exchangers are used extensively in military radar, commercial radar and industrial microwave systems. They can be designed to operate at any desired frequency or band of frequencies. They are capable of handling large amounts of average and peak power ranging from a few watts to hundreds of kilowatts. Their normal function in radar and industrial systems is to absorb energy reflected back to the transmitter by the antenna or load. A circulator is inserted between the transmitter and the antenna. Its purpose is to prevent reflected power from reentering the transmitter where it could damage the microwave power tube. Reflected power enters the circulator at port 2 and is delivered to the water load connected to port 3.

In military and commercial applications the fluid used to absorb the microwave energy is typically water or, where low temperatures might be encountered, a water/glycol mixture. However, any fluid capable of absorbing microwave energy may be heated, for all practical purposes, with 100% efficiency. High conversion efficiency assumes that a good impedance match exists between the microwave transmission line and the microwave heat exchanger.

There are some applications in food, pharmaceutical, and other industries in which it is necessary to heat pumped fluids. The purposes vary but frequently include sterilization and pasteurization. Process criteria include the following.

Conventional forced convection tube–type fluid heat exchangers heat a fluid by passing it through a fin/tube heat exchanger. Whether the flow is laminar or turbulent, the maximum heat flux appears at the wall of the tube where the fluid velocity is at a zero. Turbulent flow improves mixing of the fluid, enhances the

---

* In this discussion the term *fluid* refers specifically liquids, not gases.

heat transfer coefficient, and reduces the thickness of the stagnant film boundary layer. However, either mode of heat transfer can result in overheating and damage of the fluid and degraded heat exchanger performance. Higher Reynolds numbers can improve the heat transfer coefficient, but achieving them is difficult especially with food substances that have high viscosity and "lumpy" consistency.

Microwave heat exchangers, in theory at least, have the potential for overcoming some of objections to conventional heat exchange methods. The advantages of using a microwave heat exchanger can include the following:

> Rapid heating to the desired temperature (heating rates determined by applied RF power)
> Elimination of "fouling" (instantaneous internal heating)
> High efficiency

Unfortunately, using microwave heat can be as much as 10 times more expensive to use than conventional heat exchangers. It follows that microwave heat exchangers should be used only on very high value products and/or on products that can't be processed any other way. To illustrate this point, a 300,000 (100 kW) Btu/h gas-fired fin/tube heat exchanger, manufactured by Raytheon, for heating cooking oil in KFC stores sold in the 1980s for about $5,000. An equivalent microwave system consisting of a 100 kW transmitter and a microwave heat exchanger would have a cost of about $100,000.

In spite of this difficulty at least two companies, Charm [36] in Somerville, Massachusetts, and Questron [37] in Florida have identified niche markets for microwave-based fluid heating. Charm is involved in the pharmaceutical industry and is using microwave energy for viral inactivation in very high value fluids.

5. Combination Cooking

Microwave heating, in combination with conventional heating, offers benefits that include higher throughput per linear foot of oven length, more uniform heating, shorter cook times, and improved yields.

Recently, one company, without outside help, started using microwave energy in combination with their existing process for producing a meat product. The market is institutional where, admittedly, the user has limited influence on menu selection. The manufacturers felt that adding microwave energy to the process was the only way that to meet an increased production demand without resorting to "bricks and mortar."

Others, most notably in the poultry industry, have tried, retried after Litton, and rejected combination microwave heating. Recent process attempts generally involved sequential microwave-fry or fry-microwave-fry of breaded bone-in poultry or breaded poultry nuggets. A major benefit of the process was deep cooking at the bone and elimination of uncooked blood at the bone surface. Negative factors included concerns about toughness, system complexity, yield, and lifting of the breading (heating rate sensitivity). While it is clear that combination processing can result in shorter line lengths, the economic and quality benefits, if any, are yet to be established and documented.

It is recognized that many other microwave combination processes have been proposed, evaluated, and reported on in the literature. Notable examples include microwave freeze-drying studies led by R. Decareau at the U.S. Army Research Labs in Natick, Massachusetts. Another is the vacuum drying process proposed and marketed by Mivac [38] in the United States. None of these processes other than those cited has, to the authors' knowledge, achieved general market acceptance. Our own experience indicates that industrial microwave process development has the best chance for success when it is based on a partnership or joint venture between the microwave manufacturer and the food processor.

V. POWER AND EFFICIENCY CONSIDERATIONS

A. Consumer Ovens

In the United States the household utility power available is 120 V (60 Hz), nominal and 15 amps max (fused or circuit breaker) with a three-hole socket per the latest National Electrical Code. Therefore, the power input to a countertop microwave oven is limited, at least in the United States. About 15 years ago Amana attempted to market a higher-power consumer oven, which would require a 20 amp socket. Since this socket is not commonly installed in houses, the marketing goals were not met.

This limitation may be overcome in the United States eventually. In Europe, with a 220 V supply (50 Hz) as standard this limitation does not exist. Despite this, most microwave ovens around the world have rated powers that are compatible with the U.S. standards of available power at 120 V. This power is roughly in the range of 750–1000 W.

There has been great confusion in the past about what the rated power of a microwave oven should be. Buffler [39] has reviewed all the factors that influence the measured value including, the size and shape of the container for a water load, the initial water temperature, size of water load, location of load, influence of variation in line voltage, effects of the warming up of the magnetron, etc.

Buffler has shown that variation of line voltage by just $\pm 6\%$ can cause a variation of output power of more than $\pm 15\%$ in some ovens. Even worse, if line voltage should drop to as low as 95 V during a "brownout," the power could drop close to zero in an extreme case.

Although the effect of line-voltage variations is highly variable from oven to oven, the effect of tube warmup is rather universal for microwave ovens. Since all inexpensive magnetrons use ferrite magnets, when the tube anode warms up,

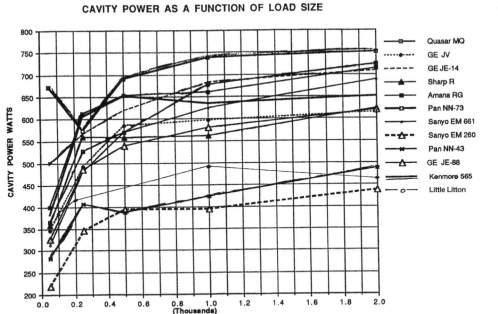

Figure 14 Cavity power as a function of load size for a variety of microwave ovens. (Courtesy of Charles Buffler, Ph. D.)

e.g., to 250°C, the magnet is warmed enough so as to drop magnetic field, voltage, and power by as much as 10%.

Furthermore, the measured oven power is a function of load size. The data shown in Fig. 14 for a variety of ovens is typical, showing a significant spread between ovens of supposed similar ratings. Also, we see that oven power drops rapidly as the load size drops below about 400 g water. One reason for this is the finite losses in the walls and waveguides. The more fundamental reason is the improbability of matching the magnetron to arbitrary small loads. In fact, the usual design procedure is to match at a large load, e.g., 1000 g water, and then to let the match at small loads fall to chance.

After many years of deliberation there has evolved a standard method of power measurement, the IEC 705-88 standard [40]. This standard details precisely how the test should be made including the specification of initial water temperature as $10 \pm 2°C$. The water load in precisely described container is 1000 ± 5 g and it is heated for a time which results in an increase of $10 \pm 2°C$. The oven must have been nonoperating for at least 6 h before the test. This ensures that the tube anode is close to room temperature. This requirement also helps avoid the decrease of power accompanying longer-term warming of the tube. It is not surprising, therefore, that the nominal oven power as measured by the standard IEC method is somewhat higher than some of the other methods used in the past.

In the discussion on tubes we mentioned the effect of cold start and the starting time delay on effective power. In some schedules of on/off tube switching for variable power the on time could be as small as a few seconds. In such a case the turn-on time in cold start will be a significant factor in determining the oven power under such conditions.

All these considerations encourage the user to conduct his or her own power tests for the conditions that interest him or her. It must be recognized, however, that any water load is not an exact equivalent to some food product so that eventually probe measurements in a real food are required for a full description of the heating of a food in a particular oven.

The standard power tests is of some use for food manufacturers in attempting to provide appropriate user instructions in any oven among the world's great variety.

A few years ago the U.S. Department of Energy (DoE) conducted serious discussions on a proposed rule for increased efficiency standards for microwave ovens. During those discussions it was the consensus that existing ovens typically yield a net efficiency—wall plug to food (equivalent water load)—of about 50%. It was the desire of DoE to see this increased by at least 5–10%. This would have required considerable redesign and increased cost in the absence of a fundamental increase in tube efficiency. This was not forthcoming in the past, but as indicated here, there may be available a modest increase (e.g., +5%) in the future for cooker tubes. It was pointed out to DoE that the operating time of consumer ovens is quite small per year compared to other appliances. After due consideration the DoE decided to drop the idea.

B. Commercial Ovens

The power and efficiency considerations for commercial ovens are basically the same as for consumer ovens except for the effects of the increased usage of commercial ovens and the possibility of short time on schedules in some fast-food applications.

The commercial ovens are designed to withstand abuse more than the typical consumer oven. Also, tube selection is more critical relative to tube life. It is clear, however, that commercial ovens will by and large be operating with warmed-up tubes with the attendant penalty in power due to magnet warming. And because of the more frequent use, the user will experience phenomena characterizing end of life more frequently, e.g., the increased turn-on time during cold start and the possibility of intermittent moding, both of which will lower effective power. Hot-start ovens will get around the cold-start delay problem but it will not solve the eventual moding problem. Even though intermittent and affecting power only slightly, its presence may become recognized because of another phenomenon—interference.

The moding phenomenon in the magnetron is an abnormal state in which no power is generated at the intended frequency, 2.45 GHz, but instead a small amount of power is generated at a spurious frequency. It turns out that most cooker magnetrons when moding generate frequencies in the range of 4.2–4.7 GHz. In this range there do operate some earth-to-satellite links for cable TV. As a result, there have been cases of such interference which required remedial action by the oven manufacturer. The simplest solution is for the tube manufacturer to alter its moding frequency, which can be done. The longer-range solution is to develop much longer-life tubes.

C. Industrial Ovens

Power and efficiency are always important in high-power installations since electricity costs play a major role in operating costs. Process efficiency is the product of the ac to dc, dc to RF, and the cavity coupling efficiency.

High-power systems operating at 915 MHz are blessed with better than 90% power supply efficiency and 85% magnetron efficiency. Waveguide transmission line losses between the microwave generator and the applicator, even for long distances, are generally negligible. Parasitic losses are those that occur in the cavity walls and in suppression tunnels if they are the absorption types. Additional losses can be attributed to power reflected back to the source. In a well-designed and properly operated system these losses can be kept to within 5–10% of the energy delivered to the cavity. This means that overall process efficiency is strongly affected by the load being heated. In conveyor- and batch-operated systems, product-coupling efficiency is usually plotted as a direct function of belt loading. In some cases where dielectric properties vary from product to product or as a function of temperature, a family of curves may be plotted. This collection of this data is usually an empirical process that is only used in the development stage. This type of information is rarely, if ever, referred to again after successful process implementation. Throughput information gained from the development process and early production is usually reflected in manufacturer claims regarding throughput and payback.

VI. UNIFORMITY CONSIDERATIONS

From the very beginning of microwave heating to the present the uniformity of heating has been a continuing concern. Still there have been no clear insights gained regarding which guidelines, if, any, in microwave-oven design yield superior uniformity in general. In one sense this is not surprising because there are so many factors that could influence uniformity that it is hard to generalize among various foods and ovens. These factors include:

> *Food properties:* Size, shape, dielectric properties, presence of multicomponents of differing absorption properties. . . .
> *Container properties:* Dielectric loading, influence on corners of foods
> *Cavity properties:* Mode phenomena, closeness of metal walls
> *Randomizing elements:* Turntables, moving conveyor belts, stirrers, rotating antennas, frequency diversity [41]

The classical approach has been to randomize in view of this complexity on the simple principle that by randomizing one will achieve an average and at least avoid singularly bad patterns that are intensified by being time invariant. (There is another side benefit with this randomizing in terms of other performance factors such as leakage.) More recently there has appeared proposals [42] for a radically different approach using applicators with known and controlled modal properties and using them not only in microwave ovens but also in industrial machines, perhaps in an array of such applicators. In some cases the design approach has the goal of preferentially exciting TM modes as seen impinging on the food surface or plane. A number of proprietary [43] concepts using both circular and rectangular guides have been recorded. It is claimed that the TM modes "match" into the food with less reflection and less edge heating. Furthermore, circular polarization is introduced in some [43] of these embodiments. In this way some of the benefits of a randomizer are restored; i.e., the polarization rotates with time in a circularly polarized field.

This has also led to the proposal of a new consumer oven design, the "Thermaliser" [44]. In this device, which is small and compact, the "cavity" is a small circular waveguide acting as an applicator for the food near the bottom of the cavity. Besides the properties claimed in general for this applicator, the Thermaliser also has an induction-heating provision at the bottom of the device and a halogen lamp at the top of the device. The device thus is claimed to provide ideal heating by appropriate sharing of these three modalities. It is not known how close to practical realization this concept is.

Associated with these proposals is a growing reliance on the effectiveness of computer modeling [45] not only for concepts using a few modes but in general. Designers of ovens and industrial systems still must rely mostly on empirical optimization of designs. See Chapters 1 and 2 for a review of modern computer modeling and its status relative to design of microwave heating systems.

VII. CONTROLS AND SENSORS

This subject, in its generality is treated elsewhere in this book. From the limited view of designers of ovens and industrial equipment, a few observations can be made. In the past, considerable use of temperature probes in ovens was made even though they were field perturbing, i.e., metallic prong probes containing thermocouples or thermistors. In addition, there have been fleeting attempts to use weight sensors and infrared probes (sensing surface temperature). For various reasons they have disappeared essentially in practical oven products and also in industrial equipment. Instead the trend has been to use moisture sensors that detect the early stages of doneness of foods, allowing with proper calibration intelligent control of subsequent heating.

There is no question of the value of nonperturbing fiber-optic temperature probes. These are ideal for research on microwave heating but not as an accessory to practical products.

Outside of cooking controls, the microwave oven does use temperature sensors for simple safety considerations. The magnetron has a thermocouple mounted on its base to sense anode temperature and to cut off the oven power supply if the temperature becomes excessive, e.g., because of failure of the cooling fan. Furthermore, most ovens have a sensor of either the oven wall temperature near the exhaust vent or the exhaust air temperature. Again if this temperature exceeds a predetermined limit the oven power supply is shut off, including that to fans, etc. In this way a fire can be detected and snuffed out.

In industrial systems arcing around foodstuffs occasionally occurs and there are arc sensors that sense the arc and shut off the main power supply to the magnetron tube. This not only prevents the growth of a fire, but also prevents an arc from traveling back toward the tube and damaging the protective circulator or the associated load (usually a water load). These sensors operate through detection of visible light. In addition, in some systems, there is a microwave detector near the exits of conveyor tunnels that will shut down the system if excessive microwave energy is detected, signifying a failure of the tunnel suppression means.

We have already mentioned recent advances in inverter power supplies. They appear practical for industrial systems and there is a report of at least one oven manufacturer introducing them into oven products. These supplies truly offer new dimensions in power control, i.e., continuous variation with change in anode current operating point of the magnetron and not just variation of duty cycle. There will, however, be an associated change in out-of-band emission with such a power supply. Earlier studies indicated that these changes were somewhat adverse. It remains to be seen to what degree the new power supplies show the same phenomenon.

Although out of the scope of this chapter it is appropriate to make brief mention of ovens modified with new controls and sensors for the purpose of evaluation of moisture content in samples of foods or other materials. Systems using conventional ovens modified for this purpose are available from CEM Corporation [46]. More recently a compact oven using the design principles in Refs. 42 and 43 has become available [47]. It is claimed that this smaller oven performs the evaluation faster than the conventional device. In all such cases, it is clear that more stringent control of the oven is required if the device is to reliably carry out its function of precise and accurate measurement.

VIII. TRENDS AND OUTLOOK

A. Available Frequencies

In Table 9 we show the existing frequency allocations for ISM equipment. This includes microwave ovens and other heating equipment which utilize radio fre-

Table 9 Frequency Allocations for ISM Applications

Frequency (MHz)	Region	Conditions
6.765–6.795	Worldwide	Special authorization with CCIR[a] limits; both in and out of band
13.553–13.567	Worldwide	"Free" radiation bands
26.957–27.283	Worldwide	"Free" radiation bands
40.66–40.70	Worldwide	"Free" radiation bands
433.05–434.79	Selected countries in region 1[b]	"Free" radiation bands
	Rest of region 1	Special authorization with CCIR[a] limits
902–928	Region 2[c]	"Free" radiation band
Frequency (GHz)		
2.4–2.5	Worldwide	"Free" radiation band
5.725–5.975	Worldwide	"Free" radiation band
24.0–24.25	Worldwide	"Free" radiation band
61.0–61.5		
122–123	Worldwide	Special authorization with CCIR[a] limits; both in and out of band
244–246		

[a] CCIR, International Radio Consultative Committee of the International Telecommunications Union (ITU).
[b] Region 1 comprises Europe and parts of Asia; the selected countries are Germany, Austria, Lichtenstein, Portugal, Switzerland, and Yugoslavia.
[c] Region 2 comprises the western hemisphere.
Source: Ref. 48.

quency energy. The frequencies in the table are from the World Administrative Conference of 1979 [48]. They have not been altered since. The ITU requested the CISPR organization to investigate the subject of possible in-band limits for all the bands including the so-called free radiation bands, i.e., where there is at present no limit on radiation except for health and safety reasons. The CISPR report in the mid-1990s merely listed the probable maximum levels of leakage radiation that exist today in such bands without any recommendation for in-band limits. This may change; see Sec. E.

Although the potential problem of in-band limits is troubling and potentially burdensome for equipment manufacturers and users, the basic problem limiting the choice of operating frequencies is simply the availability of power sources— meaning magnetrons for most practical purposes.

It is true that the development of tubes such as magnetrons becomes more difficult as frequency increases. We already note that while the high-power tubes at 915 MHz exhibit efficiency above 85%, the cooker tubes exhibit only 70% at present. It is conceivable that magnetrons at efficiency somewhat less could be developed at 5.8 GHz. But this requires investment, which is not probable without substantial government intervention or other major economic events. The higher frequencies are more problematical and await either breakthroughs in tubes or power sources or new applications that justify expensive equipment.

B. Available Powers

At 915 MHz, inexpensive tubes are not available at power levels of 1 kW or lower. At 2.45 GHz, inexpensive tubes for the 100–200 W level are not available. In principle, the 1 kW tube at 915 MHz is easily developed with better efficiency than at 2.45 GHz. The problem again is the economic events required to justify the investment for low-cost manufacture. There have been attempts to manufacture low-voltage (e.g., 400–600 V) tubes at 2.45 GHz. There has not been any consistent success although laboratory results show feasibility. It should be noted that the original tube made by GE for the 915 MHz GE oven operated at 600 V but was expensive to manufacture.

Again, in principle, lower-power tubes at these frequencies should be feasible and eventually they probably will become available. Tubes at higher frequencies will require much more development.

For years the technological community has been awaiting the arrival of efficient solid-state sources, especially for use in low-weight compact microwave ovens for use in cars, etc. As pointed out above this promise has not yet been realized. In principle it should be feasible but it appears the economic investment for large-scale manufacture is the stumbling block.

C. Attainable Efficiencies

The efficiencies achieved in magnetrons used for microwave heating, although high, are always below the theoretical limit by 5–10%. One avenue to higher efficiency is to operate at higher magnetic field, with other parameters held constant. It would appear that there is possible a 5% improvement at 2.45 GHz following this approach, while at 915 MHz in the higher-power tubes it is less likely that there remains much more improvement in this direction.

One explanation for the discrepancy in theoretical and experimental efficiency of magnetrons is the effects that derive from axial electron energies [49]. Again, in principle, it should be possible to overcome such effects but with significant development.

If one looks to oven design for increased efficiency there are several avenues that can improve efficiency but they involve cost or other penalties, e.g., restriction in use conditions. It is conceivable that with future advances in modeling more efficient coupling to one specific load will be achieved but the design is not likely to work for all loads.

D. Reduced Costs

The achievement of the low-cost cooker magnetron is such a remarkable milestone that for those skilled in the art of tubes, or solid-state devices, it is difficult to imagine any further reduction, other than that which comes from radical change in raw material prices or political influences on manufacturing practices.

In the high-power area reduction of cost is possible when and if the market for such tubes greatly expands.

E. Influence of Use of ISM Bands for Communications on Future Cost and Availability of Microwave Heating Systems

This is a subject that only recently has become important, maybe even critical, for the future of all ISM applications of microwave power. The problem is reviewed in technical and political detail in recent publications [50] of the International Microwave Power Institute (IMPI). In recent years there has been a trend for the use (unlicensed) of ISM bands for communications purposes, e.g., wireless LANs (local area networks) for industry, office, and the home. Already there are reports of serious interference of such systems in industry and the office. Still theorists believe interference can be overcome through digital techniques and spread-spectrum techniques.

On the other hand, the consortia for developing the widespread systems that will eventually go into all homes have advised the FCC that emissions from proposed microwave lighting systems cannot be tolerated by future wireless systems operating in the 2.45 GHz band. We have shown that if this is true then interference will also be caused by microwave ovens. We believe that the eventual result of these trends could be the political desire to eliminate microwave oven emissions. This could be done, presently, at great cost and inconvenience.

Similar attacks on the rights of ISM users in these bands are occurring at other ISM bands across the spectrum. Therefore, IMPI has mounted a campaign to oppose this trend and to restore the rights of ISM users for "free" radiation rights in ISM bands. This has been the predicate for inexpensive and practical ISM in the past. It remains to be seen if the IMPI campaign will preserve the rights of future ISM band users.

REFERENCES

1. J. M. Osepchuk. A history of microwave heating applications. IEEE Trans. MTT-32(9), pp. 120–1224 (September 1984).
2. FCC Docket No. 6651, by order dated May 15, 1947.
3. J. M. Osepchuk. Microwave technology. In Kirk-Othmer Encyclopedia of Chemical Technology, vol. 16, 4th ed., 1995, pp. 672–700.
4. E. C. Okress. Microwave Power Engineering, vol. 2, Academic Press, New York, 1968.
5. John P. O'Meara. Why did they fail? J. Microwave Power 8(2), July 1973.
6. Robert A. Peterson. A 915 MHz folded waveguide applicator. Microwave Energy Applications Newsletter vol. II, May–June, 1969.
7. Franklin J. Smith. Microwaves–hot air drying of pasta, onion, and bacon. Microwave Energy Applications Newsletter XII(6), 1979.
8. Robert F. Schiffman. Applications of microwave power, J. Microwave Power 8(2), July 1973.
9. A. Bezanson and R. Edgar Microwave tempering in the food processing industry. Electronic Progress XVIII(1), Raytheon Company, Spring 1976.
10. J. M. Osepchuk, J. E. Simpson and R. A. Foerstner. Advances in choke design for microwave oven door seals. J. Microwave Power, 5(3):295–302, November 1973.
11. Ira C. Magaziner and Mark Patinkin. Fast heat: How Korea won the microwave war. Harvard Business Review, January/February 1989, pp. 83–92.
12. A. S. Gilmour, Jr. Microwave Tubes. Artech House, Inc., Dedham, MA, 1968.
13. J. M. Osepchuk, The cooker magnetron as a standard in crossed-field research, Proc. 1st Int. Workshop on Crossed-Field Devices, Ann Arbor, University of Michigan, 1995.
14. G. A. Solomon. Analysis of magnetron failure modes versus power level. Proc. 30th Microwave Power Symposium, pp. 63–65, 1995.
15. Derek Chambers and Cliff Scapellati. New High Frequency High Voltage Power Supplies for Microwave Heating Applications. Spellman High Voltage Electronics Corporation, Plainview, NY.

16. M. J. Schindler. GaAs power PHEMT technology & applications. Proc. 31st Microwave Power Symposium, IMPI, Manassas, VA, 1996, pp. 139–142.
17. R. Yokoyama and A. Yamada. Development status of magnetrons for microwave ovens. Proc. 31st Microwave Power Symposium, IMPI, Manassas, VA, 1996, pp. 132–135.
18. R. Decareau. Microwaves in the Food Processing Industry. Academic Press, New York, 1985.
19. Charles R. Buffler. Microwave Cooking and Processing. Van Nostrand Reinhold, New York, 1993.
20. A. Metaxas and R. Meredith. Industrial Microwave Heating. Peter Peregrinus, London, 1983.
21. A. C. Metaxas. Foundations of Electroheat. John Wiley, Chichester, England, 1996.
22. Roger Meredith. Engineers' Handbook of Industrial Heating. The Institution of Electrical Engineers, London, 1998.
23. Ref. 20, pp. 286–289.
24. J. M. Osepchuk. Microwave heating. In Wiley Encyclopedia of Electrical and Electronics Engineering, vol. 13, John Wiley New York, 1999, pp. 118–127.
25. J. P. Quine and S. M. Bakanowski. Impedance and heating characteristics of microwave ovens. In Industrial Appl. Microwaves. Digest of Workshop, IMPI, Manassas, VA, 1984.
26. J. M. Osepchuk. A review of microwave oven safety. J. Microw. Power 13(1):13–26, 1978.
27. J. M. Osepchuk; U.S. Pat. Re.32,664, May 10, 1988; reissue of U.S. Pat. 3,767,884, Oct. 23, 1973.
28. J. M. Osepchuk. Debunking a mythical hazard. Microwave World 16(1):16–19, Summer 1995.
29. U.S. Dept. of HEW, FDA, Bureau of Radiological Health, Regulations for the Administration and Enforcement of the Radiation Control for Health and Safety Act of 1968, 1030.10, Microwave Ovens, pp. 36–37, DHEW Publication no. (FDA) 75-8003 (July 1974). Also see website: http://www.fda.gov/cdrh/radhlth/index.html.
30. IEEE. IEEE Standard for Safety Levels with Respect to Human Exposure to Radio Frequency Electromagnetic Fields, 3 kHz to 300 GHz, C95.1–1991, 1999 Edition, IEEE, Piscataway, NJ, 1999.
31. Federal Communications Commission. "Industrial, scientific and medical Equipment. Code of Federal Regulations, Title 47, Part 18, 1991.
32. International Electrotechnical Commission, Special Committee on Radio Interference (CISPR). Limits and Methods of Measurement of Radio Interference Characteristics of Industrial, Scientific and medical (ISM) Radio Frequency Equipment, Publication 11, rev. 1999, Geneva, Switzerland.
33. Underwriters Laboratories, Inc. Standard for Safety: Microwave Cooking Appliances, UL 923, 3rd ed., 1990.
34. K. Kawamura and S. Hiraoka. Microwave oven for passenger Airplane. Toshiba Review 156:37–41, Summer 1986.
35. C. Buffler, private communication.
36. Charm Bioengineering, Inc., 36 Franklin Street, Malden, MA 02148.
37. Questron Corporation, 4404 Quaker Ridge Road, Mercerville, NJ 08619.

38. Pitt-Des Moines, Inc., Grogan Mills Road, The Woodlands, TX 77380.
39. C. Buffler. An analysis of power data for the establishment of a microwave oven standard. Microwave World 11(3):10–15, 1990.
40. IEC. Methods for measuring the Performance of Microwave Ovens for Household and Similar Purposes, CEI IEC 705, 2nd ed. Bureau Centrale de la Commission Electrotechnique Internationale, Geneva, Switzerland, 1988.
41. R. J. Lauf, D. W. Bible, A. C. Johnson, and C. A. Evenleigh. A 2 to 18 GHz broadband microwave heating system. Microwave J. 36(11):13–26, 1993.
42. P. O. Risman, T. Ohlsson, and B. Wass. Principles and models of power distribution in microwave oven loads. J. Microwave Power 22: 173–181, 1987.
43. P. O. Risman and C. Buffler. Cylindrical microwave heating applicator with only two modes, U.S. Pat. 5,632,921; P. O. Risman. Rectangular Applicator, U.S. Pat. 5.828,040; P. O. Risman. Tubular microwave applicator, U.S. Pat. 5,834,744.
44. C. T. Buffler and G. W. Leimer. 'The Thermalizer™ combination cooking appliance, a new concept in appliance technology. Presented at the 1997 Appliance Manufacturer Conference, Nashville, TN. See also V. D. Chase, Creating a new nuke. Appliance Manufacturer, p. 8, December 1997.
45. M. Sundberg, P. Risman, P. S. Kildal, and T. Ohlson. Analysis and design of industrial microwave ovens using the finite-difference time-domain method, J. Microwave Power and Electromagnetic Energy 31(3):142–157, 1996.
46. CEM Corporation, 3100 Smith Farm Road, Matthews, NC 28106.
47. Denver Instrument Co., 6542 Fig Street, Arvadam, CO 80004.
48. ITU. Final Acts of the World Administrative Radio Conference (1979), vols. I and II, International Telecommunications Union, Geneva, Switzerland, December 1979.
49. J. M. Osepchuk. Axial energy phenomena in cooker magnetrons. Proc. of the 1998 Int. Conf. on Crossed-Field Devices and Applications, Northeastern University, Boston, MA, June 1998.
50. J. M. Osepchuk. The bluetooth threat to microwave equipment. Microwave World 20(1):4–5, May 1999.

8
Measurement and Instrumentation

Ashim K. Datta
Cornell University
Ithaca, New York

Henry Berek
FISO Technologies
Rockford, Illinois

Douglas A. Little
FLIR Systems, Inc.
Portland, Oregon

Hosahalli S. Ramaswamy
McGill University
Montreal, Quebec, Canada

I. INTRODUCTION

Electromagnetics, heat transfer, moisture transfer, and food quality changes are the components of a microwave heating process. Thus, the parameters that may need to be measured during a microwave heating process include dielectric properties, electromagnetic field strength, thermal properties, temperature, mass transfer properties, moisture content and time-temperature histories (as indicators of quality). Dielectric property measurement was discussed in Chapter 3. Thermal and mass transfer property measurements are outside the scope of this handbook and the reader is referred to books such as Ref. 1. This chapter will discuss measurement of process parameters such as electromagnetic field strength, temperature, moisture, and integrated time-temperature history. Measurement of other process parameters such as pressure, refractive index, strain are possible in the microwave environment using specialized systems, but these are not discussed here.

Measurement of parameters such as temperature and moisture during microwave heating requires special considerations, primarily due to the interference of the measuring systems with the microwaves. For example, temperature measurement during microwave heating is difficult with thermocouples since they can perturb the field and cause powerful electrical discharges. Measurement of electric field is needed as an indicator for heating rates or to measure leakage from a microwave oven. This chapter summarizes the theory and practice of instrumentation relevant for microwave process monitoring and control, both in research and in production uses.

II. MEASUREMENT OF ELECTRIC FIELD

Measurement of microwave E field or power density is needed for at least three applications in food processing—stray fields (leakage), applied and reflected power in an oven and point values inside an oven (or food).

A. Simple Measurement

A very simple to way to have a quick estimate of electric field at any location or averaged over a small volume of material is by equating the energy absorption to temperature rise in absence of any diffusion. This leads to the equation

$$E_{rms} = \sqrt{\frac{\rho C_p}{\omega \epsilon_0 \epsilon_{eff}''} \frac{\Delta T}{\Delta t}} \tag{1}$$

where E_{rms} is the electric field strength corresponding to the location where a rise in temperature of ΔT is measured over time Δt. It is emphasized that Eq. (1) is not a good approximation over long periods of time since it ignores diffusion and the effects of thermal boundary condition on the heat transfer process.

Several indirect measures of electric field patterns (that relate to heating uniformity) have been used [2]. These include fax and teledeltos paper, thermoset polymer matrix, liquid crystal sheets [3], and cobalt chloride (either as colorant in silica gel dessicant crystals or on absorbant paper). The fax paper blackens in the range 100–150°C, and has been used to provide a two-level (black/white) measurement. The liquid crystal sheets provide a range of colors depending on temperature, but over a limited range. Use of a nonenzymatic browning reaction to map the electric field distribution has also been reported (cited in Ref. 4). Another method to obtain uniformity of field patterns (heating potentials) in an empty oven is to place small amounts of water in an array of cups of nonabsorbing materials like styrofoam (e.g., Ref. 4). Measurement of temperature rise for each cup of water will provide the spatial variation in the heating rates and the electric field following Eq. (1).

Another relatively simple, nonperturbing method for measuring of electric field in a microwave oven was described by Ref. 5. It is based on measurement of the minimum and/or maximum pressure at which a selected gas (Helium) in a small cell breaks down. As microwave breakdown in gases is well understood, pressures measured can be easily converted to absolute values of field strength.

B. Measurement Using Probes

Review of various standard antennas for measuring electric and magnetic fields can be seen in Ref. 6. Design and fabrication of an E field probe for measurement of point values inside an oven or food has been discussed at length in Refs. 7 and 8. This probe was designed for the range of frequencies 1–10 GHz, with special emphasis on 2.45 GHz [8]. A schematic of the probe is shown in Fig. 1. The probe is based on a fiberoptic temperature sensor [9] which is used to measure the temperature of a resistive element when exposed to electromagnetic field. A second sensor is used to measure the ambient temperature, and the difference between the two measurements is the temperature rise of the resistive element. This type of probe was referred to as thermooptic probe [8]. The theoretical background of the probe is discussed in Ref. 8, whereas development of a commercial probe is discussed in Ref. 7. The probe has been claimed [7] to be capable of mapping field and power distributions in an oven or process chamber and monitoring on-line power changes in process applications. Power densities up to 1000 W/cm^2 can be measured at 2.45 GHz. It was also claimed that while the probe is predominantly for the high power range, its sensitivity can be extended down into the mW/cm^2 range, where it overlaps the region covered by conventional antenna-type stray field meters such as those used to check radiation levels for health hazards. Drawbacks include the size of the capsule (>3 cm) which can introduce geometry con-

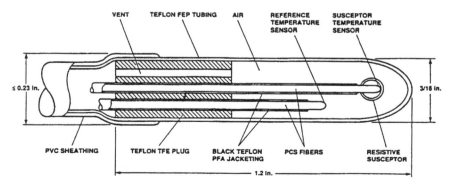

Figure 1 Schematic of an E-field probe. (From Ref. 7.)

cerns when working with small liquid or semisolid food loads and the temperature limitations of the capsule materials. The probe is rated for 250°C operation but the thin wall Teflon used can easily become soft at 200°C. Additionally, the thin metal oxide sphere inside the probe is fragile and will crack if heated too quickly or vibrated too strongly. Very few studies are reported [10] of the use of this probe to measure electric field inside a food sample heated in a microwave oven.

III. POINT MEASUREMENT OF TEMPERATURE

General discussions of temperature measurement can be found in books such as [11]. Specific applications of temperature measurement to heating of foods in a microwave oven can be found in Refs. 10 and 12–14.

A. Thermocouples

Thermocouples and other metallic probes are generally unsatisfactory for precision temperature measurements in microwave ovens for several reasons:

1. Metallic probes reflect and absorb energy in response to incident microwaves, and require special grounding and installation to withstand microwave operations. Electrical discharges have been reported [14].
2. Electromagnetic field disturbances caused by metallic probes create localized changes in heating patterns in the material that can induce variability in overall heating patterns.

There have also been attempts to quantify the errors of using thermocouples in a microwave environment [15].

However, use of thermocouples in microwave heating is discussed here since they are quite common otherwise and are perfectly suitable when the microwaves are off, as at the end of heating. They can also be modified with some success for use in a microwave environment and they are inexpensive.

1. Thermocouples

Thermocouples are perfectly suitable for measuring temperatures immediately *after* microwave heating, but not necessarily *during* microwave heating. Thus, if the final temperature is desired, thermocouples will work just fine, if inserted immediately after the heating. It (obviously) will not have the usual limitations of signal perturbations from microwaves. Some researchers [16] have used thermocouples to measure temperature at discrete times during heating by turning the microwaves off. Likewise, thermocouples can obviously be used to monitor the temperature of liquid in continuous flow through a microwave oven by mounting them just outside the inlet and outlet ports of the microwave oven [17].

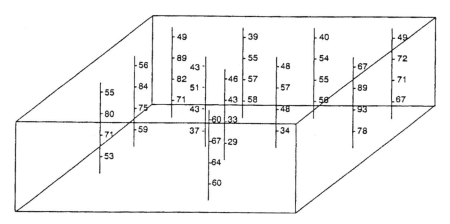

Figure 2 An example of temperatures obtained using an array of thermocouples. (From Ref. 18.)

The benefit of using inexpensive thermocouples for temperature data acquisition become quite obvious when temperatures at multiple locations are needed simultaneously. An array of thermocouples can be mounted in a fixture known as "a temperature hedgehog." Such a fixture would normally be made to replace the lid of the sample cup or dish after it is heated in the microwave oven. The fixture would then be placed on the sample so that the thermocouples penetrate to the predesignated positions and using a rapid data acquisition device the temperature data is gathered in a normal way. Since the hedgehog is placed on the sample only when the microwave oven is "off," it will not have any of the signal interference problems discussed earlier. The skill required is to have the temperature logging initiated a little while before the end of heating, and to replace the normal lid of the container with the hedgehog immediately after the oven is turned off. Small amount of inaccuracies will be there due to the short time lag, but a reasonably accurate distribution of temperatures can be achieved. Such a technique has been used [18] to measure temperature distribution in several studies, an example of which is shown in Fig. 2.

2. Thermocouples Modified for Microwave Oven Use

Thermocouples have been modified to make them work in a microwave environment [19, 20]. A thermocouple which was shielded using an aluminum tube and grounded to the microwave frame was designed and found to work well (to within 1°C) without discharging or signal perturbations. It had limited shield heating, even with no load in the cavity.

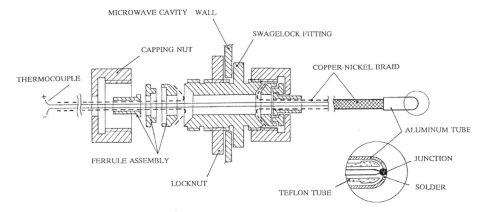

Figure 3 Schematic of a shielded thermocouple for microwave oven. (From Ref. 20.)

To construct the probe (Fig. 3), a 30 AWG copper-constantan thermocouple [21] with welded junction is shielded in an aluminum tubing (3.65 mm OD and 4 cm long). The thermocouple junction is sealed about 3 mm inside the tip of the tube by crimping the end with smooth pliers (crimped junction probe), rounding with a file and working with an emery cloth to form a smooth tapered tip. The thermocouple wires from the other end of the aluminum tube is shielded using nickel coated copper braid which is passed through brass Swagelock adaptors attached to the cavity wall using a retaining nut. This entire attachment is grounded to the metal frame of the microwave to prevent discharges within the Swagelock orifice or in the microwave cavity.

The probe was used satisfactorily with a variety of samples [14, 19] such as water, sulfuric acid, ground meat, methanol, coconut oil, linseed oil, sunflower oil. Shield heating of the thermocouple probe is a problem; however, the heating rate of the shield was about an order of magnitude less than that of a 100 ml sample of water being heated. A probe isolating the thermocouple tip from the shield material showed a marked improvement in the acquired data and provided a heating curve essentially parallel (with a small lag) to that obtained using a fiber optic probe. Generally speaking, however, thermocouples are not used in any significant extent to monitor temperatures during microwave heating.

B. Fiberoptic Thermometry

Fiberoptic thermometry is commonly used and widely accepted today to measure the temperature during microwave heating. This is primarily due to the nonmetallic and electrically nonconducting nature of the fibers and sensors it uses. There are two types of fiberoptic thermometers. One type measures temperatures along the length or at points along the length of a fiberoptic cable. The second type of fiberoptic thermometer measures temperature at just one point at or near the end of a fiberoptic cable. In these systems the fiber acts as a transport medium for light to and from the sensors and the signal conditioner or meter. Commercially available systems of the first type exist. One existing company has had moderate success selling a distributed temperature sensing fiberoptic thermometer to monitor temperature as a function of position in large electrical systems. Commercially available models of the second type of fiberoptic thermometers utilize several different technologies including fluorescence, spectral modulation [22], and interferometry. Two of the more interesting of these technologies will be discussed in more detail below.

1. Fiberoptic Thermometers Based on Fluorescence

The fiberoptic thermometer with the greatest commercial success uses the fluorescence of magnesium fluorogermanate activated with tetravalent manganese (MFG), as shown in Fig. 4 [9]. A tiny amount of this material shaped into a disk is glued to one end of a fiberoptic cable, the other end of which is connected in turn to a signal conditioner. This conditioner employs a light source rich in ultraviolet or blue light to activate the MFG which emits red light in response. All fluorescent materials have some afterglow or persistence whose rate of decay (Fig. 5) turns out to be temperature dependent. MFG, for example, has a decay rate which varies from 5 ms at $-200°C$, to about 0.5 ms at $450°C$. By pulsing the exciting light source and averaging the decay times from the emission signals, commercial units have been able to demonstrate an accuracy of better than $0.2°C$ within a span of $50°C$ centered around a single calibration temperature point. The range of these instruments is typically between $-300°C$ and $+450°C$ and most can measure at least 4 times a second. Drawbacks of these units have mainly been the high cost of signal conditioner and sensors and the fragility of the sensors.

2. Fiberoptic Thermometers Based on Interferometry

Another fiberoptic thermometer makes use of a novel adaptation of an older Fabry-Perot interferrometric (FPI) approach [23], as shown in Figs. 6 and 7. A FPI is constructed out of two mirrors facing each other and the gap between the mirrors is called the cavity length. Light reflected in the FPI is wavelength modulated in exact accordance to cavity length. This new commercial thermometer makes the FPI out of tiny pieces of polished fiber. Most of the fiber is quartz, but one of the pieces is made from a type of glass that has a noticeable thermal expansion coefficient. Thus, when the temperature changes, the cavity length changes measurably. This particular thermometer adds a unique white light cross correlator by passing the signal from the FPI through a prismlike device before it goes to a CCD detector. This gives two independent measures of the cavity length, one from the

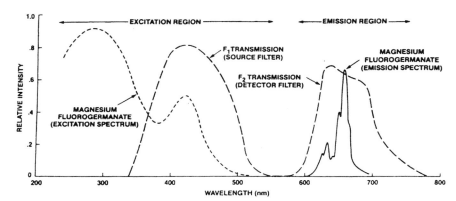

Figure 4 Emission and excitation spectra for manganese-activated magnesium fluorogermanate. (From Ref. 45.)

wavelength modulation of the FPI, and the second from the spatial variation of the cross-correlator produced lateral position on the CCD pixels. Low-cost commercially available units can be obtained with an accuracy of 0.3°C or better. Generally the range of these units is −50°C to 350°C with an operating rate of 10 Hz. Slightly more expensive units are available with better accuracy (0.1°C), higher

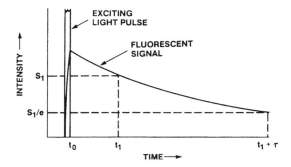

Figure 5 One method of measuring the decay rate of the fluorescence signal. An initial reading of the detector voltage S_1 representing the intensity of the fluorescence is taken at time t_1. S_1/e is then computed and established as a reference voltage. When this crossover point is detected, the time difference from the initial reading to the crossover reading is the decay rate. (From Ref. 45.)

Measurement and Instrumentation

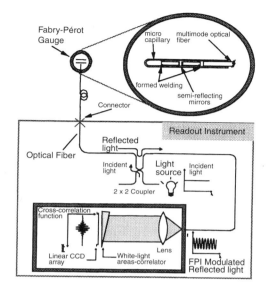

Figure 6 An idealized schematic of a Fabry-Perot interferometer with FISO white light cross correlation.

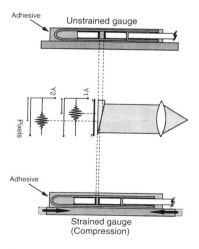

Figure 7 Details of the Fabry-Perot gauge.

speed (200 kHz), or larger number of channels (4, 8, 16, or 32). While these units allow single and multipoint calibrations, probes for these units come precalibrated. Drawbacks to this system seem to be in the nature of the FPI construction. The FPI cannot be stressed or bent during measurement (as the device also works as a strain gauge) and care must be taken when surface temperatures are made to ensure good thermal contact [24].

IV. MEASUREMENT OF SURFACE HEATING PATTERNS

For obtaining surface temperature patterns, a common method is to use thermography. Thermography is the use of a thermal imaging camera to take a "heat picture" of a target. Thermographic cameras view infrared (IR) energy (as opposed to visible light energy) and display the resultant temperatures as shades of gray or colors. The cameras available today are typically fully calibrated to not only provide a thermal picture but also accurately measure target temperatures. Thermography has been used extensively to study heating patterns in microwave product and process development [25].

A. Background to Thermography

Basics of infrared thermography can be seen in texts such as Ref. 11 and is not repeated here. Thermography technology is not new; actually the first thermography systems were introduced in the mid 1960s. These systems produced extremely low resolution images yet at the same time were extremely large and bulky. The application for these earlier systems were limited to gross thermal problems, not to smaller and more detailed targets. As thermographic systems evolved and matured, a new detector technology called focal plane array (FPA) was introduced in the early 1990s, opening up the applications for thermography.

First-generation IR systems used a single IR sensor; a mechanical scanning device was required to create the thermal image. This scanning mechanism would mosaic a thermal image one pixel at a time. These mechanical scanning devices increased the physical size of the systems as well as their weight and power consumption. Also, because the IR detector in a scanning system is constantly "flying" across the image, the thermal sensitivity is compromised.

The focal plane array, or FPA, is an IR detector that incorporates rows and columns of individual sensors eliminating the requirement for scanners. In FPA systems, each IR detector "stares" at the target and creates its own pixel of information. An FPA system typically utilizes a 256 × 256 or 320 × 240 infrared detector that has over 65,000 infrared sensors. This is quite an improvement over the scanning systems that only had one sensor! The new sensor design allows camera size to be significantly reduced so that modern thermal imaging systems now resemble camcorders. In fact, the cameras are now referred to as ThermaCAMs.

Previously, all IR systems required some sort of cooling of the detector typically using liquid nitrogen or gaseous argon. Eventually, closed-cycle Stirling cooling systems replaced the gaseous or liquid cryogenic cooling. However, recently a new technology in IR detectors eliminated the need for any type of cryogenic cooling.

B. Innovations in Infrared Technology

Recent innovations in a century-old technology have provided the next generation of thermal imaging systems: "thermal detectors" that operate at room temperature. As a result, this technology is also referred to as "uncooled" detector technology.

Among the most exciting thermal detector technologies to evolve recently is commercial uncooled technology, notably the microbolometer, and ferroelectric or pyroelectric detectors. The resistive microbolometer is the only uncooled thermal detector technology yielding high response that is DC restored, and therefore does not require a mechanical chopper to modulate the detector response. This excellent linearity is better than most cooled detectors, and translates directly into low spatial noise over a very broad temperature range. Isolation from neighboring pixels ensures negligible pixel crosstalk, resulting in high thermal sensitivity. In side-by-side measurement tests, the uncooled microbolometer has proven itself more accurate and repeatable than comparable cooled camera.

Modern silicon processing and micromachining techniques are used to create an array of microbolometers, each less than 47 mm wide. These elements, each with highly uniform characteristics, are formed into an array of detector cells, typically in a 320×240 format to match standard display devices. In monolithic designs, each microbolometer element is a thermally isolated microbridge suspended over a silicon substrate containing the readout integrated circuit and analog to digital converter.

ThermaCAM systems can produce high-resolution images. This is a result not only of the increased spatial resolution, but also a result of increased sensitivity; this combination also allows the camera to detect small targets. In the past, it was common to have low-resolution thermal images which appeared blurred or out of focus. Many images of small targets might have been misinterpreted because of low resolution. Optical design advances for commercially available systems allow powerful magnification down to 25 mm targets.

C. Technical Challenges

Technical challenges confront thermographers who test, measure and analyze microwave temperature. Most food items introduced into a microwave vary in shape and composition. Each of these materials emit infrared energy differently. As a result, the thermographic camera can have difficulties properly identifying true and correct temperatures.

Infrared analysis and reporting software captures and records the temperatures of the varying food shapes and materials that are placed into the microwave for cooking. Algorithms are then calculated and can then be reproduced as a color histogram image (Fig. 8, see color insert).

Developments to allow rapid storage of thermal images have enhanced camera performance as well. Previously in older generation systems, thermal images were created quite slowly and dynamic events were not able to be captured. Now, the FPA systems can create up to 60 images per second (and more). Dynamic storage systems available allow up to 60 thermal images to be digitally stored in 1 s, greatly enhancing the capability of the systems.

Figure 8a–d show typical time-lapsed agar samples being heated in a microwave oven. The results clearly show the nonuniformity of microwave heating. Thermographic testing like this can lead to, not only improved ovens, but better quality control for packaged foods.

Visually comparing thermal images of microwave packaging in a serial manner, one by one, can be difficult. Typically there are subtle temperature changes which cannot be discerned easily by the naked eye. Again, software has come to the rescue! Thermal image processing software includes a feature called image subtraction. Basically two thermal images can have the temperature of each pixel subtracted from each other. If the temperature between two targets is equal, there will be a 100% subtraction. The resultant image will only show temperature differences between two resultant targets.

Thermal imaging and measurement systems provide an effective means for identifying and resolving thermal related problems (hot and cold spots) in microwave food product and process development by giving a direct measurement of the actual target. The imaging systems give an idea of the range of temperatures present during heating of a food as well as some information on the fraction (surface area wise) of area at any particular temperature. Localized hot spots are difficult to locate using point measurement of temperature (such as discussed in the previous section).

D. Accuracy and Related Issues

The resolution and accuracy of one of the systems is discussed here and the author would need to refer to respective manufacturers for their own specifications. For example, the FPA microblometer infrared camera is a thermal detector and is not wavelength specific. However, because they rely on the radiation emitted by the target object to cause a change in the microbolometer's bridge structure temperature, they are optimized to operate in the 7.5–13 μm region, or spectral range. Additionally, the uncooled microbolometer detector must collect as much heat energy as possible in order to effectively capture normal room temperature targets with better than 0.1°C resolution.

Using FLIR Systems ThermaCAM 595, as an example, we can see some performance features. The camera's temperature measurement is accurate to ±2% of range, or 2°C and can effectively view and analyze temperatures from −40°C up to 1500°C (−40–2732°F). The infrared measurement accuracy is always a function of emissivity. Emissivity is a fractional representation of the amount of energy from the target material versus the energy that would come from a perfect emitter (blackbody). In ThermaCAM 595, for example, the emissivity effect is considered through automatic emissivity correction that is loaded into each camera system and comprises adjusted values from predefined emissivity tables.

V. MEASUREMENT OF INTERNAL TEMPERATURE PROFILES USING MAGNETIC RESONANCE IMAGING

Comprehensive internal temperature distributions can be obtained using magnetic resonance imaging (MRI) technique. Outside of food processing, intended for medical applications, this technique has been used to measure temperature profiles in gel phantoms and muscle tissue using microwave sources at 915 or 2450 MHz compatible with MR [26, 27]. In food applications, although MRI has been used to measure temperature profiles [28, 29], only one study has been reported [30, 31] where temperature profiles shortly after microwave heating have been measured. This study is now discussed in detail as an example of a possible technique for internal temperature measurement during microwave heating. Note that it is a complex and expensive technique that is currently suitable for research.

The MRI study [30] was based on phase mapping which uses the sensitivity of water proton chemical shift to temperature. A gel made of TX 151 (an organic hydrophylic polymer) and water was made in glass jars (100 cm^3), as shown in Fig. 9. A jar was sectioned at $\frac{1}{4}$, $\frac{1}{2}$, and $\frac{3}{4}$ heights so that the gel surfaces at these levels could be exposed to a thermal imaging system to obtain thermal images for comparison with MRI temperature images.

A waveguide system terminated with a circular applicator was used for microwave heating at 2450 MHz, as shown in Fig. 9a. A sample holder built into the radio frequency probe of MRI allowed the jar to be taken out of the magnet bore to be heated in the microwave applicator, and then accurately repositioned in the probe, thereby ensuring that the room temperature phase image and that obtained after microwave heating could be compared. After acquisition of MRI heated image, the surface was exposed to infrared thermal-image camera.

Figure 10a (see color insert) shows two-dimensional multislice MRI temperature maps at the three sections shown, for an intact jar (without the sectioning). Figure 10b shows the temperature maps obtained from infrared thermal images of the surfaces exposed by separating a split jar. The MRI temperature maps and the infrared thermal images are in good agreement qualitatively. The actual temperatures from the thermal images were about 10°C lower than those derived from MRI, which the authors attributed to the time lapse between MRI and thermal images, rapid cooling of surface (for thermal image) and averaging effect over 5 mm slice (for MRI scan). The authors plan to improve this method to make it a quantitative measurement procedure.

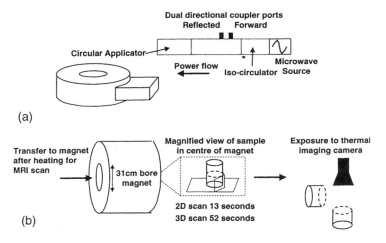

Figure 9 Schematic diagram of (a) waveguide circuit showing circular applicator; (b) temperature measurement by MRI temperature mapping and infrared thermal imaging. (From Ref. 30.)

VI. MEASUREMENT OF COOK OR STERILIZATION VALUES (TIME-TEMPERATURE HISTORY EFFECTS)

Changes in a food during heating, such as sterilization, cook values, nutrient retention, etc., are functions of not just temperature but also the time-temperature history. Measurement of such changes are needed for purposes such as process verification (as in sterilization) or process optimization (as in cook values). If time-temperature histories can be measured, sterilization or cook values F_O, measuring extent of reaction at a location, can be calculated following first-order kinetics from the following equation (or some variations of it)

$$F_0 = \int_0^t \exp\left[-\frac{E}{R}\left(\frac{1}{T} - \frac{1}{T_R}\right)\right] dt \qquad (2)$$

where t is the duration of heating, T is temperature at the location any time. T_R is the reference temperature, E is the activation energy, and R is the gas constant. Direct measurement of such integrated time-temperature history effects are referred to as time-temperature indicators (TTIs) that can be based on biological, physical, or chemical mechanisms. For a comprehensive overview of TTIs, see Ref. 32. Examples of chemical TTI are the "intrinsic chemical markers" in the food [33]. These intrinsic chemical markers are compounds that are formed due to reaction between protein precursors and ribose or glucose. By analyzing the spatial distribution of the amount of marker formed after a heating process, the extent of thermal treatment (temperature-time history) at different spatial locations in the food can be achieved and can be related to the spatial distribution of sterilization and cook values. This has been applied to microwave sterilization [34, 35], an example of which is shown under the sterilization section (VIIIB) in Chapter 4.

VII. SET-POINT TEMPERATURE CONTROL IN A MICROWAVE OVEN

Microwaves generate heat constantly and therefore the temperature keeps increasing; unlike in surface heating where the food attempts to reach the heating medium temperature. To maintain a set-point temperature in a microwave oven requires additional control aspects. This has been attempted by researchers in the contexts of controlling and enhancing chemical reactions, evaluating destruction kinetics of chemicals, enzymes as well as microorganisms, and for controlled establishment of thermal processes such as pasteurization, sterilization, thawing, drying, etc. Several methods have been used by researchers to achieve this purpose. Feedback mechanisms using temperature gathering devices (thermocouples, thermisters, infrared or fiberoptic sensors) to achieve on/off control of domestic microwave ovens appears to be the simplest approach in this direction. Similar mechanisms can be used to adjust the power level of the variable power source in the more expensive industrial ovens for continuous operations.

A. Domestic Microwave Oven

Commercial probes used in domestic microwave ovens are generally thermistors located inside stainless steel tubes and a copper braid. Like the resistance temperature detector (RTD), the thermistor is also a temperature sensitive resistor. While the thermocouple is the most versatile temperature transducer and the RTD is the most stable, thermistors are more "sensitive" [21]. Amongst the three sensors, thermistor exhibits by far the largest parameter change with temperature. They are composed of semiconductor materials, with their resistance generally decreasing with temperature. The thermistor is an extremely nonlinear device, often difficult to calibrate and is less reliable than thermocouples and RTDs. These probes require insertion into a microwave absorber (sample being heated) to isolate them from the microwave field. They are generally interfaced with the microwave on/off control circuitry to achieve some control of set-point sample temperatures and/or specific product heating patterns such as program cycles for defrosting, cooking, etc.

B. Research Microwave Heating Applications

While the commercial probes discussed in the previous paragraph could be adequate for the consumer, they are not of sufficient accuracy for maintaining desired temperature in a research environment. Studies in maintaining constant temperature have often been in connection with investigations into possible nonthermal effects associated with microwaves. The general idea in these studies is to remove the generated heat to maintain a constant temperature. For example, a modified Liebig condenser was used for maintaining set-point temperature of test samples subjected to microwave heating by using a kerosene (microwave transparent) heat exchanger for removing the produced heat [36]. Using such an equipment, the authors reported the sample temperatures could be easily maintained at sublethal levels for evaluating nonthermal effects. More recent studies have also involved such heat exchange devices as stainless steel (e.g., Ref. 37) or silicone tube [38] heat exchanger which is completely immersed in continuously flowing liquid samples for efficient removal of microwave heat generated in the sample. A slightly different concept has been used in industry to maintain a low temperature in frozen samples subjected to microwave heating for tempering purposes. In these applications, cold air ($-20°C$) is blown over the frozen sample to keep its surface frozen, permitting the microwaves to penetrate deep into the sample and heat more uniformly. Surface thawing would result in microwaves mostly absorbed near the surface and shield the inside.

Improvement in maintaining a desired temperature lead to a microwave kinetics reactor apparatus (MWKR), as shown in Fig. 11 [39]. The MWKR consisted of a feedback temperature control system and a specially designed reactor vessel. The feedback control system used a fiberoptic temperature probe and temperature control software for turning the microwaves on/off. The MWKR consisted of plastic vessel (500 cm^3 volume) and a completely sealed mechanical mixing assembly with ports for drawing samples, inserting fiberoptic probes, pressure gauge, and an

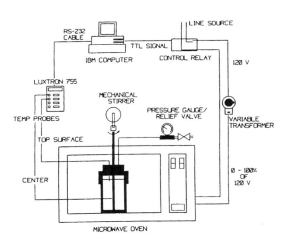

Figure 11 Schematic diagram of the microwave kinetics reactor apparatus (MWKR). (From Ref. 39.)

Measurement and Instrumentation

overpressure safety release valve. Set-point sample temperature accuracy of ±0.2°C was reported. An interesting variation in the feedback on/off control of microwave ovens has been reported for temperature control in small volumes [40]. Since small volume test samples (≈5 cm^3) heat up very rapidly, a heat sink in the form of 250–350 cm^3 water was placed below the test tube containing the sample to prevent temperature overshooting. The control was based on the temperature signal from a shielded thermocouple probe placed in the sample.

VIII. MEASUREMENT OF MOISTURE LOSS AND MOISTURE PROFILES DURING MICROWAVE HEATING

Measurement of total moisture loss typically involves measuring changes in weight. This can be done as a continuous measurement during the heating or drying process or stopping the microwaves at intervals and making a quick measurement of the weight and continuing with the heating process. In continuous measurement, the sample sits on a pedestal (e.g., Ref. 41) that is connected to the outside for weighing. Such weighing systems can be built by making a small hole

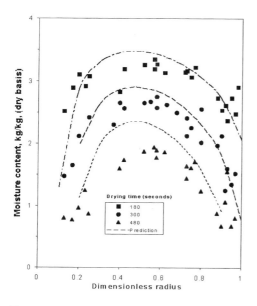

Figure 12 Moisture distribution inside a potato sphere of diameter 6 cm at three different drying times obtained by slicing the sphere and measuring the moisture content of the slices using the standard hot air oven technique. Initial moisture content is 3.95 kg of water/kg of dry matter. (From Ref. 43.)

in a typical domestic oven, or these can be built into laboratory microwave ovens such as the ones by Cober Electronics, Stamford, Connecticut.

Measurement of internal moisture profiles in a moisture transport process (conventional or microwave) is quite difficult. An ideal setup for measuring such moisture profile during microwave heating is to use the MRI described in Sec. IV, but it has not been accomplished so far.

Simpler methods [42, 43] of measuring moisture profiles depend on withdrawing the sample from heating and its rapid cooling to arrest further moisture transport, often using liquid nitrogen. The sample is then sliced and the moisture content of each slice is measured using the standard oven method (e.g., Ref. 43) or other methods such as the near infrared (NIR) spectrospic technique [44, 42]. An example of moisture profiles obtained by slicing is seen in Figure 12.

REFERENCES

1. M. A. Rao and S. Rizvi, Engineering Properties of Foods. Marcel Dekker, 1995.
2. S. Bradshaw, S. Delport, and E. van Wyk, Quantiative measurement of heating uniformity in a multimode microwave cavity, Journal of Microwave Power and Electromagnetic Energy, vol. 32, no. 2, pp. 87–95, 1997.
3. A. Hiratsuka, H. Inoue, and T. Takagi, Observations of electric field intensity patterns in microwave ovens, Journal of Microwave Power, vol. 13, no. 2, pp. 189–191, 1978.
4. B. J. Bobeng and B. D. David, Identifying the electric field distribution in a microwave oven: A practical method for food service operators, Microwave Energy Applications Newsletter, vol. 8, no. 6, pp. 3–6, 1975.
5. C. S. MacLatchy and R. M. Clements, A simple technique for measuring high microwave electric field strengths, Journal of Microwave Power, vol. 15, no. 1, pp. 7–14, 1980.
6. M. Kanda, Standard probes for electromagnetic-field measurements, IEEE Transactions on Antennas and Propagation, vol. 41, no. 10, pp. 1349–1364, 1993.
7. K. Wickersheim, M. Sun, and A. Kamal, A small microwave e-field probe utilizing fiberoptic thermometry. Journal of Microwave Power and Electromagnetic Energy, vol. 25, no. 3, pp. 141–148, 1990.
8. J. Randa, Theoretical considerations for a thermo-optic microwave electric-field-strength probe, Journal of Microwave Power and Electromagnetic Energy, vol. 25, no. 3, pp. 133–140, 1990.
9. K. A. Wickersheim and M. H. Sun, Fiberoptic thermometry and its applications, J. Microwave Power, pp. 85–94, 1987.
10. J. Mullin and J. Bows, Temperature measurement during microwave cooking. Food additives and contaminants, vol. 10, no. 6, pp. 663–672, 1993.
11. L. Michalski and E. Eckersdorf, Temperature measurement. Chichester, England: John Wiley & Sons, 1991.
12. T. Ohlsson, Methods for measuring temperature distribution in microwave ovens, Microwave World, vol. 2, no. 2, pp. 14–17, 1981.

13. R. Y. Ofoli, Thermometric consideration for rf and microwave research in food engineering, Journal of Microwave Power and Electromagnetic Energy, vol. 21, no. 2, pp. 57–63, 1986.
14. H. S. Ramaswamy, J. M. Rauber, G. S. V. Raghavan, F. R. van de Voort, and T. Kudra, Evaluation of some shielded thermocouples for temperature measurement in microwave environment, ASAE Paper, vol. 90–6601, pp. 1–11, 1990.
15. W. E. Olmstead and M. E. Brodwin, A model for thermocouple sensitivity during microwave heating. Int. J. Heat and Mass Transfer, vol. 40, no. 7, pp. 1559–1565, 1997.
16. X. Zeng and A. Faghri, Experimental and numerical study of microwave thawing heat transfer for food materials, Journal of Heat Transfer, vol. 116, pp. 446–455, 1994.
17. T. Kudra, R. F. V. D. Voort, G. S. V. Raghavan, and H. S. Ramaswamy, Heating characteristics of milk constituents in a microwave pasteurization system, Journal of Food Science, vol. 56, no. 4, pp. 931–934, 937, 1991.
18. H. S. Ramaswamy and T. Pillet-Will, Temperature distribution in microwave-heated food models, Journal of Food Quality, vol. 15, pp. 435–448, 1992.
19. F. R. van de Voort, M. Laureano, J. P. Smith, and G. S. V. Raghavan, A practical thermocouple for temperature measurement in microwave ovens, Canadian Institute of Food Science and Technology Journal, vol. 20, no. 4, pp. 279–284, 1987.
20. T. Kudra, G. Raghavan, F. van de Voort, and H. S. Ramaswamy, Microwave heating characteristics of rutile, ASAE Paper 893542, 1989. Presented at the ASAE International Winter Meeting at New Orleans, LA. Published by ASAE, St. Joseph, Michigan.
21. Omega, The Temperature Handbook. Stamford, CT: Omega Engineering Co., 1995.
22. E. W. Saaski, J. C. Haartl, and G. L. Mitchell, A fiber optic sensing system based on spectral modulation, Advances in Instrumentation, vol. 41, no. 3, pp. 1177–1184, 1986.
23. C. Belleville and G. Duplain, White-light interferometric multimode fiber-optic strain sensors, Optics Letters, vol. 18, no. 1, pp. 78–80, 1993.
24. H. E. Berek and K. A. Wickersheim, Measuring temperatures in microwaveable packages. Journal of Packaging Technology, vol. 2, no. 4, pp. 164–168, 1988.
25. J. Bows and K. Joshi, Infrared imaging feels the heat in microwave ovens, Physics World. vol. 5, no. 8, pp. 21–22, 1992.
26. H. Schwarzmaier and T. Kahn, Magnetic resonance imaging of microwave induced tissue heating, Magnetic Resonance in Medicine, vol. 33, pp. 729–731, 1995.
27. I. A. Vitkin, J. A. Moriarty, R. D. Peters, M. C. Kolios, A. S. Gladman, J. C. Chen, R. S. Hinks, J. W. Hunt, B. C. Wilson, A. C. Easty, M. J. Bronsil, W. Kucharczyk, M. D. Sherar, and R. M. Henkelman, Magnetic resonance imaging of temperature changes during interstitial microwave heating: a phantom study, Medical Physics, vol. 24, pp. 269–277, 1995.
28. X. Z. Sun, J. B. Litchfield, and S. J. Schmidt, Temperature mapping in a model food gel using magnetic resonance imaging, Journal of Food Science, vol. 68, pp. 168–172, 181, 1993.
29. R. Ruan, K. Chang, P. L. Chen, and A. Ning, Simultaneous heat and moisture transfer in cheddar cheese during cooling. II. Mri temperature mapping, Drying Technology, vol. 16. no. 7, pp. 1459–1470, 1998.
30. K. P. Nott, L. D. Hall, J. R. Bows, M. Hale, and M. L. Patrick, Three-dimensional mri mapping of microwave induced heating patterns, International Journal of Food Science and Technology, vol. 34, pp. 305–315, 1999.

31. K. P. Nott, L. D. Hall, J. R. Bows, M. Hale, and M. L. Patrick, MRI phase mapping of the temperature distribution induced in food by microwave heating, Magnetic Resonance Imaging, vol. 18, pp. 69–79, 2000.
32. M. Hendrickx, G. Maesmans, S. D. Cordt, J. Noronha, A. V. Loey, and P. Tobback, Evaluation of the integrated time-temperature effect in thermal processing of foods, Critical Reviews in Food Science and Nutrition, vol. 35, no. 3, pp. 231–262, 1995.
33. H. J. Kim and I. A. Taub, Intrinsic chemical markers for aseptic processing of particulate foods, Food Technology, vol. 47, no. 1, pp. 91–99, 1993.
34. H. Zhang and A. K. Datta, Experimental and numerical investigation of microwave sterilization of solid foods, American Institute of Chemical Engineers Journal, 2000.
35. T. Wig, J. Tang, F. Younce, L. Hallberg, C. P. Dunne, and T. Koral, Radio frequency sterilization of military group rations, in Presented at the AIChE Annual Meeting (Dallas, Texas), pp. T3017–66B, 1999.
36. R. V. Lechowich, L. R. Beauchat, K. I. Fox, and F. H. Webster, Procedure for evaluating the effects of 2450 mhz microwaves upon *Streptococcus faecalis* and *Sacharomyces cerevisiae*. Applied Microbiology, vol. 17, pp. 106–110, 1968.
37. M. Kozempel, O. J. Scullen. R. Cook, and R. Whiting, Preliminary investigation using a batch flow process to determine bacteria destruction by microwave energy at low temperature, Lebensmittel Wiss. u. -Technology, vol. 30, pp. 691–696, 1997.
38. C. H. Tong, Effect of microwaves on biological and chemical systems, Microwave World, vol. 17, no. 4, pp. 14–23, 1996.
39. B. A. Welt, J. A. Street, C. H. Tong, J. L. Rossen, and D. B. Lund, Utilization of microwaves in the study of reaction kinetics in liquid and semi-solid media, Biotechnology Progress, vol. 9, pp. 481–487, 1993.
40. H. S. Ramaswamy, F. R. van de Voort, G. S. V. Raghavan, D. Lightfoof, and G. Timbers, Feedback temperature control system for microwave ovens using a shielded thermocouple, Journal of Food Science, vol. 56, no. 2, pp. 550–552, 555, 1991.
41. P. Bhartia, S. S. Stuchly, and M. A. K. Hamid, Experimental results for combinational microwave and hot air drying, Journal of Microwave Power, vol. 8, no. 3, pp. 245–252, 1973.
42. L. Zhou, V. M. Puri, R. C. Anantheswaran, and G. Yeh, Finite element modeling of heat and mass transfer in food materials during microwave heating—model development and validation, Journal of Food Engineering, vol. 25, pp. 509–529, 1995.
43. L. Lu, J. Tang, and L. Liang, Moisture distribution in spherical foods during microwave drying, Drying Technology, vol. 16, no. 3–5, pp. 503–524, 1998.
44. R. S. Yeh, R. C. Anatheswaran, J. Shenk, and V. M. Puri, Determination of moisture profile in foods during microwave heating using vis-nir spectroscopy, Lebensm.-Wiss. u. Technol., vol. 27, pp. 358–362, 1994.
45. K. A. Wickersheim and M. H. Sun, Fluoroptic thermometry, Medical Electronics, vol. 2, pp. 84–91, 1987.

9
Microwave Processes for the Food Industry

Robert F. Schiffmann
R. F. Schiffmann Associates, Inc.
New York, New York

I. INTRODUCTION

The use of microwave energy for industrial processing was of interest soon after the discovery of its ability to heat foods. In fact, beginning in 1950 and for over 20 years thereafter, the pundits focused upon the great opportunities to be had in industrial processing with microwaves [1–3]. A special issue of the *Journal of Microwave Power,* published in 1973, surveyed the status of industrial microwave processing in the United States, Europe, and Japan [4–6]. Very few persons believed that consumer microwave ovens would be the sensational appliance they have become. How wrong the pundits were is easily seen in Table 1, which illustrates the approximate relative sizes of the two markets. (*Note:* While all the numbers in the table are approximate, a recent article presented a similar set of numbers for the installed megawatts of microwave power for consumer ovens and industrial systems [7].

How could the experts be so wrong? First, they failed to see that convenience (which has no real meaning in industrial processing) is very important to consumers, especially at a time of enormous social change. But on the industrial side they did not foresee the intense resistance to change from the conventional means of doing things to using microwaves. Therein lies a great lesson: What matters in industry is not how something is done, but that it be done well and with the greatest return on investment. Microwave equipment is expensive, as is the generation of electricity. All things being equal, microwaves usually can't compete with natural gas or steam. To be successful the microwave system must provide some unique benefit and have an economic return that justifies its installation.

Table 1 Comparison of the Consumer and Industrial Microwave Markets in the United States

	Consumer ovens	Industrial systems
Number of units installed	170 million	< 1000
Total sales value	$25.5 Billion[a]	$500 Million[b]
Installed megawatts	100,000[c]	100

[a] $150 per oven
[b] $5000/kw installed
[c] 600 w per oven

A. Criteria for a Successful Microwave System

Over the years a number of experts have produced lists of criteria required for the successful adoption of a microwave system. One of the earliest and most quoted is Jeppson:

Criteria for successful adoption of a new microwave application
Product cannot be produced any other way
Cost is reduced
Quality is improved
Yield is higher
Raw material can be cheaper

Source: From Ref. 8.

On the other hand, Krieger [9] has suggested a different set of criteria, shown in Table 2. Krieger's list is especially useful for this text, and for anyone considering the use of microwave energy. He is a leading industrialist and manufacturer of numerous industrial microwave systems, and understands those attributes most important to industrial utilization of microwaves. However, even if the major criteria are met, the chances of the adoption of the microwave process are small and some of the reasons for this failure are presented later in the chapter.

Chapters 2 and 4 provide a good assessment of the mechanics of microwave heating. The single most important thing about microwave and RF (radio frequency) heating is the unique opportunity to create heat within a material—the volumetric heating effect—not achievable by any other conventional means. This volumetric heating creates another important phenomenon—a relatively cool surface. This is the cause for much consumer frustration with the inability to brown and crisp foods in domestic microwave ovens, unless hot air or special packaging materials are used. However, for industrial processing this can be a welcome phe-

Table 2 Criteria for Successful Adoption of New Microwave Applications

1. Energy transfer
   a. High speeds
   b. Short line lengths (less floor space)
   c. Less scrap
   d. Energy conservation because only the material is heated
2. Deeply penetrating heat from molecular agitation
   a. Better quality from more uniform heating
   b. Greater yields
3. Selective heating based on dielectric properties
   a. Unique solutions to problems
   b. Ability to heat within a package
4. An electronic mechanism
   a. Automation
   b. Control and data logging
   c. Environmentally clean and quiet
   d. Quick set up and fast changeover to new products

Source: Ref. 9.

nomenon since it avoids overheating the surface of products. It has been successfully adopted for precooking bacon and sausages, proofing dough, tempering meat, and many other processes. In other words, when properly applied, often with auxiliary heat forms, microwave heating can produce highly controllable thermal profiles at speeds not attainable by other means. To be successful, an industrial process uses this advantage to produce the desired product with the desired quality, and with a return on investment equal to or better than any other competing process.

The following pages examine a number of microwave food processes, some quite successful, some successful for a time and no longer employed, and some which never went beyond the pilot stage but which could be adopted successfully today. Not included are those things whose chances of successful commercialization, in the opinion of the author, are unlikely, despite laboratory success. The literature and patent lists are filled with such things—interesting from an academic point of view but not meeting the economic criteria.

B. Early History of Industrial Microwaves

The first patent granted for a microwavable food product was in 1949 to Percy Spencer, for popcorn [10]. In 1952, Spencer patented the first means for conveying materials through a microwave oven [11]. However, it awaited the development by Cryodry, in 1962, of a conveyorized microwave oven with large openings for the product path, before this technique became practical [12]. What began with water-filled chokes slowly developed over the years to various types of more practical absorptive, reflective, and capacitive choke designs. Products as thin as bacon or as large as loaves of bread or 100 lb cartons of frozen meat may be transported through open-ended tunnels without personnel hazards due to microwave leakage.

Also critical to the growth of microwave processing was the development of multikilowatt sources capable of delivering sufficient microwave power to process foods at hundreds or thousands of pounds per hour [13]. Today, microwave generators exist capable of delivering anywhere from less than one to nearly 100 kW and at either 2,450 or 915 MHz. (RF generators operating at 27 and 40 MHz are capable of delivering hundreds of kilowatts from a single tube.) Another critical advance in microwave hardware was the introduction of high-power ferrite circulators in the mid 1970s [14]. Prior to this, severe mismatch in the form of high reflected power could dramatically shorten the magnetron or klystron tube life. This problem caused the shutdown of at least one large industrial application.

It was in 1960s that the availability of high-power tubes and safe conveyor tunnels (but without ferrite circulators) led to the beginning of industrial heating applications. The first big success was the microwave drying of potato chips, which resulted in over 100 operating systems in the early 1970s. Other highly successful early applications included precooking chicken and bacon; proofing and frying donuts; tempering frozen foods; and drying pasta. Most of the early systems are no longer operational, for a host of reasons described below. A detailed discussion on microwave ovens is presented in Chapter 6.

The companies involved in these early days were large, although their industrial microwave groups were usually small divisions. The early players included Raytheon, Litton Industries, Armour-Cryodry, and Bechtel. Of these, only Amana (an off-shoot of Raytheon) and Cryodry (a small independent company: Microdry) are still in existence. There are several other companies today as well, all of them small, or small parts of much larger companies.

C. Overview of Food Industry Applications

Microwave processing has been successfully applied on a commercial scale to

Meat and poultry cooking
Bacon precooking
Tempering meat, poultry, fish, butter, fruit
Baking and proofing dough
Drying snacks and vegetables and
Pasteurizing ready meals and pasta

Note that smaller or proprietary applications do not appear in this list.

While tempering and bacon cooking together account for hundreds of operating systems, many of the others have disappeared. Of the remainder, most are single customer installations. An in-depth discussion of all these systems follows.

II. MEAT AND POULTRY PROCESSING

The use of microwaves in processing meat accounts for some of the earliest installations of microwaves in the food industry, as well as the applications accounting for most of the current operational systems. The two most successful uses of microwave energy for food processing, in terms of numbers of operating systems, are tempering meat, poultry, fish, fruit, and butter and the precooking bacon. Each hour, 300 major food companies temper a total of approximately 1000 tons of bulk frozen food for further processing. At the same time, processors in the North America and United Kingdom precook about 2,500,000 slices of bacon each hour [7].

There are other applications in the meat industry whose success is limited to very few systems: precooking chicken, sausage, and meat patties. The driving forces for the adoption of microwaves to meat processing vary, but rely upon the controlled internal heating which provides speed, yield, and end product quality not achievable by conventional techniques.

A. Precooking Chicken

This was one of the earliest industrial processes, beginning in the 1960s. At one time there were three large systems operating, capable of processing several thousands of pounds of poultry per hour with a single microwave precooker. None of these is operating any longer for various reasons described below.

Precooking poultry, especially chicken, is a major operation in the food industry. Millions of pounds of breaded precooked chicken are sold annually, usually in frozen form, in retail and food-service distribution. The two most common methods of precooking are by live steam or hot water immersion. The latter is largely a batch operation, while steam cooking lends itself to continuous operation. Cooking times vary between 45 min and 2 h. In a typical processing plant, the slaughtered, cleaned poultry are cut into pieces: breasts, legs, thighs, backs, and wings. After cold-water chilling, the pieces pass through a batter/breading operation and then are either deep fat fried for retail and some food service distribution or are simply frozen—primarily for food service, where the pieces are likely to be fried just prior to serving.

As earlier described, pieces are segregated by size and cooked together, i.e., breasts and thighs; and wings, backs, and legs. Because these parts vary in size, it is not simple to ensure uniform cooking. A common problem is "bloody bone" where meat close to the bone is undercooked and/or bloody marrow comes out of the bone, discoloring the meat. For this reason, it is often necessary to overcook the chicken thereby reducing the yield since there are higher cooking losses. Note that in the poultry industry yield is determined by comparing the finished cooked weight of the breaded chicken to the weight of the raw chicken without breading. Thus, it is possible to have yields exceeding 100%. However, yields are commonly on the order of 90% and higher. Beyond the cooking losses, another problem is that these systems, especially water immersion, may be somewhat unsanitary.

The first full-scale precooked poultry operation was designed and built by the Litton Industries Atherton Division and installed in 1966. It was also the longest running installation. It consisted of two conveyorized stainless steel microwave tunnels, located side by side, cooking poultry parts. One tunnel received the larger pieces—breasts and thighs—and operated at 80 kW via thirty-two 2.5 kW generators, operating at 2450 MHz. The other tunnel, cooking the smaller wings, legs and backs, used twenty 2.5 kW generators for a total of 50 kW. Both units operated with an atmosphere of saturated steam to prevent drying out of the chicken. The pieces were transported on a silicone-coated fiberglass conveyor belt. The microwave units were located at the top of the oven and fed the microwave energy into the cavity via flexible waveguide. The production capacity was 2500 lb/h.

A second system using two 30 kW klystrons for one tunnel, and one 30 kW klystron for the other, followed. This also had a production capacity of 2500 lb/h. However, problems soon arose which forced its shutdown. Steam condensing on the microwave windows, at the entry of the waveguides into the cavity, caused severe mismatches resulting in high reflected power, which damaged the very expensive klystrons. Losing even one klystron would shut down the processing line. This is a good argument for the use of multiple smaller power sources rather than a single large source.

A third system employed a single 25 kW generator at 915 MHz. Here the precooked poultry was produced only for the retail market. It employed a short hot-oil cook to brown and crisp the breading, and the product was frozen but later sold from the refrigerated meat case.

What is surprising is that none of these systems is still in use, despite the success of the first processing system. That system operated for nearly 20 years producing tens of millions of pounds of high quality precooked chicken. The reasons for the lack of further growth and the disappearance of these systems is only partially explainable. The klystron system was shut down because of the tube damage that could be avoided today through the use of ferrite circulators. However, the multiple generator systems worked well. Perhaps, it is because these systems were so early in the microwave heating technology, at a time when domestic microwave ovens were not common that there was ignorance about microwaves and reluctance to use these "mysterious rays." This author has had similar reactions to a number of systems in the baking industry installed in the 1960s and 1970s. It is also likely that the improvements in continuous live-steam systems—a conventional process which engineers and plant superintendents could understand—made the microwave/system process less attractive.

A 1969 analysis by May [15] projected savings over batch cooking of 1.2¢/lb for continuous steam cooking and 3.0¢/lb for the microwave/steam process. Batch systems are very labor intensive, so large savings are usually possible with continuous systems. These savings might make the microwave/steam process more attractive today, but it is hard to economically replace steam in any process, unless other significant benefits are achievable.

B. Precooking of Bacon

One of the two most successful applications of microwave heating to food processing is the microwave precooking of bacon. There are now well over 30 continuous microwave bacon cookers operating in the United States, most of large microwave power and product output. Microwave heating is the ideal way of accomplishing this process and currently no other process matches or exceeds its advantages. Bacon is a food that is inconvenient to prepare in an era of convenience foods. One of the few foods that consumers cook in their microwave ovens is bacon. In food service, bacon is a food that is not only inconvenient but messy. We've all seen the short order cook grilling bacon with hot oil spitting into the surroundings, and the grill coated with burned on meat that the cook scrapes off with a spatula. What a difference there is when the restaurant or fast food establishment is supplied with precooked bacon that can be put under infrared lamps, or in an oven, or even put quickly on a grill to be reheated prior to serving.

Microwave heating is an ideal way to cook bacon, much preferred over grilling. The reason has to do with the effect of heat upon meat protein. When proteins are heated they denature, congulate, and shrink, causing much of the water bound to the protein to be expressed, or squeezed, out of the structure. When heat is applied externally by grilling, severe water loss does occur and the entire structure shrinks. There is some evaporation, but the squeezing out of fluid dominates. Bacon is essentially a two-component food: lean and fat. When heated, the fat melts (renders) and runs out of the fat cells. On a grill, the extremely hot surface causes the thermal breakdown of the rendered fat, lowering its quality.

Microwave cooking of bacon causes less physical shrinkage, so the structure has greater final volume and better control of the final shape and dimensions than when grilled. This is largely due to the use of a hold-down belt, which keeps the bacon flat and allows it to be cooked to the length and/or shape specified by the processor. For example, bacon prepared this way for the McDonalds® chain must fit a specific bun or sandwich size. Thus, microwave precooked bacon can be cooked to meet a processors specifications for:

- Length
- Yield (typically 28–32% by weight)
- Water activity
- Appearance
- Flavor, based upon the processors specified cure

The bacon is also evenly heated, thereby avoiding local overheating and burning which is common with grilling and infrared cooking.

The fat rendered from the bacon can be collected in a state in which it has not been thermally abused and, so, has much higher economic value. One microwave equipment manufacturer indicated that the money received from the sale of fat is often sufficient to pay for the cost of electricity to operate the system [16].

The economic importance of the more gentle cooking accomplished with microwave is influenced by way bacon is sold: by count, i.e., the number of slices per pound. For example, "18 count bacon" has 18 slices per pound. With conventional cooking, the slices of 26 count bacon are thinner than 18 count slices. However, with microwave precooking the 26 count slice may be as thick as an 18 count slice that is grilled! This means the processor can sell more slices per pound, but with higher consumer satisfaction.

For the processor, there are other important economic and quality factors, as well:

- Central process control resulting in uniform high quality
- Less shipping weight vs. bulk raw bacon
- Central recovery of fat for sale and other uses
- Central recovery of ends and pieces, which can be sold as high profit bacon bits
- Greater customer satisfaction resulting from elimination of in-store cooking; better maintenance of flavor; no fat disposal problems

In addition to sliced strips of bacon, round slices with spices are also cooked this way. This process is likely to continue its growth in the United States, Canada, and some European countries. In the United States, there are currently three major suppliers of bacon processing systems: Amana Industrial Processing, Ferrite Components, Inc. and Microdry. It is estimated that the cost of the microwave precook is 0.3¢/slice, or 6–10¢/lb. If the value of the fat rendered is equivalent to the electricity cost, this reduces the pre slice cost to 0.015¢/slice! [16].

Two early bacon cookers were identified in an article by this author in 1973 [4]. These used microwaves at 915 MHz in combination with hot air, with cooking times of 3–4 min—similar to conventional grill cooking. However they had advantages in product quality, ease of cleaning, and minimal damage to the rendered fat. Today, most systems no longer use hot air but still employ 915 MHz largely because of the high power requirements. When warm or hot air is used, it serves to efficiently remove the water vapor liberated by the microwave cooking.

Typical systems are 50–100 ft long, and utilize 400 or more kilowatts, processing as much as 50,000 slices of bacon per hour, the largest one being 600 kW. Commonly, slices of bacon, from specially designed high-speed slicers located in a refrigerated room, are placed on a conveyor belt usually made of polypropylene in the form of a linked or articulated belt. These pass through the stainless steel applicators in 45 s, during which the bacon is cooked and the fat is rendered. (The combination of rendered fat and evaporation reduces the raw bacon weight by about 60–70%.) During its transport through the applicator, the bacon is kept flat by a hold-down belt, similar in design to the conveyor belt, for the reasons described earlier. The steam is exhausted out of the ovens while the fat is collected by draining from the bottom and pumped to storage tanks for later sale as a valuable commodity. The bacon is then conveyed to the refrigerated packaging room [7]. A typical commercial bacon cooker is seen in Fig. 1.

An interesting feature of the bacon cookers is that the applicators are often built as modules—each with two microwave inputs, which are then bolted together, with short microwave ducts, to make the system of desired size and throughput. For example, a 420 kW system would be built from three applicator modules, each with two 75 kW power inputs, or a total of six microwave generators. The generators are usually located remotely from the cooker—often in a room directly overhead. Waveguide is used to bring the microwave power from the generators, through the floor and into the applicators—usually through the top. Commonly, some type of mode-mixing device is used to couple energy from the waveguide input into the applicator. Amana and Ferrite Components Inc. use ro-

Figure 1 Continuous microwave bacon cooker. (Courtesy of Amana Appliances Commercial Products.)

tating 3-port waveguide antennas; Microdry splits the output power from each transmitter to feed a pair of cross-polarized "waveguide dump" feeds, with a rotating mode stirrer placed between each pair of feeds.

C. Cooking Sausage Patties

While numerous laboratory systems were developed for the microwave processing of sausages, none of these became commercial. Two successful, but differently designed, microwave systems were employed for the processing of meat patties. The first was developed in the early 1970s in Sweden for Indra Foods. The microwave process replaced an earlier deep-frying process for beef patties. In the Indra process, the patties are grilled between electrically heated grilling plates in close contact with the meat surface, which first has a 1.5 g dot of margarine placed thereon. This causes surface browning and seals the surface to reduce loss of meat juices. After this the patties are transported through a 30 kW 2450 MHz microwave applicator wherein the internal temperature is raised to 70°C (158°F). The equipment processes approximately 16,000 patties per hour [17]. The main advantages of the microwave process are speed and improved economics. The deep-fat-frying process requires continuous fat filtration and replacement of the frying fat, an expensive process.

The second system is proprietary and was developed by this author for precooking pork sausage patties. Prior to the installation of the microwave system in 1983, the processor was grilling the patties between gas-fire heated grill plates. This created several problems. The operation was messy and the rendered oil led to frequent fires—at least once a day—causing the line to be shut down and the cooker and grill plates to be cleaned before the cooking could continue. At the end of the day, two persons spent an entire shift cleaning the grill cooker. Finally, the grilling created a thin patty, somewhat distorted, with a low yield; approximately 65% (i.e., 65 lb of finished cooked patties per 100 lb of raw sausage meat).

In the microwave process, the patties were extruded by a patty former onto a conveyor belt and passed through a microwave system employing six 6 kW generators to generate a total of 36 kW at 2450 MHz. This raised their internal temperature to 38°–43°C (100–110°F). From there they proceeded directly to an infrared cooker which browned the outside and finished the internal cooking to 70°C (158°F). Following this the patties were frozen for retail or food service distribution. The production rate was 30,000 1 1\2 oz patties per hour.

Enormous advantages were realized by this processor. First, the patty quality was much improved; the patties were of greater volume and more tender. Second, the yield was raised to 80% or more, an enormous economic advantage. Third, the overall process was much simpler, safer, and more sanitary: fires were eliminated, as was the cleanup associated with them. Daily cleanup of the entire processing system could be done in less than 1 h by one person. This led to the payback of the entire cooking system, including the patty former, in less than 6 months!

An interesting problem encountered here and other microwave meat processing systems is that of the large amounts of water used constantly in meat processing plants. Clearly, this is incompatible with the high voltage electronics in microwave generators. For this reason, the generators are usually located remotely, in this case in a separate room. The microwave energy is fed to the applicator via waveguide from the generator.

The sausage precooking system is no longer in operation. However, the processor still speaks highly of it and how it supplied the millions of pounds of product that captured the lions' share of the retail market in which they were distributed. After 10 years of trouble-free operation the system was shut down and replaced with a new all-steam and gas grill system. These dominate the production of all types of meat patties in the United States.

III. TEMPERING

Without doubt, this is the single most successful application of microwave heating in industry. There are at least 400 tempering systems operating in the United States alone, primarily for tempering of frozen meat, but also for fish, poultry, vegetables and fruit. In the United Kingdom there are several large systems, up to 200 kW, utilized for tempering of frozen beef, as well as butter. 915 MHz tempering systems, batch and continuous, are sold worldwide, including the United States, Europe, United Kingdom, China, Japan, Korea, and Australia. Section III.S of Chapter 4 includes further discussion on microwave tempering.

A. Meat and Fish

In the early days of microwave processing, the 1950s to the early 1970s, a great deal of effort was expended trying to use microwave heating to thaw frozen foods. A typical system viewed by this author in the early 1960s was at the U.S. government fisheries laboratory in Gloucester, Massachusetts. When fisherman catch fish, the fish must be frozen aboard the boat, unless the boat returns to port the same day. Prior to further processing, the frozen fish must be thawed. In its experiments, the frozen fish blocks were cut to 60 lbs and placed on the entrance conveyor of a microwave tunnel. Ten minutes later a mess of water, ice, frozen fish and cooked fish exited from the other end of the tunnel. This was a perfect illustration of the problem with microwave thawing. The loss factor (ϵ'') of water is approximately 12; while that of ice is approximately 0.003 [18]. The transition from ice to water produces an enormous increase in the loss factor. Ice is almost entirely transparent to microwave energy. However, there is always some unfrozen water present, and that acts to capture the microwave energy, heating preferentially instead of the ice. As this continues, the water gets hotter and melts some ice; by itself, the ice is not heated by the microwaves. This leads not only to uneven heating, but "thermal runaway" causing ever-hotter water, resulting in boiling water in a block of ice.

The above illustrates the problem that has always been encountered in the attempts to microwave thaw products on an industrial scale. However, tempering can be done highly effectively. In tempering, the frozen product is warmed by means of microwaves from hard frozen [<0°C (< −18°F)] to tempered [−5 to −2°C (23–28°F)]. At this temperature, the meat or other food product tissue is softened so that it can be sliced, diced, ground, or otherwise further processed. And this can be done very quickly: 5–10 min for 30–80 lb (15–40 kg) blocks of frozen meat, meat, poultry, fruit, vegetables, etc. and at processing rates up to 10 tons/h. In the conventional methods of meat tempering, the frozen blocks of meat, usually in cartons, are placed in "tempering rooms," which may or may not be climate controlled, and where they may stay for 1–4 days as they temper. During this period they may suffer severe drip loss, i.e., the loss of blood and other high protein containing fluids, which reduces the weight of meat by up to 10% and is a serious economic loss. Also, this leads to unsanitary conditions, and an unreliable and unpredictable source of meat. There is also the problem of inventory control and the cost of inventory. A 4 day tempering process requires tying up the capital, represented by the stored meat, which could be otherwise invested.

The microwave tempering systems reverse these disadvantages and provide more reliable products of higher quality, in only a few minutes. Meat that has been shipped in cartons, usually 60–100 lb (30–60 kg) each, such as raw deboned beef, can be tempered within the carton, thus saving labor and handling. This enables the tempered meat to be easily matched to operations such as patty formation. Today's microwave tempering ovens operate as either batch or continuous systems. They can range in power from 50 to many hundreds of kilowatts. 915 MHz is commonly used because of its longer wavelength and larger depth of penetration. Even when meats and other foods are frozen, there is still considerable free water present, containing dissolved salts and other meat constituents, which affect the penetration depth. The use of the 2,450 MHz frequency has, therefore, been of limited success in this application. This could be a problem for mainland Europe and other countries that do not allow the use of 915 MHz. However, 915 MHz batch systems, which meet the CISPR emission limits are being sold in Europe [40 dB (μv/M) at 30 m].

The fish processing industry has also adopted tempering in a large way. Interestingly, it was the development of the Ross Slicer, which cleanly slices fish blocks at 22–24°C with minimum loss, that led to the growth of this application. Following slicing, the fish sticks or other shapes are battered and breaded; after which they may be frozen or par-fried and then frozen.

Microwave Processes for the Food Industry 311

A typical continuous microwave tempering system, as seen in Fig. 2, is 25–50 ft long, exclusive of feed and discharge conveyors, with the applicator constructed of stainless steel. It may employ four or more 75 kW 915 MHz generators, remotely located from the conveyor. The articulated conveyor itself is made of polypropylene, which is easy to handle and clean. At the entry and exit ends of

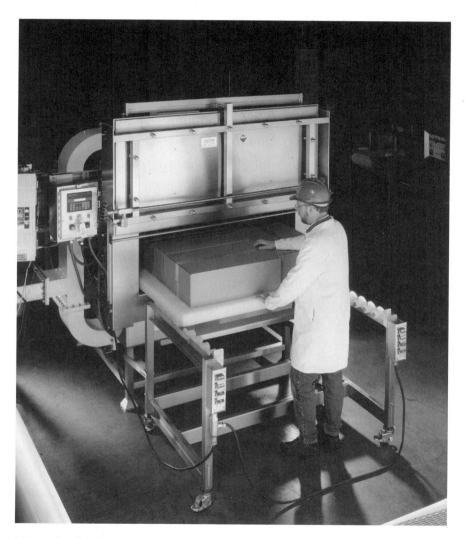

Figure 2 Continuous microwave tempering tunnel. The operator is loading frozen cartons of meat onto the conveyor. (Courtesy of Amana Appliances Commercial Products.)

the applicator there are large leakage suppression tunnels with openings as large as 24 in wide by 12 in high. There is continuous monitoring of the leakage. These systems are housed in refrigerated meat inspection rooms, and meet the USDA requirements for meat fish and poultry plant use. They are on the list of approved USDA equipment.

Modular construction of applicators, similar to that described for bacon cookers; higher power generators; sophisticated control systems utilizing PLC controls; better controlled and constructed conveyor belts; and lower costs of many of the components has made microwave tempering more attractive than ever. The actual cost of tempering has been significantly reduced since its inception. Eves [7] indicated that a tempering cost of 0.25 to 0.5¢/lb. assuming an electricity cost of 10¢/kWh and a 7 year amortization of the capital cost.

B. Butter

There are at least four large (well over 100 kW each) systems operating in England for the tempering of butter. In order to avoid the development of rancidity, bulk butter is stored in a frozen state. It must be reduced in size and formed into the proper shape to prepare the butter for retail sale as wrapped sticks, $\frac{1}{2}$ kilo or 1 lb blocks, etc.; or as patties for use in restaurants. In order to do this efficiently, with the maximum yield, the butter is first tempered, prior to slicing or shaping.

IV. BAKING

If there is any industry in which one would expect the widespread use of microwave processing, it is for baking. In the late 1980s to mid-1970s there were numerous successful bakery plant installations. Today, there continues to be a great deal of interest and some R & D activity. However, there are no known operating microwave systems operating in the baking industry, except for pasteurization and finish drying, which will be discussed below.

The various unit operations involved in baking, especially proofing and baking, lend themselves to microwave processing because the heat transfer problems encountered by conventional means are easily overcome by microwave heating. Baking, in all cases except unleavened products, involves the creation, expansion, and setting of edible foams through the use of heat. Cakes, bread, doughnuts, muffins, biscuits, etc. are all examples of edible foams. As they undergo changes of state from wet flowable batter or dough to semisolid foam structures, they all become better and better thermal insulators. All baked products form some sort of crust which acts as an ablative shield making heat even harder to reach the inside. This is the reason that conventional baking can take up to an hour or more.

The addition of microwaves to baking processes can increase the speed dramatically—3, 4, or more times faster. (For microwaves, foam is not an obstacle. In fact, since air is transparent to microwaves, the foam allows the microwaves to penetrate easily to the center of the product.) When properly applied, usually combined with a conventional heat source, the microwaves seek out the wetter interior, while the external heat bakes the outer layers of the product. When conventional heat and microwaves are used to bake the product, it is easier for microwaves to penetrate and heat the center.

There are two broad classes of products that have been studied for microwave applications: yeast-leavened (or yeast-raised) and chemically leavened products. A third type, steam-leavened products, such as angel food cake, is of lesser interest, although they have been studied. There is also a class of unleavened products such as cookies, matzos, and crackermeal, where there has been some application of microwaves for finish drying. (There seems to be an increasing interest in baking cookies with microwaves as can be seen in the recent patent literature. However, there does not seem to be any commercial activity to date.)

A. Yeast Leavened Products

The prime examples are bread, sweet dough products such as Danish pastry, and yeast raised donuts. These are products that depend upon the metabolism of yeast to create carbon dioxide that causes the expansion of gas cells to create the foam. Proofing is the step of causing the dough to rise and precedes the final baking or frying in the case of donuts. Sometimes proofing is preceded by a step known as "setting a sponge" in which part of the flour is mixed with the yeast, water, and other ingredients and the sponge dough is allowed to stand, usually a slightly elevated temperature, for several hours as the yeast begins its metabolism. Later, the rest of the ingredients are mixed with the sponge and the entire mass proofs. This is a traditional method of making sourdough and many other breads. An alternative to sponge dough is straight dough, in which all the active ingredients are combined at once, including yeast foods, which hasten the metabolism of the yeast. After this the products enter the proofer.

1. Microwave Proofing Yeast-Leavened Products

When products proof they assume their basic final shape, although further expansion will occur during baking or frying. Proofing involves the metabolism of yeast and the release of carbon dioxide, but also the dissolution of dissolved gases, resulting in the formation of gas cells or bubbles. These cause stretching of the flour protein—gluten—that forms the cell walls and acts as the mortar that holds the starch "bricks" in place during baking. Also, the flavors associated with baking occur in this phase. All of this occurs through the application of heat to the dough, usually in a carefully controlled intermediate humidity atmosphere. However, since the thermal death point of yeast is approximately 57°C (135°F), the temperatures in the proofer can only be a few degrees above this, or the surface of the dough will become hard, and dry and crack, which allows gas to escape, resulting in poor loaf volume. As a result, proofing times are usually quite long: 45 min to 2 hs.

An early application of microwaves to the baking process was described by Fetty [19], as part of a proof-baking system. Bread dough was first proofed for approximately half its conventional proof time, and then finished, proofed, and baked within the same microwave applicator. A field kitchen developed for the U.S. Army also utilized a partial proofing in a conventional proofer prior to finish proofing and baking in the microwave thermal oven [20]. In England, Chamberlain [21] reported on microwave bread baking at 896 MHz, using low-protein flour to improve gas-holding capability. A patent by Schiffmann et al. [22] describes microwave proofing and baking of bread in metal pans. This technique utilizes partial proofing in a conventional proofer followed by proofing in a microwave proofer utilizing warm, humidity-controlled air and reduces the proofing time by 30–40%. This was then followed by microwave baking in a separate oven, contrary to the previously described procedures that completed proofing and baking in the same oven.

A highly successful commercial proofing procedure was developed by DCA Food Industries [23]. This process replaced the usual, slow, inefficient batch and continuous conventional proofers with a straight-line microwave conveyor, which reduced the total proofing time from 45 to 4 min! Unique advantages of this proofing system were the excellent process control, uniformity of proofing, and equipment sanitation, matched by excellent product quality, none of which had ever been achieved by conventional means. These systems operated at 2450 MHz and varied in output power from 2.5 to 10 kW for production rates of 400–1500 dozen doughnuts per hour. They could also be used for honey buns and other yeast raised products. At one time, there were more than 24 doughnut proofers using microwaves in operation in the United States, United Kingdom, and Europe. Today, none of these proofers remain, although, once again, this has nothing to do with the performance of the equipment, or the quality of the finished product.

The development of the microwave doughnut proofer is an illustration of the technical problems that may arise when microwaves dramatically shorten the time of a process. By trying to achieve in 4 min that which normally takes 45 min, the researchers discovered that no known dough system could provide an adequately proofed and stable product. The dough might expand under the influence of microwave heating it, but quickly collapsed as it exited the microwave applicator. It took the researchers over 1 year to create a microwave proofer dough mix, in effect creating a whole new science of dough technology. By comparison, the development of the microwave equipment was a trivial problem. It is, perhaps, this situation which has prevented the development and installation of other microwave-proofing systems.

2. Microwave Baking Yeast-Leavened Products

Bread baking by means of microwave energy was first reported in the literature by Fetty [19]. The process was studied at Litton Industries and was referred to as microwave proof baking since it involved both proofing the dough to about 50% of normal volume by conventional means and then finish proofing and baking in one operation in a microwave oven. This work was done at 2450 MHz and involved the use of thermoformed low-density polystyrene trays, since at that time it was felt that metal bake pans were not practical. Only microwave energy was involved so there was no development of the typical browned crust. Therefore, these products were of the brown and serve type.

Decareau and Peterson [24], in an article on potential applications of microwave energy to the baking industry, calculated the energy requirement for bread baking as 150–200 Btu/lb, and showed that microwave baking would require only 200 Btu/lb; however, this did not include browning and crusting. At that time, they used a figure of 500 Btu/lb as a typical amount in conventional bake ovens. Lorenz et al. [25] conducted a large number of experiments on microwave baking using Plexiglas bake pans especially designed for these tests. Since white breads baked in microwave ovens do not develop crust color and for that reason are not very appealing, they concentrated their work on breads baked from relatively dark dough: rye, whole wheat, Boston brown, and high-protein breads. These require very little, if any, additional heating in a conventional oven for browning. Also, the study included brown and serve rolls and cookies that were made from relatively dark dough. The overall conclusion of this work was that no changes in formulation of bread and rolls were required, although fermentation and proofing time needed to be decreased. Cookie dough to be processed by microwave energy needed to be slightly stiffer than those processed conventionally. To assure uniform baking, they should not be refrigerated before baking or contain nuts or raisins since the latter are quite absorptive of microwave energy.

In 1967, Decareau [26] noted the possibility of combining microwave energy and hot air to produce typically brown and crusted loaves of bread in a shorter time than by conventional baking methods. His review included estimates of costs, and suggested that economics of such a process would depend on savings in time, space, and equipment. New plant construction might use only one-third of the space normally required. The microwave proof-baking system became the basis of a mobile field bakery that was being developed by the U.S. Army in the late 1960s [20]. The loaves were baked in a 6 kW 2450 MHz microwave oven and then later browned in a forced convection oven. A thermal microwave oven was later developed in the program including controlled oven temperature and microwave power. Proofing, baking, and browning were accomplished simultaneously. This system included the use of metal bake pans. During this time a great deal of effort was being placed upon the potential application of microwaves for baking bread in the United Kingdom.

Chamberlain [21] reported on work being done at the Flour Milling and Baking Research Association in order to allow the utilization of British wheat. This wheat, which is low in protein, also has high amylase content and is not satisfactory for bread baking. Dough made from such flour is too permeable to gas and gives poor volume, among other deficiencies. The activity of amylase has its maximum rate between the temperatures at which starch begins to gelatinize, (about 55°C) and the temperature at which the alpha amylase is finally inactivated (about 90°C) in the conventional hot air oven, it takes approximately 10 min for a 2 lb bread loaf to traverse that temperature range and a great deal of starch breakdowns occurs in that period. The use of microwave energy at 896 MHz allowed the product movement through this temperature zone to be made in much shorter time, and the results were very encouraging, leading to bread which was in many ways reasonably compared to bread baked conventionally, with non-British flours. In order to accomplish this work, the researchers employed special pans made of phenolic resin-bonded asbestos, called Durestos, which is both heat stable and microwave transparent. Again, these researchers avoided the use of metal bake pans.

A major advance in the baking of bread is described in the series of patents by Schiffmann et al. [22, 27, 28]. These patents describe procedures for the baking of bread utilizing metal pans and, in some cases, also provided for partial proofing of the bread in the pans. This was done for both conventional white bread and firm white bread. The conventional baking of soft white bread normally takes 18–22 min in a hot air oven. In the procedure described in the aforementioned patents, this time could be reduced to 6–8 min, producing thoroughly browned and crusted loaves of comparable volume, grain structure, and organoleptic properties. The process involved the simultaneous application of microwave energy and hot air to both bake and brown the bread. The use of metal in microwaves has always been controversial; however, there is no reason for not using metal if the system is property designed. The large opening of the metal pans allows good penetration of the microwave energy, while the dough near the walls, where there is little or no microwaves can successfully bake by the hot air. It was found that the use of either 915 MHz or combinations of 915 and 2450 MHz were quite effective in baking a loaf of bread.

The baking of a firm white bread is described as a two-step procedure [27] in which the first stage employs conventional ovens with the dough in standard, lidded baking pans, but for a substantially reduced time. Oven spring and set are accomplished with the dough being confined at the top by the lid on the pans. At this point, the lids are removed and baking is finished during second stage with simultaneous application of conventional heat and microwave energy. The entire process requires only half the conventional bake time for firm bread, which is usually one hour, thereby allowing a doubling of throughput.

Neither of these procedures ever became commercial, although economic forecasts for the process indicated a substantial return on investment. Whereas the calculated energy requirement to bake a pound loaf of bread is approximately 250 Btu/lb (580 kJ/kg), Schiffmann and coworkers estimated that microwave process would require 400 Btu/lb (930 kJ/kg). The actual microwave energy requirement was very low on the order of 75 Btu/lb (170 kJ/kg). (Note that most bread loaves are 1 lb (454 g) in the United States.) At the same time, a study of the actual energy requirement in commercial baking ovens for the same bread was found to be 700–1400 Btu/lb (160 to 3250 kJ/kg), depending upon the particular bakery. Hence, the energy savings would have been substantial if microwave baking was implemented.

One surprising obstacle to the adoption of this unique system was the increased throughput, which would have required the purchase not only of a new microwave/hot air baking oven, but the mixers, the molder dividers, proofers, coolers, packaging lines, etc. more than doubling the capital requirements. It should also be noted that it would be difficult to retrofit an existing baking oven with a microwave system primarily because of problems of microwave leakage. Today no microwave bread baking systems are in operation, to this author's knowledge.

3. Brown-and-Serve Products

The work described earlier by Fetty [19], Lorenz et al. [25], and Decareau [26] indicated the potential for microwave baking of brown-and-serve products. A patent by Schiffmann et al. [29] describes a procedure for brown-and-serve baking utilizing microwave energy. The advantage of these systems would be to allow the baking of the product in the container in which the product may be finished baked in the home, and which may then also serve as a serving container. This process may also include overwrapping so as to prevent surface contamination by mold and therefore extend shelf life. Numerous successful tests at 915 and 2450 MHz have been described for use in these procedures, although no successful commercial applications are known.

B. Chemically Leavened Products

These products depend upon the interaction of baking soda, sodium bicarbonate, with a leavening acid such as tartaric acid, or various compounds of phosphoric acid (sodium acid phosphates, sodium aluminum phosphate, etc.). Often, two or more leavening acids are used in dry mixes for cakes or doughnuts, each dissolving at different rates, in order to react at different times with the baking soda. For example, a fast dissolving "bowl" leavener may be used to create gas cell nuclei when water is added to the dry mix and mixing of the batter begins. One or more slower reacting leavening acids may act later in the mixing cycle or when the batter is exposed to heat in an oven or deep fat fryer. All these serve to controllably form the gas cells in what is effectively a gas emulsion in the batter—a foam.

Baking or frying of a batter is an extremely complex system analogous to blowing up a balloon while creating the walls of the balloon. Blow it up too quickly and the walls won't be strong enough and the balloon will pop. Blow it up too slowly and the walls will be too rigid and the balloon won't be big enough. In other words, the expansion of the cell walls must occur while the walls are elastic but strong and then the increasing gas pressure must cease while the cell walls set. Baking is a complex physicochemical reaction in which all the events must be carefully timed and must occur in a well-defined sequence.

The addition of microwaves to the baking or frying of batters has great potential applicability, since we are again dealing with ever expanding foams. However, the microwave application must be carefully controlled or heating and expansion will occur too quickly, and while the cake may look fully expanded and baked, it will collapse to a pancake when the microwave energy is removed.

Another thing that distinguishes yeast-raised dough from chemically leavened batter is that the former has a well-defined structure and shape prior to the final heat setting treatment: baking or frying. The latter is flowable and amorphous in shape and, therefore, requires some sort of shape defining structure to be present—such as a cake pan, or the rapidly formed crust of a donut. In nearly every case, some auxiliary heat form is required along with the microwave energy—a hot air oven for cakes, or deep fat fryer for donuts.

1. Cake Baking

Although numerous studies have been done on the baking of cakes in microwave ovens, both with and without auxiliary heat [30–32], it appears that no industrial system exists. The reasons for this are not clear. The economic advantage for such a system may not be attractive enough. Companies are reluctant to change their conventional baking procedures without a clear reason, especially in terms of return on investment. What might perhaps be of commercial interest would be specialty cakes baked within their serving containers, perhaps in a unique form, with toppings and fillings, to be eaten directly from the container. Such products have been produced on a laboratory scale, but have not reached the marketplace.

2. Doughnut Processing

One microwave baking process that was quite successful for several years was the microwave frying of doughnuts [23]. Once again, the unique interior heating properties of microwaves were able to overcome the otherwise difficult heat transfer conditions normally found in frying doughnuts. During the frying of doughnuts, the dough rapidly expands, forming an open-celled structure with an ablative crust on its surface, thereby restricting heat transfer from the hot fat to the interior of the doughnut. Doughnuts are normally fried first on one side and then flipped over to be finish fried on the second side. After turning, the remaining unexpanded dough is now captured in an immovable crust and is unable to expand further, thereby forming a dense area within the doughnut, commonly called a core. This dense mass usually represents two-thirds of the weight and only one-third of the volume and is that portion most likely to stale first. By applying microwave energy on the first side of frying, the doughnut is allowed to expand to almost its full volume before being turned, at which point a brown crust is formed, but no further expansion is necessary. Microwave doughnuts, therefore, reach much larger volumes, in shorter frying time. Frying times of approximately two-thirds normal time are possible with 20% larger volumes, or 20% less doughnut mix required for standard volume. Fat absorption can be 25% lower than conventional. These doughnuts have longer shelf life, better sugar stability and excellent eating quality. The larger volume and lower fat absorption provided high profits for the bakery, due to mix and fat reductions. This system was developed at DCA Food Industries in the late 1960s and at least 12 commercial microwave doughnut fryers were installed. These fryers were successful for quite some time during the 1970s. However, after several years they disappeared from the scene. The reasons are quite complex, have primarily to do with the internal operations of the manufacturer, and have little or nothing to do with their performance or the quality of the doughnuts. The financial benefits to the baker were clearly evident and consumer satisfaction was high, so their disappearance is rather surprising. It is also interesting that this process was developed by a bakery equipment/mix manufacturing company to satisfy the needs of its customers, rather than originating from a microwave engineering company. A lesson to be learned from this is that it is important that the actual user of the equipment becomes intimately involved in its development in microwave processing.

In conclusion, one can only wonder why the seemingly ideal application of microwave heating—baking of edible foams—has not been more successful. Probably because the baking industry is extremely slow to adopt new technology. There is an "If it ain't broke—don't fix it" attitude. Also, baking ovens are expensive and represent major capital investments. Since it is almost impossible to retrofit an existing baking oven with microwaves, it is only possible to install a microwave baking oven or proofer when a new line is being installed. At that time, the purchaser will buy what he or she is most comfortable with—and that is rarely a microwave system, except for bacon precooking or tempering.

V. DRYING

A. Background and Advantages

Microwaves and RF lend themselves well to speeding up almost any drying process in which the liquid being evaporated is neither explosive nor flammable. (The latter can be accomplished with microwaves under very special conditions such as microwave-vacuum drying in which the exclusion of air prevents otherwise catastrophic effects. This is done routinely in the pharmaceutical industry for the microwave-vacuum drying of tablet granulations.) The great advantage of microwave drying is speed, often allowing the drying of a material in 10% or less of the normal drying time. However, in no application, other than laboratory analytical drying systems, are microwaves used to dry a product alone. Always there is the use of additional heat—hot air, ambient or forced circulation; infrared; or some combination of these. In fact, microwave heating, properly applied, usually represents a minor part of the total heat energy required for drying, the reason being cost. Section IV.G of Chapter 4 discusses the heat and mass transfer associated with microwave drying.

Drying with microwaves and RF is radically different from any other conventional drying means. When drying conventionally, with hot air or infrared, the speed of drying is limited by the rate at which water or another solvent diffuses from the interior to the surface from which it is evaporated. The diffusion occurs by capillarity and the longer or more difficult the diffusion path, the slower the drying. Increasing the ambient temperature, thereby evaporating surface water faster, can sometimes speed up drying. However, this, too, is limited by the rate at which the interior water can reach the surface. It is usually not a good idea to try to dry too quickly since the surface may overdry, case harden, or crack because the interior water hasn't reached it quickly enough. Also, as drying progresses, the path for diffusion of the interior water becomes longer and more difficult and the drying rate slows dramatically, as seen in Fig. 3 [33]. All drying curves look alike and usually two-thirds or more time is required to remove the last one third or less of the water.

Microwave drying employs a completely different mechanism. Because of the internal heat generated by the microwave field, there is an internal pressure gradient, which effectively pumps water to the surface. When judiciously applied, this water is easily removed by hot air or infrared heating. In fact, the hot air now serves a different purpose—to transport the water out of the dryer—so it can be maintained at a much lower temperature, say 121–150°C (250–300°F) instead of 175–230°C (350–450°F), just above the dew point temperature. High-velocity air movement is usually helpful in this regard.

The usual means of applying microwaves to a drying process is at the end of the falling rate period (Fig. 4), in which case this is referred to as finish drying. A good example of this is the finish drying of cookies and biscuits. Microwaves

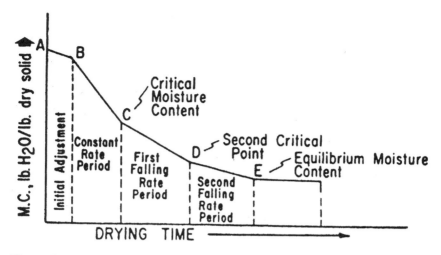

Figure 3 Typical moisture drying curve, showing moisture content vs. time in the dryer, and the various stages in the drying process. (From Ref. 33.)

can also be applied throughout the drying process at a low level: "booster drying," constantly pumping water to the surface. This technique is used for the microwave drying of pasta. There is also the possibility of applying microwave heating prior to hot air drying, thereby preheating the product to the drying temperature. This was tried with cake batter many years ago. However, the only commercial appli-

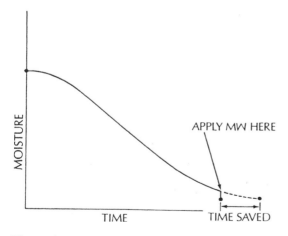

Figure 4 Microwave finish drying. Conventional hot air drying is employed first and microwave energy is added near the end of the falling rate period to rapidly remove the last few percent of moisture.

cation of which the author is aware is the preheating of rubber blanks prior to vulcanizing of tires.

The advantages of microwave drying are listed below:

1. *Efficiency:* In most cases the energy couples into the solvent, not the substrate.
2. *Nondestructive:* Drying can be done at low ambient temperatures; no need to maintain high surface temperatures. This leads to lower thermal profiles.
3. *Reduction of migration:* Solvent often mobilized as a vapor thereby not transporting other materials to the surface.
4. *Leveling effects:* Coupling tends toward wetter areas.
5. *Speed:* Drying times can be shortened by 50% and more.
6. *Uniformity of drying:* By a combination of more uniform thermal profiles and leveling.
7. *Convevorized systems, less floor space, reduced handling:* No need for batch processing in most cases.
8. *Product improvement in some cases:* Eliminates case hardening, internal stresses, etc.

B. Microwave Drying Potato Chips

The first great industrial microwave heating success was a process employing microwaves for finish drying of potato chips. This process, started in the 1960s, led, at one time, to more than 100 operational microwave systems, only to disappear totally by the mid-1970s. One of the most instructive papers ever published in this field is by John O'Meara, titled, "Why Did They Fail?" [34], which examines the reasons for the initial, enthusiastic adoption of this process and its later failure.

The finish drying of potato chips with microwaves satisfied important process requirements. Potato chips are a large part of the multibillion dollar snack food market. Their appearance, especially color, is extremely important to the consumer. When a potato chip is fried it dehydrates and develops its characteristic light brown color. The color is a result of Maillard reaction browning—the interaction of amino acids with reducing sugars. However, the sugar content of potatoes can be very variable, especially as a result of the storage conditions.

In the production of potato chips it is essential that the final moisture content of the chips be no more than 2% or they will become moldy in the bag; and crispness is dependent on extremely low moisture content and water activity ($A_w = 0.1$). However, due to high sugar content of potatoes at that time, frying to such a low moisture content was likely to cause overbrowning or burning of the chips. The basic concept of the microwave potato chip dryer was to fry the chips to a desired color, at which point they might have a moisture content of 6–8%, remove them from the fryer, and finish dry them with microwaves to the desired 2% or lower moisture content. While the all-fried chip operation required costly varieties of potatoes that had not developed too much sugar in storage, the microwave finish drying allowed the use of cheaper, higher-sugar potatoes. Storage of potatoes could be done at lower temperatures, which, while causing higher-sugar content, reduced weight loss and avoided sprouting of potatoes, thereby giving higher yield.

The microwave process was based upon the patent of Lipoma and Watkins [35]. By 1967 over 800 kW of installed microwave power was being used to finish dry potato chips [36]. All of the equipment for finish drying of chips operated at 915MHz (or 896 MHz in England). The U.S. equipment was manufactured by Cryodry (now Microdry), and used two 25 kW magnetrons coupled to a multimode resonant conveyorized cavity, handling approximately 1500 lb of chips per hour. The British equipment used three 25 kW magnetrons with a split-folded waveguide applicator [37]. Larger systems were eventually developed using up to 300 kW of microwave power.

Eventually, however, all these systems (more than 100) were shut down and replaced with improved fryers, many of much higher capacity. The reasons for failure included technical problems combined with major changes in the potato chip industry. On the technical side were unexpected equipment problems: the low dielectric loss of nearly dry potato chips caused difficulties in achieving the desired moisture contents, and differences in drying rates of supposedly similar potatoes. Fires due to arcing were not uncommon and required the installation of fire detection and control systems. (A similar problem exists with microwave bacon cookers). On the other hand, the industry was changing—large cooperatives and larger manufacturers gained better control of the chipping potato market and this, combined with improvements in storage techniques reduced and practically eliminated the problem of high-sugar potatoes. Also, fryer technology and construction improved dramatically. The end result was the elimination of the need for microwave process [33].

In the past several years, a new application appeared for microwave drying of potato chips. This was used to produce nonfat and low-fat nonfried potato chips. This patented process [38] was proprietary to a company, Louises' Inc. The sliced potatoes were conveyed through a hot air predryer, then through a "high-intensity microwave field" that rapidly converted the potato moisture to steam, causing the slice to puff and the surface roughened so it resembled that of a traditional potato chip. Then the surface was coated with flavoring and a small amount of oil and finally through a second microwave applicator (low intensity), where they are dried to the final desired moist content and crispness. (Note that specific information on power levels and applicator design is not available.) While 100 or more processing units were claimed, it seems that only up to about half that were actually in operation at any time. It appears that all of these have been shut down due to bankruptcy of the company. In retrospect, one wonders how such a system could have been successful. Microwaves are an extremely expensive way to evaporate water by comparison to frying, high-velocity hot air, or infrared. While there was a demand for reduced fat potato chips the processing cost had to be high, thereby making this an expensive product. Also, the market for low fat products has seriously declined.

C. Pasta Drying

During the 1970s and 1980s, 25 or more microwave pasta dryers were installed in the United States. The process was developed by Cryodry, and is limited to "short goods" (macaroni), as opposed to "long goods" (spaghetti). Pasta is extremely difficult to dry because it is so dense that moisture moves slowly to the surface. The drying temperature must be kept low—approximately 35°C (95°F) in order to avoid case hardening of the surface. If case hardening occurs, the evaporating moisture may cause the surface to split, in which case the starch exuded during cooking will cause the pasta to stick together. As a result, the conventional drying times are long: 8–12 h. Also, the warm, moist conditions are an ideal breeding ground for microbial growth, especially *Salmonella*—a serious problem in the production of pasta containing egg as an ingredient.

The microwave process has three stages:

Stage 1: Hot air predrying at 82°C (180°F) for 35 min,
Stage 2: Microwaves and hot air at 82°C (180°F) for 12 min,
Stage 3: A final equalizing stage in which there is no heat or air movement

The wet pasta enters the first stage at 30% moisture; enters stage 2 at 18% moisture and drops to 13–13.5%; and exits at approximately 12% moisture. An important part of the process is humidity control—low: 15–20% in the first two drying stages and high: 70–80%—in the final equilibration stage.

The total microwave process takes 1 1\2 h compared to the 8–12 h conventional processes. There are major floor space savings, improved sanitation, product quality, and less energy requirement. The equipment uses 915 MHz, through large multimode resonant cavities. Unfortunately, this is another example of a microwave process that is falling into disuse. There are few, if any microwave pasta dryers still in operation.

D. Snack Drying

At least one large snack product microwave dryer is in operation in the United States. It is a finish dryer utilizing microwaves in combination with high velocity hot air. Figure 5 is a photograph of an operational snack food finish dryer. The operation of this system is proprietary; however, it utilizes many state-of-the-art sys-

Figure 5 A microwave finish dryer for snacks. It controls the final moisture in the snacks. Note the open choke assembly, which makes cleaning easier. (Courtesy Cober Electronics).

tems, including PLC control, internal fire detection and extinguishing systems; in-place cleaning; easy access chokes (as shown in the figure); and more.

E. Finish Drying Biscuits and Crackers

This is a process used extensively in the cookie, cracker, and biscuit industries. It is not unusual to see ten or more of these dryers in operation in a bakery. However, they are all RF systems rather than microwaves, although they finish dry equally well with the latter. The reason is that the conveyors tend to be quite wide—1 or more meters—and it is much easier to achieve uniform drying over this width with RF. What is truly impressive is that the baking oven might be 20 or more meters long, but the finish dryer is only 2 or 3 m long.

F. Cereal Cooking and Drying

At least one large continuous system is currently being used for the cooking and finish drying of breakfast cereal. It is installed overseas and is proprietary. It replaces a difficult to control and labor intensive batch system with a state-of-the-art microwave continuous system, with a production rate of nearly 1 ton/h.

VI. PASTEURIZATION AND STERILIZATION

From the viewpoint of technical ability, using microwaves to pasteurize and sterilize foods seems to be an excellent opportunity. Yet, in the United States, there have been few applications. There are, currently, perhaps two or three pasteurizers operating on a commercial scale and no known food sterilizers, in the United States. However, there are quite a few pasteurizers in Europe and, perhaps, some sterilizing retorts.

Over the years, microwaves have been tested for the pasteurization and sterilization of foods as diverse as milk and milk products; fruit and vegetable products such as fruit juice, jam, oranges in syrup; solid vegetables and meats; bread; beef; pasta; pâté; and whole meals. Only the latter have had some commercial success, although all of the former were technically successful in achieving the desired reduction in the microorganism population. What prevented commercial success was the poor economic return in the use of microwaves, especially since most of these processes are easily achieved in high-temperature water, steam or pressurized steam. Bacterial destruction associated with microwave heating is discussed in Chapter 5 and food safety–related issues are discussed in Chapter 14.

A. Pasteurization vs. Sterilization

It is important to clearly differentiate between these two processes.

- Pasteurization is a thermal process, which provides a partial sterilization of substances by inactivating pathogenic microorganisms, notably vegetative cells of bacteria, yeast, or molds. The effectiveness with which such organisms are deactivated is a measure of the adequacy of pasteurization. Pasteurization is usually carried out at a temperature of 100°C or lower (39) and the products have to be refrigerated.
- Sterilization processes are designed to inactivate microorganisms or their spores which would grow and cause spoilage or health hazards under the normal conditions of storage. Thermal sterilization is usually done at temperatures in excess of 100°C which means they are usually done under pressure [39].

B. Microwave Pasteurization

As illustrated in Fig. 6, all thermal food pasteurization processes depend on:

- Raising the temperature of the food to a "pasteurization temperature" at which the desired level of pasteurization can be achieved ("come-up time")
- Holding that temperature for the required time to complete the pasteurization ("hold time")
- Cooling the product to room temperature or lower as required for storage and/or shipping ("cool-down time").

Assuming that the microwave process uses the same pasteurization temperature as a conventional process (which is the usual case), the only thing which changes is the come-up time, which can shorten to a small fraction of the time used by the conventional process.

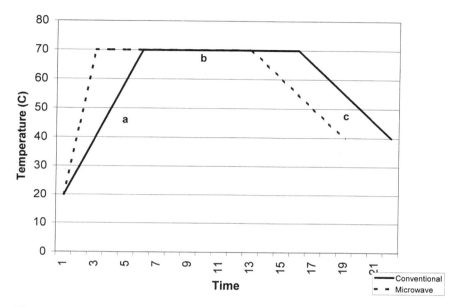

Figure 6 A typical pasteurization curve showing (a) come-up time; (b) hold time; (c) cool-down time. In using microwave heating to pasteurize only the come-up time is shortened.

1. Microwave Pasteurization of Ready Meals

The reduction of process time is not the main impetus for the microwave processing of ready meals. Rather, it is the ability to process the food in hermetically sealed microwavable trays to increase the shelf life of product. These products are typically distributed and sold in the refrigerated state. This is an ideal process for refrigerated entrees or whole meals packaged in sealed plastic or paperboard trays, usually with a see-through plastic film lid. The meals may then be rethermalized in a microwave oven and then eaten directly from the tray, if desired. This process is successful in Europe where refrigerated or "ready" meals are quite popular and of good quality. Since the distribution lines are short in most European countries, the product can be prepared in a plant and delivered by refrigerated truck directly to the supermarket within a day. Such products usually have shelf lives of 3–5 days, after which they are removed from the shelves and discarded. The process is also used to produce meals for airline feeding. One plant located near Schiphol Airport in the Netherlands uses microwave pasteurization to produce the food used on many commercial airplane flights.

While there has been great interest in using this process in the United States, the barriers to its adoption are such that no similar products are in commercial distribution. The two main barriers are:

- *Distribution network:* The United States is geographically very large and most supermarket distribution is via warehouses. Even with several manufacturing plants located strategically around the country, products must have 30–60 day shelf lives in order to match the potential distribution times. It is extremely difficult to provide this long a shelf life in any refrigerated product and still maintain quality as well as food safety.
- *Lack of satisfactory refrigeration throughout the supermarkets, groceries, and other food vendors:* Kraft and General Foods attempted to sell lines of refrigerated products by direct store delivery, bypassing the warehouse distribution network with its long shelf life demands. However, they found that the refrigerated storage in the various stores varied widely in temperature. This resulted in product going bad very quickly in some stores. Attempts at providing separate refrigerated kiosks for their products also failed for numerous reasons, and both companies withdrew their products.

Where microwave pasteurization of meals is done, several unusual demands are made upon the processing equipment:

- All products must enter the microwave pasteurizer at the same temperature. Since they will all be exposed to microwaves for the same length of time, and it is important that they all reach the desired pasteurization temperature at the same time, a common initial temperature is essential. This may entail the use of preconditioning (nonmicrowave) oven to provide the uniform temperature [40].
- In multicomponent meals—meat/starch/vegetable, for example—it may be necessary to expose the various meal components to different amounts of microwave power. This is because the microwave loss characteristic of each meal component is likely to be different and will heat at different rates unless different levels of microwaves are employed. In one operational system, multiple power sources feed multiple waveguide horns located directly above the meal trays as they are conveyed through the applicator. In this way, individual meal components receive the required amount of microwave power to complete the pasteurization. The entire process is computer controlled [41].

After this, the microwave heated meals pass into a nonmicrowave hot air tunnel for the hold-time period, and then to the cooler. Microwaves should not be used to hold a constant temperature—it is an expensive and difficult way to do something that can be better conventionally.

2. Microwave Pasteurization of Fresh Pasta

Pasteurization of fresh pasta (i.e., not dried) in the package has had commercial success in the United States, especially when combined with controlled atmospheres. However, the extent of use is low. It is employed to a greater extent in Europe and Japan.

3. Microwave Pasteurization of Bread

This was in wide use in Germany as a result of the change in food additive laws that forbade the use of preservatives. By using microwave or RF pasteurization, whole loaves of bread could be treated in their sealed plastic sleeves to extend the shelf life without preservatives. However, the high cost of these systems has either forced their shutdown or the bakers themselves have constructed cheaper systems, often using domestic microwave oven power supplies and tubes.

4. Microwave Pasteurization of Brown-and-Serve Products

A microwave process for baking and pasteurizing and, thereby, extending the shelf life of brown and serve products was developed by this author [29]. The normal shelf life of 7–10 days was extended to 21–28 days. The process never became commercial, however.

C. Microwave Sterilization

The sterilization curve for food products is similar to the pasteurization curve, except that the temperatures required are higher and the hold times are usually significantly longer. For example, typical sterilization temperatures in the product may be 121–129°C (250–265°F) with hold times of 20–40 minutes depending upon such factors as container size and shape, food content, and viscosity. These conditions don't change for a microwave sterilizer or retort; only the come-up time may be significantly reduced. It has been the hope of investigators that this reduced come-up time would provide greater product quality. However, studies by several major U.S. food manufacturers were unable to support this claim, or to see enough benefit to warrant the additional expense of the microwave system.

Another problem associated with microwave retorting is meeting the regulatory requirements to insure that the entire contents have been at the sterilization temperature for the required amount of time. This requires sophisticated thermometry utilizing nonperturbing fiberoptic probes, and, since microwaves heat nonuniformly, it has not been possible to meet these requirements to date. Chapter 7 discusses issues related to measurement of temperature during microwave heating of foods. Recent work done at the U.S. government research facility in Natick, Massachusetts, on biological indicators for describing temperature distribution during microwave heating, has some promise [42].

Other problems, such as maintaining package-seal integrity, seem not as large now as they did 10 or more years ago.

One possible area of technology to which microwave processing could be applied is HTST (High temperature, short time) processing. In this case, the sterilization temperature may be at or near 150°C (300°F), in which case the sterilization time is on the order of seconds. Microwaves could be used to rapidly increase temperature, in a minute or less, for the last 50°C, approximately. However, the pressure requirements, approximately 50 lbs/in^2 above atmosphere, make unusual demands upon the process equipment and the package. Temperature uniformity also becomes more problematic. To date, no known successful industrial microwave system utilizing this principle has been installed.

D. In-Pouch Sterilization

At about the same time, Natick Laboratories in the United States and Alfa-Laval AB in Sweden developed systems for sterilizing foods in pouches. The former work was done in an attempt to prepare shelf-stable individual rations for military applications. While such rations had already been developed using foil-laminated pouches, the microwave process was developed as a way of shortening process time and improving quality. Polyethylene-laminated polyester and polypropylene/polyester pouches were used. The pouches were first conventionally preheated to 98°C, then exposed to microwave heating to raise the temperature (in a pressure vessel) to 121°C, followed by a holding period in water at 121°C, to achieve sterilization, and finally rapid cool-down with water. Such products as chicken fricassoe and beef stroganoff were processed. However, negative results with inoculated pouches indicated that uneven microwave heating prevented uniform sterilization in all areas of the packages.

The Multitherm Process of Alfa-Laval AB also microwaved food in pouches that were immersed in a fluid medium having a dielectric constant at least one-half that of the product. Adjusting the temperature of this circulating medium could then control the temperature of the product surface. The work was dropped in the late 1970s due to lack of a suitable packaging material. However, such materials are now available.

VII. FUTURE OF MICROWAVE PROCESSING IN THE FOOD INDUSTRY

For the manufacturers of industrial microwave equipment, the food industry has been a mixed success. Their objective is to manufacture and sell as many microwave systems as possible, for as much profit as possible. The best way to do this is to manufacture many systems of the same kind. While this has happened several times, for an industry that is 50 years old the overall results are disappointing:

Process	Installed systems	Status
Potato chip drying	>100	None operating
Pasta drying	>20	Very few operating
Donut proofing	>20	None operating
Donut frying	5–10	None operating
Meat tempering	>400	Very successful
Bacon cooking	>25	Very successful
Low fat potato chips	>100	None operating

However, single installations of particular applications are more common: sausage cooking, granola cooking and drying, snack food finish drying, etc. There may be hundreds of these isolated systems operating, most under a cloud of secrecy.

A. Barriers to the Successful Adoption of Microwave Processes

These are numerous and sometimes quite subtle. However, a few stand out.

- Manufacturers of microwave processing systems are small companies or small divisions within large companies. Most have annual sales of only a few million dollars. These companies are reluctant to invest their own money to develop new processes. Rather, they rely on users to do so, or focus on one or two well-developed successful processes. This means users must purchase microwave test equipment, devote time, money, and people power to developing the system, often with employees having little or no microwave background and a management reluctant to try something new. On the other hand, the equipment manufacturer may or may not have equipment for testing, but is unlikely to understand the food technology or food science.
- Companies which sponsor or perform microwave process development wish to reap the benefits themselves and are unlikely to spread the process to the outside and competition. In other words, there is a basic conflict the equipment manufacturer would like to sell as many units of the same system to all prospects, while the food company/sponsor wishes to maintain a proprietary hold on the process. So one, two, or three machines may be built and operated, but no more. It is only when the microwave equipment manufacturer invests the money to develop the process, or copies such an in-house developed product that there can be many installed systems. But this, also, requires that the equipment manufacturer build a sales team to sell to the food industry, or form a strategic alliance with a food equipment company or processor who would

benefit from many operational microwave systems. This technique has been successfully employed in the development, sales production, and installation of meat-tempering and bacon-cooking systems.
- The lack of understanding of microwave heating leads potential users to have unrealistic expectations of what can be done. It is not a question of how to do something using microwaves, but does it make sense, from product performance and a return on investment point of view, to use microwaves. This question should be studied in depth before spending a great deal of effort and time. That means that a thorough economic analysis and projection be done as early as possible. In nearly every case, the results will indicate the commercial impracticality of the microwave process. That is because most foods are generally of low intrinsic economic value and cannot carry the extra cost burden put upon them by the microwave process. As an example of this, a recent review of microwave and radio frequency research in *Food Technology* [43] described projects on drying of apples, carrots, potatoes, blueberries, and more, which, while of academic interest, this author believes have no chance of commercial success. It is simply too expensive to dry these items with microwaves, and many less expensive and satisfactory methods exist.

B. Improving the Likelihood of Success

In order to maximize the chances for success and avoid costly failures of microwave processing applications it is important that processors and equipment manufacturers realistically assess the potential applications. Of utmost importance are in-depth examinations with meaningful answers to the following questions [44]:

- *What is the real problem?* This is the most basic of all questions and should be followed by examining why other solutions to the problem have been rejected. The approach should be "Why microwaves?" rather than "Microwaves!" The problem must be clearly be understood and approached.
- *What are the true economics of the situation?* As a rule of thumb, an industrial microwave system usually has an installed cost of $4,000–$7,000/kW, which includes the generator, applicator, conveyor, and control system. (This cost range is for new systems, which often require extensive engineering and my only lead to a single installation. Mature applications such as bacon cooking have much lower costs. This is shown in Fig. 7 [7].) The required amount of microwave power for the process is easily estimated by standard engineering heat balance equations. To illustrate a case involving drying: 1 kW of microwave power

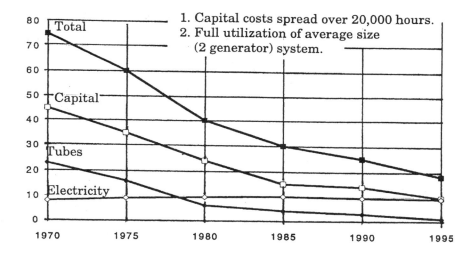

Figure 7 Industrial microwave processing costs in constant 1996 dollars. This shows the decline in capital and operating costs realized as a result of multiple process system installations, and the use of re-built magnetron tubes. (From Ref. 7, courtesy of G. Eves and *Microwave World*.)

will dry $2-2\frac{1}{2}$ lb of water in an hour. So if a drying operation needs to remove 250 lb of water in an hour, it will require about 100 kW at a capital cost between $250,000–$500,000, or more. It is apparent that high capital cost is why this energy source is never used alone to dry huge quantities of product.

- *Is the effect of the microwaves in the system truly unique?* Microwave tempering and bacon cooking are successful because no other energy source can provide all the unique benefits that microwaves can. Also the intrinsic value of meats, fish and bacon are high, and can bear the added cost of the microwave process. But, most often, the proposed systems could be done as well or nearly as well with another cheaper energy source.
- *Would direct and indirect labor changes on a given process be reduced by the addition of microwave power?* With ever-increasing labor costs and need for improved profits, microwaves provide a unique opportunity among other energy producing processes. They may make it possible to significantly increase the throughout of a present process with little or no increase in floor space and a net decrease in labor on a pound of product produced basis. Increases of 50–100% above the present throughput are

within reason. There may be other significant labor savings in cleanup or product handling.
- *What is the best combination of energy sources for the new applications?* Most applications use microwaves in combination with hot air, steam, infrared, or other heating means. (Note that bacon cookers and tempering systems nearly always use only microwaves.) This will usually reduce the cost of the system since these heating forms are usually far less expensive than microwave energy and the combination is often synergistic.

C. Potential Applications for Future Success

The most likely future for microwave food processing is the continued development of unique single systems such as the sausage and breakfast cereal precookers. Very few large multiple-machine markets such as tempering or bacon cooking are likely. Those applications which may offer multiple system opportunities, if the unique combination of users and manufacturers can be brought together, are:

- Baking cakes, breads, and other dough products
- Cooking of meats and poultry
- Pasteurization and retorting of shelf-stable meals

Note, however, that all of the above have been tried, mostly with limited commercial success. This does not mean that they cannot be achieved; rather they have to be done better and/or less expensively, and not necessarily from a microwave point of view.

VIII. CONCLUSION

Having been personally involved in the development, design, and installation of a number of successful microwave systems, I remain a cautious optimist about the future of this field. But it remains a very exciting processing tool, unmatched by any other technology when there is a proper match between microwave heating and an application. If such a match appears likely, the development path to be followed should be as follows:

1. Do an approximate economic analysis of the capital and operating cost of such a system. If it is satisfactory, proceed to the next step.
2. Using a laboratory microwave oven, run a series of tests to produce the desired product. Evaluate combinations of microwaves with other heat sources. Compare the product produced with the conventional product. (Note that the author discourages the use of domestic microwave ovens

for this purpose. They have too much power for most applications, which usually require 50–100 W per unit, not the 800–1000 W available in home ovens, which cannot be successfully reduced in power by ordinary means.)
3. Do a careful energy balance based upon the test results and redo the economic calculations. Take into account all the auxiliary equipment that may be required, such as vacuum, auxiliary heat, conveyors, pre- and post-microwave-processing equipment, etc.
4. Discuss the results, economic analysis, process needs, etc. with manufacturing, marketing, and all pertinent personnel.
5. Contact a manufacturer of microwave equipment to discuss your requirements and extend your test program.

Following this path does not assure success, but it does make it possible to reach a rapid conclusion, with minimal expenditure of funds, whether or not the process idea has a reasonable probability of success.

REFERENCES

1. A. E. Supplee, High Power Microwave Systems, J. Microwave Power, 1(3):89–96 (1966).
2. Industrial Applications of Microwave Energy, Proceedings of the First Varian Symposium on Microwave Energy, July 1968.
3. R. F. Schiffmann, The Industrial Applications of Microwave Energy, Proceedings of the IMPI Short Course for Users of Microwave Power, 74–89 (1970), International Microwave Power Institute, Manassas, VA.
4. R. F. Schiffmann, The Applications of Microwaves to the Food Industry in the United States, J. Microwave Power, 8(2):137–142 (1973).
5. N. Meisel, Microwave Applications to Food Processing and Food Systems in Europe, J. Microwave Power, 8(2):143–148 (1973).
6. T. Suzuki and K. Oshima, Microwave Applications to Food in Japan: Industrial Applications, J. Microwave Power, 8(2):149–159 (1973).
7. G. Eves, Industrial Microwave Food Processing, Microwave World, 17(1):15–18 (1996).
8. M. R. Jeppson, J. Microwave Power, The Evolution of Industrial Microwave Processing in the United States, 3(1):29–38 (1968).
9. B. Krieger, Commercialization: Steps to Successful Applications and Scale-Up, Microwaves: Theory and Applications in Materials Processing. III. Ceramic Transactions Volume 59, The American Ceramic Society, Westerville, OH, 1995, pp. 17–21.
10. P. L. Spencer, Prepared food article and method of preparing, U.S. Patent 2,480,679 (1949).
11. P. L. Spencer, Means for treating foodstuffs, U.S. Patent 2,605,383 (1952).
12. M. R. Jeppson, The Evolution of Industrial Microwave Processing in the United States, J. Microwave Power, 3(1):29–38 (1968).

13. E. Shelton, Devices for the Generation of Microwave Power for Industrial Processing, J. Microwave Power, 1(1–3):21–27 (1966).
14. F. Okada, et al., The Development of a High Power Microwave Circulator for Use in Breaking of Concrete and Rock, J. Microwave Power, 10(2):181–190 (1975).
15. K. N. May, Applications of Microwave Energy in Preparation of Poultry Convenience Foods, J. Microwave Power, 4(2):54–59 (1969).
16. G. Eves, Private communication with the author, 2000.
17. R. V. Decareau, Microwaves in the Food Processing Industry, Academic Press, New York, 1985, pp. 141–142.
18. A. R. Von Hippel, Dielectric Materials and Applications, The M.I.T. Press, Cambridge MA, 1954, p. 361.
19. H. Fetty, Microwave baking of partially baked products. Proceedings of the American Society of Bakery Engineers, Chicago, IL, 1966, pp. 145–166.
20. A. L. Dugan and M. Fox, The SPEED microwave oven in integrated cooking. J. Microwave Power 4(2):44–47 (1969).
21. N. Chamberlain, Microwave energy in the baking of bread, Food Trade Review 43(9):8–12 (1973).
22. R. F. Schiffmann, Microwave proofing and baking of bread utilizing metal pans, U.S. Patent 4,271,203 (1981).
23. R. F. Schiffmann et al., Applications of microwave energy to doughnut production. Food Technology 25:718–722 (1971).
24. R. V. Decareau and R. A. Paterson, Potential applications of microwave energy to the baking industry. J. Microwave Power 3(3):152–157 (1968).
25. K. Lorenz et al., Baking with microwave energy, Food Technology; 12:28–36 (1973).
26. R. V. Decareau, Application of high frequency energy in the baking field, Baker's Dig 41(6):52–54 (1967).
27. R. F. Schiffmann, Method of baking firm bread, U.S. Patent 4,318,931 (1982).
28. R. F. Schiffmann, Microwave baking with metal pans, U.S. Patent 4,388,335 (1983).
29. R. F. Schiffmann, Microwave baking of brown and serve products, U.S. Patent 4,157,403 (1979).
30. B. E. Proctor and S. A. Goldbith, Radar cooking for rapid blanching and its effect on vitamin content, Food Technology 2(2) 95–104 (1948).
31. M. B. Street and H. K. Surratt, The effect of electronic cookery upon the appearance and palatability of yellow cake, J. Home Economics 53(4):285–291 (1961).
32. D. J. Martin and C. C. Tsen, Baking high-ratio white cakes with microwave energy, J. Food Sciences 46:1507–1513 (1981).
33. A. S. Rankell, H. A. Lieberman, and R. F. Schiffmann, Drying, in The Theory and Practice of Industrial Pharmacy, Lea & Feibiger, Philadelphia PA, p. 53 (1976).
34. J. P. O'Meara, Why did they fail? A backward look at microwave applications in the food industry, J. Microwave Power 8(2):167–172 (1973).
35. S. P. Lipoma and H. E. Watkins, Process for making fried chips, U.S. Patent 3,365,301 (1968).
36. J. P. O'Meara, Finish drying of potato chips, in Microwave Power Engineering, vol. 2, Academic Press, New York, 1968, pp. 65–73.
37. R. V. Decareau, Equipment for microwave finish drying, in Microwaves in the Food Processing Industry, Academic Press, New York, 1985, pp. 93–98.

38. D. Gaon and J. Widersatz, Process for preparing fat free snacks, U.S. Patents 5,180,601 (1993) and 5,202,139 (1993).
39. Foods and Food Production Encyclopedia (D.M. Considine, ed.), Van Nostrand Reinhold, New York, 1982, pp. 2008–2009.
40. R. B. Holmgren, High volume line turns out meals with "fresh" look, Packaging, August 1988.
41. W. Schlegel, Commercial pasteurization and sterilization of food products using microwave technology, Food Technology, 46(12):62–63 (1992).
42. A. Prakash, H. J. Kim, and I. Taub, Assessment of microwave sterilization of foods using intrinsic chemical markers, J. Microwave Power 32(1):50–57 (1997).
43. N. H. Mermelstein, Microwave and radio frequency drying, Food Technology, 52(11):84–86 (1998).
44. K. Bedrosian, The necessary ingredients for successful microwave applications in the food industry, J. Microwave Power 8(2):173–178 (1973).

10
Basic Principles for Using a Home Microwave Oven

Carolyn Dodson
Microwave and Food Specialist
Rotonda West, Florida

I. INTRODUCTION

From an engineering standpoint, both microwave heating and cooking of foods are complex. Thus, microwave cooking is certainly complex and a complete engineering description of this process that includes physical, chemical, biological, and sensory aspects is simply not available. As discussed throughout this handbook, cooking depends on such a myriad of oven and food factors. Thus, even after all the discussion of principles in this handbook, there are situations where a final product is more of an art than science. Patterns begin to emerge from persistent testing of actual food products and processes. These can lead to the development of simple, basic microwave techniques, developed from the other end of the spectrum, i.e., final product. This chapter is one person's effort [1] to provide some of these principles obtained from years of practical experience in product development using the home microwave oven. Its purpose is not to explain why things work, but rather "what works." These rules of thumb may not be unique and alternate approaches are possible, as may be seen in Refs. 2–21. The principles and rules-of-thumb presented in this chapter should complement the scientific approaches laid out elsewhere in the book. It is even possible at times that a few of these rules apparently contradict some of the principles laid out elsewhere. This is expected considering the complexity of the cooking process and its incomplete scientific understanding. Perhaps this chapter would also serve to identify some of the gaps in our fundamentally based understanding of the complex process of microwave cooking.

II. COOKING PATTERNS, POWER LEVELS, AND TEMPERATURE CORRELATION

A. Cooking Patterns and Power Levels

Microwave ovens have various cooking patterns, and it is important to determine if there are hot or cold areas which could cause uneven cooking of foods. To help ascertain where these areas are, dampen a paper towel, place it on your turntable or the floor of your oven, and heat it for a minute or two. Hot or cold spots will show up as damp and dry patterns on the towel. Sometimes these areas can actually be used to an advantage by placing certain types of foods within the areas to help cook them faster or slower. For example, a leg of lamb roast will cook more evenly when the small end of the leg is placed in the cooler area, and the larger end of the roast in the warmer area in the oven. Although newer microwaves have fewer hot and cold spots, this can be a problem in many of the older models.

Electrical power in some areas may vary throughout the day. This can explain why food cooked exactly the same way will turn out differently at times, which can be very frustrating to microwave cooks. To help determine the peak time of power usage in a certain area, it is helpful to boil a cup of room-temperature water (75°F) at three separate times throughout the day. It is best to do this test on a weekday when schedules are more similar than on weekends or holidays. At certain times throughout the day, time for the water to reach boiling point may vary as much as 1 or 2 min. Therefore, cooking times may need to be adjusted upward or downward at certain times in order to achieve the same result in the food cooked.

Microwave oven power is measured in watts, which indicates the cooking power of each new oven. Although the wattage information can usually be found in the owner's manual, actual wattage varies depending on the age of the oven and a number of other oven food factors (see Chapter 2). A simple test to determine oven wattage (not to replace the standard procedure for doing so) is as follows. Heat 6 oz of room-temperature water (approximately 75°F) at 100% power. The water should boil in approximately 2 min in a full-size microwave oven (600–750 W). For ovens of higher or lower wattage taking more or less time to boil the water, it will be necessary to lengthen or shorten cooking times as needed as these developed techniques are for full-size ovens.

B. Temperature Correlation with Conventional Heating

Experience has shown that foods cook more consistently by correlating microwave percentages of power to conventional oven temperatures. Table 1 shows this rule of thumb for microwave power levels as they relate to conventional temperatures, for a full-size microwave oven (600–750 W). Higher wattage ovens will need to lower the power or lessen cooking time. For example, 70% power on a mi-

Table 1 Approximate Conversions (Rule of Thumb) for Conventional Cooking Temperatures to Microwave Power Levels for 600–750 W Oven

Microwave power level %	Equivalent setting on conventional oven °F
100–90	425–500 (deep-fat-fry, broil, *high* stovetop)
80	375–425
70	350–375 (*medium-high* stovetop)
60–50	300–350 (*medium* stovetop)
40–30	225–300
20	200–225
10	150–200 (*low* stovetop)

crowave oven having 1000 W or more of power would be equivalent to 100% power on a 600–750 W oven. Use similar techniques as in conventional heating, such as to cover or not to cover, and then correlate microwave power levels to the proper conventional temperatures at which a particular food would cook conventionally.

Refer to power levels as 100% or 70% rather than *medium, low, simmer,* or *bake,* which may not be consistent from manufacturer to manufacturer. For example, one microwave manufacturer may have a *medium* of 50% power while another may have a *medium* of 70% power. By using percentages of power rather than inconsistent terminology, accuracy will be greater when cooking foods in a microwave.

III. WHAT AFFECTS MICROWAVE COOKING TIME?

A. Factors Affecting Microwave Cooking Time

Below is a list of factors affecting cooking times in the microwave:

1. Starting temperature of the food (cold vs. room temperature or warm).
2. Quantity of food (large or small amount). Since small pieces of food cook quicker than larger pieces, items of similar size cook more uniformly. Large foods such as turkeys or roasts, which cannot be stirred, should be turned over "rotisserie style" while cooking. This promotes uniform, even heating in the microwave just as a rotisserie cooks large pieces of meat more evenly on a charcoal grill.
3. Shape of food (thin vs. thick; round vs. irregular).
4. Shape of container in which food is cooked (square vs. round vs. ring).
5. Composition of food (high sugar-fat content heats faster than lower sugar-fat content; higher sugar-fat content heats faster than water).

Fresh vegetables contain more water and cook faster than those that are less fresh.
6. Density of food. Two foods weighing the same with different densities cook differently; a dense or compact roast takes longer to cook than a porous loaf of bread of the same weight.
7. Microwave power being used—power level (e.g., 100% vs. 30%) and oven wattage.
8. Electrical power fluctuation (power supply in your home may vary throughout the day and may affect the microwave cooking times).
9. Preference as to doneness (crisp vegetables vs. soft, tender vegetables).
10. Liquid being added to food (vegetables in sauce vs. vegetables cooked alone). Foods, such as soups, with a high liquid content will take longer to cook than those that cook from their own moisture content. Liquid ingredients slow down cooking. In other words, 1 lb of food with 50% moisture added will take less time to heat to a certain temperature than 1 lb of food with 75% moisture added. If meat is cooked in liquid, cooking time will be longer than meat not cooked in liquid; a potato cooked in water will take longer to cook than one cooked by itself. Also, if meat or vegetables are partially cooked prior to adding the liquid, cooking time will be shorter. Combine partially cooked food with hot liquid, and the cooking time will be shorter still.

B. Six-Minute-per-Pound Rule and Standing Time

With only a few exceptions, most foods (meats, poultry, vegetables, and fruit) will cook to done in a full-size microwave oven (600–750 W) at 100% power in approximately 6 min/lb plus standing time. An exception to this is fish or seafood, which will take approximately 3–4 min/lb. This "six-minute-per-pound rule," of course, will need to be adjusted based on factors discussed in Sec. A above.

Six minutes per pound at 100% power plus standing time will cook most foods to done. About three-quarters of the cooking occurs in the microwave while it is on, the rest of the cooking takes place after the microwave shuts off. Standing time allows "ongoing" cooking to complete, which is critical to good food from the microwave. If food is actually cooked until it is done, it would be overcooked after standing time takes place.

To help food finish cooking properly when removing it from the oven for standing time, cover with the appropriate cover to keep heat in, and according to the desired texture. Use a paper towel or loose fitting cover for a crisper texture, or a tight fitting cover for a moisture texture (see Sec. IV.A).

C. Adjusting for Oven Wattage

Recipes must be adapted according to size and wattage of the oven. Approximate cooking times for various ovens are shown in Table 2 for each minute cooked at 100% power. As most recipe books are written for 700 W ovens, it may be necessary to adjust cooking times by adding or subtracting necessary seconds in Table 2 for each minute of cooking. For example, if the 600–700 W oven takes 1 min to cook, a 500–600 W oven will take 1 min 15 s, and an 800 W or higher oven will take only 45 s. To allow for these differences on higher wattage ovens, you may choose to use 80–90% power rather than 100% power.

However, before adjusting the recipes it is necessary to make sure that the oven is actually producing higher watts and simply not rated by European standards, which give higher wattage numbers without actually producing faster cooking times.

D. Lowering Cooking Temperature

As in conventional cooking, foods require different temperatures to cook properly. To change cooking time when cooking food by the pound to well done in the microwave, increase 1 min/lb (starting at 6 min/lb) for every 10% decrease in the power level for a 600–750 W microwave oven. Ovens with higher wattage (800 W or more) may need less time or a lower power level.

E. Slow Cooking and Dealing with Multiples

Cooking at a lower temperature requires longer time. If adding other ingredients, such as vegetables, part way through the cooking time, the minutes per pound will increase as more volume has been added. See also discussion of meat cooking in Sec. VI. B.

Other foods containing large amounts of liquid, such as soups, stews, and sauces, should be brought to a boil at 100% power. Reduce power to 30–50% and slow-cook for about one-fourth of the conventional time, for example.

Table 2 Approximate Conversions (Rule of Thumb) of Heating Times for Ovens of Various Rated Power

Rated power	Heating time
900 W or higher	30 s or less
750–900 W	45 s or less
600–750 W	1:00 min
500–600 W	1:15 min

- *Spaghetti sauce (conventional directions):* Brown meat, add liquid and simmer for 3–4 hours.
- *Spaghetti sauce (microwave directions):* Brown meat, add liquid and bring to a boil. Reduce heat to 50% power and simmer for 45 min to 1 h.

F. Adjusting Time for Increased Quantity

The six-minute-per-pound rule should apply when cooking multiples of food by the pound. For individual food items, time to cook two pieces of the food is doubled. To increase the items by more than two, only one-half of the cooking time of the first item is recommended for each additional piece. For example, if time for one ear of corn is 3 min, for two ears of corn it will be 6 min, and for three ears of corn it is $7\frac{1}{2}$ min. When using microwave recipes for a casserole, soup, etc., if the quantity is doubled, add one-half of the time of the original amount.

IV. EFFECTS OF CONTAINERS, COVERS, AND SHIELDING

As some cookware is not good for use in the microwave, it is important to test it if uncertain. Simply place the dish to be tested and a small cup of room-temperature water, side by side, in the microwave. Turn the oven on for 1 min at 100% power. If the water gets hot and the dish remains cool, the container is safe to use. However, if the water does not get hot, but the container does, it is absorbing microwave energy and should not be used in the microwave—it could break or alter cooking times. Examples of cooking patterns of various containers are shown in Fig. 1.

A. Containers

Pick a microwave container slightly larger than needed for the conventional recipe. This will allow for expansion of the food, which occurs in some microwave foods, and for stirring. Choose a container about twice the size of the food to be cooked.

Cover the container according to the texture desired (see Sec. B below). If the food is not covered during cooking time, make certain to cover it when removing it from the oven for standing time. This will keep the "heat" inside the food so it will finish cooking properly.

B. Covers

Better results are usually achieved when cooking foods in the microwave if cooking techniques similar to those used in conventional cooking are used. For exam-

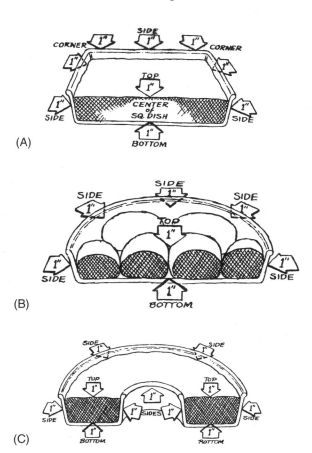

Figure 1 Cooking patterns of different containers. (A) Corners cook four ways, from two sides, top and bottom. Sides cook three ways, from one side, top, and bottom. Center cooks two ways, from top and bottom only. (B) Sides cook three ways, from one side, top, and bottom. Center cooks two ways, from top and bottom. (C) All sides cook three ways, from in/out sides, top and bottom. (Ring-shaped pans work well for cakes and other foods since they allow microwaves to reach to food from the top, bottom, inside, and outside of the ring. If you do not have a ring pan, an empty glass or cup may be placed in the center of a round dish to create a space, thus enabling the microwaves to penetrate all four ways also.)

ple, some foods require covering when cooked conventionally in order to produce a moist product. For this type food, or when converting recipes using terms such as "cover tightly" or "steaming," cover with plastic wrap or a tight-fitting cover.

Foods requiring cooking techniques which will produce neither a moist or crisp food when cooked conventionally, or when converting recipes which use terms such as "crack the lid or partially cover," cover with wax paper or a vented casserole cover when cooking it in the microwave.

Foods which require no cover and become crisp when cooking conventionally, or recipes using terms such as "crisp or uncovered," should be covered with a paper towel to help absorb moisture or left uncovered when cooking them in the microwave.

C. Shielding

Small pieces of foil may be used to cover or "shield" various parts of food to keep them from overcooking. For example, the wings and leg ends of a chicken or turkey may overcook if not protected, or "shielded," for part of the cooking time. By covering corners of square dishes, or by placing a donut-shaped foil ring over cake tops, egg casseroles, or other delicate type foods, overcooking of the corners and edges may be prevented. Never run a microwave oven empty or with too much shielding, as damage could occur to the magnetron tube. If sparks are seen, remove foil from the microwave. There must be enough water content in the food to absorb the microwaves. There can be arcing or sparks if the water content is not great enough. The food mass must be substantially greater than the amount of foil used. When shielding, place foil smoothly over the food it is protecting. Do not allow the foil to touch the sides of the microwave, which are also metal, as this will cause arcing.

V. VARIOUS PROCESSES AT HOME

A. Arranging and Stirring

Most foods cook around the edges before the centers are done with a few exceptions. For example, high sugar centers of jelly doughnuts or rolls will get hot before the actual roll does since microwaves are quickly attracted to the sugar. The egg yolk with high fat content will cook before the white is set. Figure 2 shows some food arrangements for more even cooking without stirring or rearranging. Cooking at a lower power level (50%) will help reduce overheating by allowing heat to distribute more evenly. Stirring, when possible, also distributes the heat more evenly.

B. Browning and Crisping

Meats cooked for a long enough period of time which contain fat will brown when cooked in the microwave as fat comes to the surface (protein plus sugar plus heat = brown) and will caramelize. Crisp crusts, as in deep-fat-fried foods, cannot be achieved in a microwave. Meats with high fat content brown the most. Meats

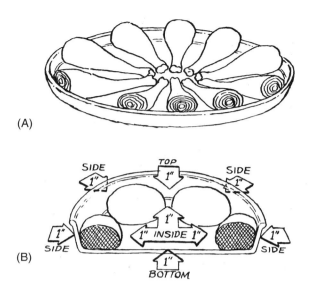

Figure 2 Arrangement of food for more even cooking without stirring on rearranging. (A) Place thicker or more dense food to outside of dish, thinner or more porous portions to the center area of dish (spoke fashion). (B) When cooking similar sized pieces of food such as rolls or boneless chicken breasts, three or more whole foods, such as potatoes or dishes such as custard cups in the microwave oven, arrange them in a circle (circle fashion) leaving about 1–2 inches between each. This pattern promotes even cooking by giving each food maximum exposure to the microwaves. "Circle fashion" allows the microwaves to reach the food from the top, bottom, and inner as well as the outer edge of the circle.

cooked 6 min or less, or with little fat content, such as fish or poultry without skin, will need spices, sauces, or browning products to make them appear "golden brown." Salt can dehydrate and toughen meats, but it will aid in browning as it attracts microwaves, which in some cases may be desired. Toasting (e.g., shredded coconut, almond) can be done by spreading thin on a glass plate and heating uncovered.

C. Defrosting

Proper time for defrosting should be dependent on the type and size of the food. Always cover while defrosting. The larger the food to be defrosted, the lower power level should be to prevent cooking. A very porous type food will defrost more easily than a dense one. When time allows, the following chart should be helpful for defrosting.

1 in thick	6 min/lb at 50% power
1–2 in thick	7 min/lb at 40% power
2–3 in thick	8 min/lb at 30% power
3–4 in thick	9 min/lb at 20% power
4 in and more	10 min/lb at 10–20% power

Defrosting frozen casseroles take approximately 15 min/qt at 30% power. Defrost before cooking for better results. Defrost breads at 30% power. Wrapping in paper towels prevents sogginess. Whenever possible, place bread on a microwave rack to allow air to circulate. To prevent food from cooking during the defrosting process, shield vulnerable areas. Lower power levels or less time may be needed for higher wattage ovens.

D. Reheating

Reheating at 100% power causes foods to heat unevenly. Reduced power levels produce better result: 70% power is good for soups, casseroles, meats, and plates of leftovers; 50% power is better for large or layered casseroles.

Choose 50%–70% power according to food type. Ovens having wattage over 750 W may take even lower power settings or less time for reheating. Arrange harder-to-heat portions to outside of plate and porous foods, which heat more easily, to the inside or center of the plate. Proper arrangement makes it unnecessary to rearrange or stir. Cover according to the texture desired. For example, moist food = plastic wrap or tight-fitting cover; dry or crusty texture = paper towel. A few frozen foods, such as pasta and soup, may be reheated without defrosting. Stir as they defrost and heat.

Vegetables and fruits for freezing may be blanched in the microwave, and thus retain more nutrients. Prepare vegetables properly by chopping, slicing, etc. Determine cooking time by using one-half of the *six-minute-per-pound rule,* or 3 min/lb. Place small amounts into microwave casserole and cook, covered, at 100% power, stirring or rearranging halfway through. Plunge immediately into ice cold water to cool. Drain, pack, and freeze.

Violent eruptions may occur when another agent such coffee, rice, etc. is placed into the microwave-heated liquid (water, broth, juice). (See superheating in Chapter 4.) To avoid this, stir while heating or before particles are added. You may also let liquid stand a minute or two in the oven after heating.

E. Softening and Refreshing

Stale rolls, cookies, chips, and crackers can be freshened by heating for a few seconds at 100% power. Microwaving frozen ice cream for 30 s at 100% power makes it easier to scoop. Tortillas can be softened by microwaving at 100% power for 20–40 s. Soften butter (1 stick) from the refrigerator by heating for 20–30 s at

100% power. To melt butter, heat for approximately 1 min at 100% power. To microwave chocolate, do not cover chocolate as it will stiffen and moisture will develop. Heat at 50% power, stirring half-way through. It takes about 2 min at this power to melt 3–4 oz of chocolate. Chocolate may retain its shape, but will change from dull to glossy and will liquefy when stirred. Times and power levels may vary for a higher wattage oven.

VI. COOKING VARIOUS FOOD TYPES

A. Vegetables

Microwave cooking preserves many nutrients lost in various conventional cooking methods. Cook vegetables at 100% power, covered, and use the *six-minute-per-pound rule* as the base for figuring cooking time. Also refer to Sec. III with regard to *what affects microwave cooking time,* e.g., oven wattage, personal taste, and size. As vegetables consist mainly of water, we should add no additional liquid to most vegetables other than that remaining from washing them. This will only slow down cooking time.

Fresh vegetables contain more water and cook faster than those that are less fresh. The shape and size of vegetables make cooking times vary. Since small pieces of food cook faster than large, vegetables of similar size cook more uniformly. Cut those that cook more slowly (dense with less water content) such as carrots into small uniform pieces. Cut foods requiring less cooking time (porous with higher water content) such as mushrooms into larger uniform pieces. Place vegetables together in a ring- or donut-shaped configuration, or arrange properly in a dish with the dense, hardest to cook toward the edge of the dish and the larger, easier to cook vegetables closer to the center of the dish. They will cook properly without stirring or over or undercooking anything.

B. Meats

As a general rule, the conventional recipe can be followed after converting times and temperatures to microwave power levels (see Sec. II.B). Irregular shapes and sizes should be avoided if possible.

1. Tender Cuts

Example of tender cuts are tenderloin or sirloin. Higher temperatures are used for tender cuts: cook large pieces at 70% power; cook small pieces at 100% power. When cooking in ovens with wattage higher than 750 W, lower power levels may be needed. No cover is required. Wax paper, a casserole cover, or paper towel may be used to help retain some moisture, or to help produce more uniform heat on larger cuts, and to prevent splatters. Use methods the same as those used when cooking conventionally. For example, in conventional cooking, a standing rib roast is cooked on a rack, uncovered, in a hot oven. In the microwave, cook on a rack when possible, at a higher power level. Cook uncovered, or covered with wax paper to prevent splatters and for heat retention.

2. Less Tender Cuts

Examples of less tender cuts are chuck roasts or brisket. Lower temperatures are used for less tender cuts: Cook at 30–50% power. Muscular roasts depend on a slow cooking process and liquid (for moist cooking) to tenderize connective tissue. Cook covered or in cooking bag. Turn large cuts (rotisserie style) at least once during the cooking period. Methods are the same as those used when cooking conventionally. For example, ask yourself, "When cooking conventionally, do I cover?" "Add liquid?" "Use high heat or low?" When this is determined, use the same method in the microwave.

3. Various Degrees of Doneness

Various stages of doneness can be achieved for meats, such as tender or semitender roasts, steaks, and chops. Timings for cooking are discussed here in reference to using a 600–750 W oven. At 100% power, "well done" can be achieved approximately at 6 min/lb, "medium" at 5 min/lb, and "rare" at 4 min/lb. A lower power setting or fewer minutes per pound may be needed for higher wattage microwave ovens to achieve the same result. High heat or power levels do not allow adequate time for the development of tenderness and flavor in many beef cuts.

To achieve various degrees of doneness while cooking at a lower power level, use the information for lowering temperatures in Sec. II.B. When meat or poultry is not cooked in liquid, 20–25% of the total cooking takes place after the oven shuts off (standing time; see Sec. III.B). Temperatures will rise a few degrees during standing time. Therefore, undercook slightly to achieve the desired final temperature. Let food remain 20–25% longer in warm oven or cover meat when removed to keep the heat inside. This allows the continuation of the cooking and helps distribute heat evenly. Meats with natural fats will brown if cooked longer than 10 min.

VII. CONVERTING DIRECTIONS FROM CONVENTIONAL HEATING

A. General Principles for Conversion

Recipes that call for "moist cooking," "stirring," "covering," "steaming," "sauces," and other such terms will convert well from conventional to microwave cooking. Stay away from those that suggest "crusty," "golden brown," etc. They may not convert well. Almost 85% of conventional recipes convert well, with many actually having a superior quality to conventionally cooked foods.

Smaller quantities are what save time in the microwave. Larger quantities will lose the time saving advantage, but may save electrical energy costs and cleanup time. As a rule of thumb, cooking time for microwaves should be reduced to one-fourth of the conventional cooking time using similar temperatures and techniques used. Eighty percent of conventional recipes convert successfully by this method. More cooking time will be required for the remaining 20%. For these, add cooking time and check product every 30 s or so until done. Always undercook when in doubt and add more time if needed. For example, if a conventional recipe cooks for one hour at 350° covered tightly, the converted microwave recipe would cook for 15 min at 70% power, covered with plastic wrap or a tightly fitting casserole cover. Food would then be allowed to "stand" a few minutes to complete cooking. Standing time of 20–25% of the total cooking time is a good rule of thumb. Take advantage of microwave cooking patterns by arranging food properly (see Sec. V.A). Rearranging and stirring will not usually be necessary. Refer to Sec. II.B for correlation of conventional temperatures to microwave power, and Sec. IV for containers and covers.

B. Adjustments for Ingredients

Reduce the "least rich" liquid ingredient in the recipe by 20–25%. Without dry heat, it will not evaporate. For example, a packaged cake mix calls for eggs, oil, and water. Water is the "least rich" of the liquid ingredients, so use 3\4 cup water instead of the 1 cup of water called for.

Cut back on sauces, spices, and herbs by 20–25%. Stronger flavors usually result in microwave cooked foods than in conventionally cooked foods, which are normally exposed for longer cooking times at more intense heat.

When converting conventional recipes, increase the leavening, such as baking powder or baking soda, by 10%. Let batter stand for a minute or two before baking. As the microwave oven cooks much more quickly than a conventional oven, this will allow the chemical reaction of the leavening to begin working so food will derive its full benefit.

Fat and oil may be reduced in most recipes, other than cakes and baked goods, by 10% or more. Adding fat causes more intense heat and can affect cooking time. Fat should be trimmed from meat, as it will heat faster with the microwaves. Eliminate butter and other oils used to fry or sauté when cooking vegetables, fruits, main dishes, and soups in the microwave. They are not needed. Instead, use a small amount to flavor only.

Water, sugar, and fat content within each food cause differences in temperature rise. Water can heat up to 212°F, sugar to about 260°F, and fat to over 300°F. The rate of temperature rise is also different, with fat heating the fastest.

Sauces, which are made with cornstarch, require less stirring and thicken more rapidly than those made with flour. It is not essential for adding air or "volume" to foods to be cooked in the microwave. Foods such as eggs and cakes will have approximately one-third more volume or "fluffiness," as no "dry crust" forms to hold down the volume as with conventionally cooked foods. Add cheese and other toppings toward the end of the cooking time so cheese and toppings will not become tough or soggy. Avoid flour coatings on meat. Flour will not brown and will get soggy when adding liquid. For pies, cook the crust before filling. Crusts will not become soggy during cooking.

See Sec. III.A for factors that lessen and lengthen cooking times. When cooking converted conventional recipes in the microwave, use techniques similar to those used for conventional cooking. People unfamiliar with conventional cooking techniques should refer to a similar food product or recipe in a conventional cookbook for directions. Simply ask yourself, "Do I cook this recipe on high heat or low heat?" Then select the suitable power level for the type of food being cooked (see Sec. II.B). Continue to ask yourself questions and compare the methods with conventional cooking methods. "Do I cook it covered or uncovered?" Use the same technique in the microwave as you would conventionally.

REFERENCES

1. C. Dodson. Definitive Microwave Cookery II. Magni Co., McKinney, TX, 1995.
2. The Microwave Cookbook. General Electric, 1998.
3. S. Brown. Cooking with a Microwave. Dorling Kindersley, London, 1995.
4. C. Bowen. The Microwave Kitchen Bible: A Complete Guide to Getting the Best Out of Your Microwave with Over 160 Recipes. Southwater, London. 2000.
5. Microwave Cookery. Dorling Kindersley, London, 1999.
6. Successful Microwave Recipes. Murdoch Books, Sydney, Australia, 1996.
7. Microwave Food Safety. The Corporation, Syndey, Australia, 1996.
8. A. Yates. The Combination Microwave Cook. Elliot Right Way, Tadworth, 1997.
9. D. Busschau. Quick and Easy Microwave Cooking. New Holland, London, 1996.
10. J. McDermott. 1992. The Ultimate Book of Microwave Hints. Bay Books, Pymble, N.S.W., 1992.
11. Microwave Meat Cookbook. Murdoch Books, Sydney, Australia, 1992.
12. A. Ward. Vegetarian Cookery by Microwave. Award Publications, Mallow, 1997.
13. W. Lee. Microwave Meals. Parragon, Bristol. 1996.
14. The Complete Microwave Cookbook. Quantum Books, London, 1996.
15. A. Yates. Microwave Cooking Times at a Glance!: An A–Z. Right Way, Tadworth, 1997.
16. J. Anderson. Micro Ways: Every Cook's Guide to Successful Microwaving. HP Books, New York, 1997.

17. Zap It! What You Don't Know About Microwaves. Learning Seed, Lake Zurich, IL, 1996. Audiovisual. Narrator, Katherine Mitchell.
18. M. A. Preston. Microwave Cooking Made Easy. 1992.
19. Microwave Vegetables Made Easy. Merehurst, London, 1994.
20. V. Hill. Microwave Tips and Techniques. Penguin, London, 1994.
21. J. Hunter. The Microwave Diet Cookbook. Magni Group, McKinney, TX, 1994.

11
Ingredient Interactions and Product Development for Microwave Heating

Triveni P. Shukla
F.R.I. Enterprises
New Berlin, Wisconsin

Ramaswamy C. Anantheswaran
The Pennsylvania State University
University Park, Pennsylvania

I. INTRODUCTION

A majority of the consumer households are using microwave ovens and microwaveable foods. The microwave oven is now widely used for reconstitution of prepared foods in the food service sector. The number of prepared foods marketed for simple reconstitution in the kitchen using a microwave oven has steadily increased to include (1) refrigerated foods such as: gravies and sauces, sausage, poultry products, baked goods; (2) frozen foods: breakfast entrees, pizza, sandwiches, dinners entrees, potato products, pot pies, vegetables; (3) shelf-stable products: popcorn, soups, entrees, soups, noodles, dessert mixes, hot cereals; and (4) specialty products: caramels, dehydrated powders, baby foods, batters, and breading products.

Food product development for the microwave oven differs very much from that for a conventional oven. Some of the characteristic features of microwave heating are unique heating characteristics of food ingredients and food composites, rapid rate of heating, impact of the wattage, and the electric field distribution on differential temperature profile within the food load. The composition of foods, in terms of microwave reactive and interactive components, limits the highest possible temperature attained within the food product (in a conventional oven the maximum temperature is dictated by the temperature setting in the oven).

An approach to product design with a good understanding of the basic principles of microwave heating, the nature of heat and mass transfer that occurs during microwave heating, and appropriate selection of ingredients and microwave reactive packaging can result in a superior microwaveable food product. Electromagnetic fundamentals, dielectric properties, and heat and mass transfer are described in Chapters 1–4. This chapter offers discussions and speculations on the extent of direct and indirect interactions of food constituents with microwaves as the basis of food product design for microwave heating.

II. MICROWAVE ENERGY

Microwaves are electromagnetic waves having an electric field and a magnetic field component. In the United States, consumer microwave ovens operate at the frequency of 2450 MHz, whereas industrial ovens operate at both 2450 and 915 MHz. The choice of 915 MHz for industrial and 2450 MHz for the consumer microwave oven is based on optimum interaction of water, salt ions, sugar, and other food ingredients at these frequencies.

The microwave energy associated with these frequencies is not sufficient to break bonds within the molecules or to remove electrons (which requires an input of 54 ev) or rotate molecules (which requires an input of 0.1–0.01 ev), unlike infrared radiation. The electric field component of the microwaves interacts with food constituents primarily through three mechanisms: (1) orientation of permanent dipoles (due to asymmetric positive and negative charge as in water molecule), (2) orientation of induced dipoles as in charged polymers, and (3) ionic migration, as with organic and inorganic salts dissolved in food.

Table 1 contains useful information needed for the calculation of various derived parameters to evaluate microwave interactions. Successful product design for the microwave oven is possible with a clear understanding of the interaction of microwaves with the food materials.

III. MICROWAVE OVEN

Microwave oven is a cavity containing standing waves and wave patterns without phase difference. Knowledge of the mode distribution or the pattern in the empty cavity is very useful in characterizing the microwave oven. Other oven-related factors need to be taken into account while evaluating new products for microwave heating. Ideally, microwaves enter the cavity from all directions and, due to reflection at the walls, a unique power density pattern is created by standing waves in the microwave oven. Part of the microwave power generated is absorbed

Table 1 Useful Data and Constants Describing Microwave Interactions

Speed of light	3.00×10^{10} cm/sec
Relative permittivity	1.00
Dielectric constant, free space	8.854×10^{-12} F/m
Dielectric constant of water	80.000
Dielectric constant of bound water	4.50
Permeability of free space	1.257×10^{-5} Henry/m
Relaxation time of water	0.25×10^{-10} s
Relaxation time of food proteins	10^{-6} s
Intrinsic impedance	
Air	377
Water	42
0.1 M NaCl	42
Food solids	188
Potato solids	200
Raw potato	50
Moist potato solids	460
Olive oil	218

Sources: Refs. 21, 26, 54, 114.

by the food, and the rest is reflected back to the feed system. The size and the absorption characteristics of food are extremely important in determining the actual power delivered during microwave heating of foods. The food load can decrease or increase both power and wavelength, and it can also influence the mode distribution within the microwave oven. It is important to note that the number of modes increases with the volume and small dimensional changes in width, depth, and height can cause large mode changes. More information on these aspects is available in Chapters 2 and 6.

IV. INTERACTION OF FOOD COMPONENTS WITH MICROWAVES

The interaction of microwaves with foods is determined by the dielectric constant and the dielectric loss factor. This is described in detail in Chapter 3. Only a fraction of power delivered to the microwave cavity is absorbed by the food product depending on the size and dimension of the food load, its dielectric properties, and oven-related parameters such as power output and distribution. The temperature rise in the food product also depends on the physical and thermal properties of food such as the density, specific heat, and thermal conductivity. The dielectric and thermal properties of foods can be modified by adjusting food composition and formulations and, therefore, are manageable within certain limits.

Some of the key microwave interactive constituents in foods are water, other polar molecules of flavors and seasonings, ionized salts, peptones and peptides, yeast autolysate, monosodium glutamate, acidulants, sugars and oligo-saccharides, high DE maltodextrins, fats and oils, emulsifiers, and high-boiling plasticizers such as glycerol and propylene glycol, and various food polymers. The list of critical and commonly used microwave reactive food components includes triglycerides, amylopectin, amylose, anionic starch phosphates, oxidized starches, polyols such as glycerine and sorbitol, monosaccharides and disaccharides, various sugars, monovalent and divalent ions and their ratios, free amino acids, protein hydrolysates, citric and propionic acids, and proteins of different charge densities and molecular weights.

Factors such as the total charge in a food molecule, asymmetry of charges, surface charges and their distribution, emulsion droplet size, volume exclusion by large polymers, dipole moments of hydrated and unhydrated charged systems such as ions, super dipoles as in free amino acids, protein hydrolysates, and peptides impact microwave heating. Factors that affect dielectric properties of water, including the presence of other interactive constituents such as hydrogen bonding resulting due to the presence of glycerol, propylene glycol, and sugar and carbohydrate-like poly-hydroxy materials will also impact microwave heating.

In the case of cereal-based foods, microwave interactions of modified starches of varying granule size and cereal proteins with varying charge density, molecular weight, and hydrophilicity occur via volume exclusion and hydrodynamic effects. Ionic strength and pH strongly affect space charge polarization. These two parameters need to be taken into account in the case of high-viscosity acidic foods with built in buffering capacity.

Salts and sugars can be used to modify the browning and crisping of food surfaces. In addition to direct microwave interactions, lipids, salt, sugar, and polyhydroxy alcohols can also raise the boiling point of water. This allows the food to reach a higher temperature needed for the development of reaction flavors, and Maillard browning reactions which is necessary for aroma, reaction flavor, and texture development common to baking operations. Glazes, icings, emulsified sauce compositions have different yet predictable microwave reactivity.

A. Dielectric Properties of Food Components

Food in general act as mixed dielectrics (mixture of proteins, carbohydrates, lipids, electrolytes, and water), whose properties need to be understood in terms of both permanent and induced electric moments. When exposed to microwaves, they act as a nonideal capacitor consisting of a mixture of low and high loss components. They can also support resonances and modes when placed within the microwave oven depending on the geometry, dimensions, and dielectric properties of the food materials. Since dielectric behavior of foods is modifiable by specific

Ingredient Interactions and Product Development

formulations, the phenomena of reflection, transmission, coupling, refraction, and absorption by food dielectrics are manageable within certain limits.

There are a variety of functional ingredients used in product development, and they should be examined in terms of their dielectric properties for designing microwaveable food products. They include water activity management ingredients (modified starches, hydrocolloids and gums, glycerol), seasonings and spices, liposome encapsulates (browning compositions, aroma and flavor compositions), aroma enhancers, controlled-release flavors, encapsulated leavening agents, nonvesicular emulsions (such as fats and oils, lecithin, triacetin), emulsified shortening, controlled HLB (hydrophile/lipophile balance) emulsifiers systems, and plasticizers (such as glycerol, propylene glycol). A compilation of data on dielectric loss, dielectric constant, loss tangent, and penetration depth of various interactive food ingredients is presented in Table 2. Additional data are available in Chapter 3.

The dielectric constant and loss are affected by the presence of free water, bound water, surface charges, electrolytes and nonelectrolytes, and hydrogen bonding in the composite food product. Most foods act like variable resistors. For example, a wheat flour dough containing 35% moisture content, 2% salt, and 10% shortening and formed into a cylinder (diameter = 1.82 in and height = 1.5 in) can be only partially gelatinized in 1.5 minutes at 205°F in a 650 W microwave oven. However, at 25% moisture content, the center of dough cylinder begins to burn in less than 3.5 min.

Formation of macromolecular complexes, emulsion, dispersions, and electrical double layers can also affect the conductivity and the relaxation time in addition to the dielectric constant and the dielectric loss. The dielectric constant and the loss affect the degree of coupling of microwave energy, the heating rate, and phenomena such as focussing. Hence there is a need to obtain reliable data on dielectric constant and dielectric loss of various foods and food ingredients from room temperature to processing temperature.

One also needs to understand the hydration dynamics of food polymers, temperature coefficient of loss factor of foods in terms of layers, phases, refractive discontinuities, and differential absorption at various penetration depths. Such an understanding is a prerequisite for designing suitable browning compositions, layered foods, and multicomponent entrees. Various proprietary disclosures provide practical information in this area [4, 7, 13, 14, 29, 30, 37, 48, 66, 109].

With minor exceptions, foods are very heterogeneous in terms of microwave transmission through them. The dielectric behavior of the continuous phase, the volume of continuous phase in relation to the discontinuous phase, number of interfaces and discontinuities, and presence of other significant phases vary from food to food. A few examples are water-in-oil emulsions, oil-in-water emulsions, foods containing emulsified microparticulates, and aerated foods.

Dielectric properties of a typical food material depend on a number of fac-

Table 2 Dielectric Properties of Composition Wise Categorized Food Materials

Food compositions (°C)	Dielectric loss (ε)	Dielectric constant (ε)	Loss tangent (ε″/ε′)	Penetration depth cm = $1.948 \times \sqrt{\varepsilon/\varepsilon'}$
Water	13.40	78.00	0.17	1.28–1.50
High-moisture foods				
Raw potato	24.00	72.50	0.33	0.90
Skim milk [30]	13.20	00.00	0.22	1.14
Cooked carrot [25]	17.90	71.50	0.25	0.92
Spinach [25]	27.20	34.00	0.80	0.41
High-protein and salty foods				
Cooked turkey [25]	26.00	36.50	0.75	0.41
Lemon	14.00	71.00	0.19	1.17
Water + 0.5 M NaCl [25]	67.00	41.90	1.60	0.20
Meat broth	21.40	73.90	0.29	0.78
Egg white	15.80	69.70	0.22	1.03
Precooked ham [20]	22.80	42.90	0.53	0.50–0.60
Pork [20]	15.70	53.20	0.29	0.90
Turbot fillet	14.10	53.60	0.26	1.01
Cooked beef [30]	9.60	30.50	0.31	1.10
Bread dough	9.00	22.00	0.41	1.00
Fats and oils				
Fat and oil ave	0.15	2.50	0.06	20.30
Peanut butter	4.10	3.10	1.32	0.80–0.95
Sunflower oil [25]	0.18	2.40	0.07	17.07
Milk fat solids	0.20	2.60	0.07	15.70
High-sugar foods				
Sucrose [25]	6.80	72.30	0.09	2.43
Sweet potato	14.00	52.00	0.27	1.00
Apple	10.00	54.00	0.18	1.43
Squash [20]	37.60	47.00	0.80	0.35
Pineapple syrup	11.60	67.80	0.17	1.38
Food solids				
Potato chips [25]	2.08	5.76	0.35	2.28
Bread	1.20	4.80	0.26	3.50

Sources: Refs. 6, 68, 69, 72, 84, 120–123.

Ingredient Interactions and Product Development 361

tors and cannot be easily predicted just based on its proximate analysis and composition. One of the reasons the composition-based predictions of dielectric property of formulated foods are difficult is that both nonelectrolyte (sugar, polyol, and oligosaccharide) and electrolyte systems play an important role in determining the dielectric properties and their interactions are complex. Details of dielectric theory and dielectric data are available in a number of publications [21, 23, 36, 43, 101] and in Chapter 3. Interactions with microwaves are reasonably well established for a number of food components, and are discussed below.

Salts depress dielectric constant and increase dielectric loss (based on Hasted-Debye model). Sodium chloride and other mineral constituents commonly present in foods, small proteins and peptides, protein hydrolysates, free amino acids, food acidulants, and other charged molecules can have similar effects. Salts in particular can have a dramatic effect on the heating rates in the commonly used range of 0.05–0.15 moles in foods depending on their ionization behavior.

Insoluble and immiscible constituents (emulsions and suspensions) reduce both dielectric constant and dielectric loss factor (based on modified Fricke model). Emulsions could be either oil-in-water or water-in-oil types with continuous phases reversed. Dissolved proteins fall into this category with a lower critical frequency of 915 MHz. In case of larger protein molecules, there could be a twisting of side chains that carries sufficient charge [1, 43, 44, 82].

Interactive constituents, such as sugar, polyols, oligosaccharides, and alcohol, exhibit their critical frequencies shifted midway between those of pure materials based on the degree of hydrogen bonding (Maxwell-Debye model). Hydrogen bonds and hydroxyl-water interactions play a significant role in high sugar, maltodextrin, starch hydrolysate, and lactoselike disaccharide-based food formulations [86].

Regions of dispersion for bound water (hydrogen bonded and with hydrated ions) exist below that for free water [43, 44]. Two similar food compositions with the same water activity at a given apparent moisture content and containing different combinations of food polymers will have different levels of bound water and thus exhibit different dispersion behavior. The effects can be very dramatic when these two foods contain food polymers of widely different water-holding capacity. A very small amount of bound water is associated with the polymer of very high water-holding capacity. The change in dispersion behavior in this case will be negligible compared with food containing mixed polymers of low water-holding capacity. Examples are foods with high-viscosity gum systems, alkylated cellulose and starches, and foods with native starches and a blend of native proteins such as soy protein isolate, casein, and whey protein concentrate.

Basic dielectric data on food ingredients, food electrolytes, and food solvents have yet to be developed. Dielectric properties of marinated food are almost nonexistent. Very little is known about the dielectric behavior and dielectric property of emulsions, relaxation of bound water, surface conductivity of low molecular weight peptides in protein hydrolysates, and oriental sauce composition. Emulsion stability and oil droplet size are very important with respect to reconstitution in the microwave oven. Given the interest in sauce, browning, and controlled-release flavor compositions, fundamental research on interaction of microwaves with food emulsions should receive a high priority.

B. Dielectric Behavior of Water in Foods

Pure water is ionic to an extent of 33% due to its dipolar nature. The values of dielectric constant and loss factor in 2450 MHz region are temperature dependent. The frequency response is characterized by a dispersion phenomenon whereby the dielectric constant decreases logarithmically with frequency from 80 in the static region to 5.5 in the optical region (Fig. 1). But the loss factor increases from a negligible value in the static region to a maximum at the critical frequency and then decreases to a negligible value in the optical region. Tables 1 and 2 in Chapter 3 have more information on complex permittivity and relaxation of water as a function of temperature.

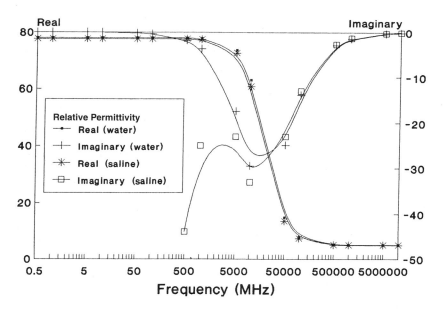

Figure 1 Dispersion behavior of water and aqueous salt solutions (ambient, 0.1 M).

The conduction behavior of water is the mirror image of the dispersion behavior. In contrast, the dispersion phenomenon is limited to 1 decade for dielectric constant, and extends to 2 decades for the loss factor. Both the static dielectric constant and the critical frequency of water decrease with increasing temperature.

Water bound to food polymers (starch, gums, and proteins) relaxes well below microwave frequencies. At 2450 MHz, bound water has almost icelike dielectric behavior. The stronger the binding forces between water and protein or carbohydrates, the smaller the value of dielectric constant and loss factor. Since the binding forces become weak at high temperatures, the water dipoles will orient more easily. The management of moisture content in foods can serve as a major controlling factor during formulation of microwaveable foods.

The relative dielectric activity of food constituents is listed in Table 3. The heating rate depends on the water and per unit volume of hydrated or unhydrated polar constituents of food ingredients such as flavorings and seasonings, emulsifiers, sugars, disaccharides, and poly-hydroxy alcohols. Orientation polarization, described earlier, is largely due to the free water [101]. Water as a permanent dipole and other charge-carrying solutes (salt, leavening agents, monosodium glutamate, organic acids, and free amino acids) interact to produce two types of dissipative effects: absorptive polarization and resistive loss. These effects are additive and are dominant within the liquid phase containing water, high boiling solvents (glycerol, propylene glycol, sorbitol, and starch hydrolysates containing di- and small oligosaccharides), electrolytes, and sugars.

A food scientist needs to understand the effects of individual ingredients on the dielectric behavior and on the changes in the state and structure of water. Various schemes of water activity control in foods need critical analysis in terms of their direct and indirect effects on microwaveability via modifications of their dielectric properties.

Table 3 Dielectric Activity of Major Food Constituents at Microwave Frequencies

Food constituents	Relative activity
Bound water	Low
Free water	High
Protein	Low
Triglycerides	Low
Phospholipids	Medium
Starch	Low
Monosaccharides	High
Associated electrolytes	Low
Ions	High

1. Effect of Nonelectrolytes in Water

Sugar is a key food dielectric. The importance of sugar as a microwave-absorbing food ingredient may supercede its conventional role as a sweetener, plasticizer, and texturizing agent and as water activity control agent in shelf-stable dry mixes, baked goods, bakery ingredients, preserved fruits, and sweet snacks. Addition of sugar increases the boiling point, and changes the dielectric properties of sugar-water mixtures (Figs. 2 and 3). Honey-based foods experience similar effects. What happens to the dielectric property of a 50:50 mixture of sugar-water or alcohol-water is only poorly understood [85].

Solutes, such as sugars, modify the dielectric behavior of water. Solutions containing mixtures of sugars are often twice as absorptive as the individual components of the binary mixture (Table 2). This interactive behavior offers a definite advantage in microwave cooking of sugar-based food formulations. The poly-hydroxy alcohols (sugars, sorbitol, glycerol, manitol, hydrogenated starch hydrolysates, low molecular weight carbohydrates, and syrups) accentuate the absorption of microwave energy due to orientation polarization of water. This is true

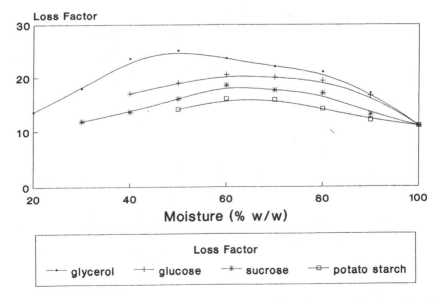

Figure 2 The loss factor dependence on concentration of sugar polyhydroxy food ingredients. (From Ref. 78.)

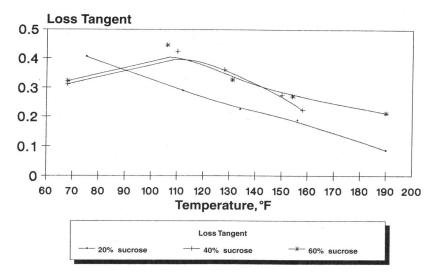

Figure 3 Variation in loss tangent (loss factor/dielectric constant) of aqueous sucrose solution (2.4 Ghz) as a function of concentration and temperature. (From Shukla and Miara, unpublished, 1991.)

of sugar-water and alcohol-water mixtures in general. The degree of microwave interaction depends on the extent of hydrogen bonding.

Aqueous solutions of sugars and alcohols are unusual in their dielectric behavior. The loss factor profile with respect to percent of sugar or glycerol in water (Fig. 2) exhibits a maximum at 50%. Also, the maximum loss factor seems to be related to the carbon number in sugar or alcohol molecules. This intriguing phenomenon has interesting implications in formulating browning and crisping compositions. The behavior, in all likelihood, depends on hydroxyl-water interactions that stabilize liquid water structure by hydrogen bonds. The observed synergy of dielectric effects between water and sugars and its dependence on a critical sugar concentration requires further studies. In all cases, the pure compounds are highly miscible in water; they shift critical frequency to the microwave region away from the critical frequency of the pure components.

Sugar can be used in formulating multilayered products for layer-by-layer selective heating. It is also effectively used for both surface heating and for creating a high loss shield that prevents the next layer of food from heating. Thus a desert topping can be designed to be heated in the microwave oven leaving the frozen deserts unheated [4, 117]. Most browning compositions use sugars and glycerol.

There are differences in the loss characteristics of monomers, dimers, and polymers. Whereas a 5% sugar solution exhibits a negative temperature coefficient for dielectric constant, the temperature coefficient for the loss factor is positive (Fig. 4). A positive coefficient results in rapid heating of the product to a high temperature [100].

In summary, the chemical composition of foods with respect to water, salts, and ions, and sugar is the major determinant of their dielectric properties. Addition of proteins, starch, and gums can further modify these properties. Collectively, these ingredients in a formula can be varied to control the rate of heating, depth of heating, and efficiency of microwave coupling. More research is needed in order to pinpoint the mechanisms underlying dielectric behavior of mixtures of sugars and complex carbohydrates in food like aqueous systems in terms of relaxation time, dipole character of the structural elements of the hydrogen bonded mixtures, and the molecular symmetry of dissolved sugars in foods.

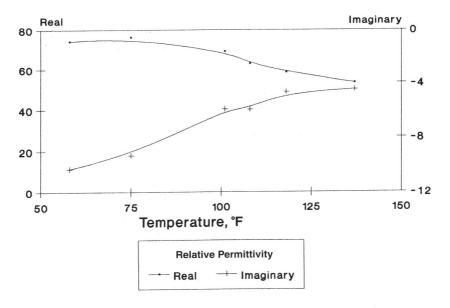

Figure 4 Temperature dependence (2.4 Ghz) of the real (dielectric constant) and imaginary (loss factor) components of complex permittivity of 5% sugar solution.

2. Effect of Food Electrolytes

Dissolved and dissociated salts impact the dielectric behavior foods. Hydrated ions, depending on their size, hydration number, and charge, depress dielectric constant of water and elevate the loss factor due to conductive effects. Hydration numbers for sodium chloride, Na, and Cl are 5.5, 4, and 7, respectively, and those of di- and multivalent ions are considerably different. Ions act as charge carriers. They increase heating rate and reduce penetration depth. Their effect on the loss factor is governed by the equivalent conductivity of ions which is both temperature and concentration dependent and varies with the frequency. In general, they increase the loss factor [43, 44, 82]. Dissolved ions bind water and the binding force depends on their size and charge. Dielectric constant is depressed by the presence of ions below that of water, and the effect is characterized by hydration number.

The ionic loss and dipole loss are the dominant components of total loss due to the aqueous phase of foods. At a fixed frequency, the dipole loss decreases with increasing temperature but the ionic loss increases. However, the total loss of food products decreases initially with increasing temperature when dipole loss is the dominant component, but then increases with increasing temperature when ionic loss becomes the dominant component [43, 82]. The higher the ionic conductivity of a food, more dominant is the ionic loss even at relatively lower temperatures. Although proteins and oil-in-water emulsions behave like salts at submicrowave frequencies, amino acids, peptides, peptones, and small molecular weight proteins are interactive in the microwave region. Even for large molecules, surface charges can be interactive because of the twisting of the side chains. Although quantitative data on the effect of water-binding agents, bound salt effects, and water emulsions are not currently available, there is considerable evidence from patented formulations as to the effects of bound water and ions in decreasing dielectric loss [1, 3, 57, 116].

Electrolyte solution enclosed by a spherical double layer, as in the case of water-in-oil emulsions, create a kind of Faraday cage where adsorbed layers of atoms on the surface can experience both polarized atoms and induced dipole moments [43, 82]. Also, emulsions and encapsulated food systems do exhibit differential heating rates when placed in a microwave oven for heating and reconstitution, e.g., sauce, butter, and browning compositions.

The conduction effects (resistive heating) are due to food electrolytes such as salts, organic acids, free amino acids and hydrolyzed proteins, leavening agents, mineral supplements, amphoteric lipids, small molecular weight glycolipids and glycoproteins, and short chain fatty acids and amino sugars. Although present only in small (0.1 molar) quantities, these substances can influence the heating rate. Since it is the dissociated or ionized form of electrolytes that interacts with microwaves, the pH and ionic strength effects become critical and significant [28, 69, 100], and hence the addition of acidulants in microwaveable foods has to be carefully looked into. Acidulants should be carefully selected on the basis of their ionization behavior in actual food systems.

3. Effect of pH and Ionic Strength

Hydrogen ion concentration of a food determines the degree of ionization. Dissociated ions migrate under the influence of an external field, and hence the pH becomes a factor in microwave heating. Ionic strength determines the collision frequency and the heat generation. Collision increases at higher concentration of ions to a point that dielectric loss factor may have positive temperature coefficient. It is for this reason ionizable materials are incorporated into browning compositions. Target pH should be below pK of amino acids, organic acids, and other ionizable groups.

4. Effect of Water-Alcohol Mixtures

Polyols (glycerol, propylene glycol, sorbitol, and manitol) are miscible and soluble in water. The mixture has a critical frequency intermediate to those of water and pure polyols. Since polyols are extensively used for water activity management and shelf-stability improvement, knowledge of the dielectric behavior of such mixtures as continuous phase is critical to the formulation of microwaveable foods and food coatings. Information presented in Figs. 9, 11, 13, 15, and 18 in Chapter 3 can be very useful for proper ingredient selection and microwaveable food formulations. Glycerol is commonly used in food formulations, and its significance goes beyond reducing water activity. For example, 4% of added glycerol in a food containing 30% moisture content results in very high microwave absorption and leads to temperatures greater than 212°F. Glycerol is widely used in browning and basting compositions.

C. Dielectric Behavior of Food Polymers and Plasticizers

1. Native and Modified Starches

The dielectric behavior of food polymers depends on their interactions with water and charged food constituents such as salt ions, leavening agents, monosodium glutamate, and free amino acids. Major food polymers are starches, proteins, and various hydrocolloids. Collectively, such polymers affect polarization of dipoles and electrolytes. The two dissipative effects (absorptive polarization and resistive loss) are additive. Low moisture, low loss, and low attenuating foods are unique in that they establish internal standing waves causing rapid heating, hot spots, and even burning. Up to 25% of incident power may dissipate by this mechanism. Also, high-fat cereal foods, such as brownies and cookies, and oil-in-water emulsions heat by similar mechanisms. Given an approximate relaxation time of

Ingredient Interactions and Product Development 369

10^{-6} s, proteins may behave like salts at submicrowave radio frequencies [82], a heating method now commonly used for proofing and baking. Viscosity of the medium in which food polymer materials relax and reorient, especially in relation to their hydration, has considerable effect on reconstitution and microwaveability.

In the case of emulsions, interactions will vary for oil-in-water and water-in-oil emulsions. Other factors such as the physical state of food and concentration of dielectrically active ingredients are not well understood in quantitative terms. Patent literature makes only qualitative references to depression of dielectric constant and increase in loss factor for various food ingredients. Least understood of all are the microwave interactions of multicellular tissues, starch and protein inclusions within cells, and cell membranes.

Prebaked bread or bread dough, when heated or reconstituted in the microwave oven, develop unacceptable texture with a rubbery, leathery, tough, and difficult-to-tear exterior and a chewy and firm interior. The dough has two polymeric materials: gluten protein and largely granular but differentially gelatinized starch embedded in the gluten network. Toughness relates to gluten protein and firmness to starch granules. Free water, through its intrinsic ability to interact with both of these polymeric materials, is indirectly involved in both of these problems. It stands to reason that two bread samples of different gluten content and different degrees of starch gelatinization will have different free moisture content, and thus will heat up differently within a microwave oven.

Baked goods are created by carefully manipulating type of wheat flour and gluten content, amount and type of fat, degree of dough development, and type and extent of leavening for a given type of heat treatment (baking, frying, steaming). Their microwaveability without the development of objectionable texture can be enhanced by manipulating the gluten protein network, size and swelling of starch granules embedded in it, and the moisture content. A lower level of moisture in the formulation should be used in order to prevent migration, reabsorption, and redistribution subsequent to microwave reconstitution. There are two events that take place at a rapid rate during microwave heating: The large molecular weight gluten proteins plasticize with additional moisture pushed outward from the interior, and otherwise minimally gelatinized starch granules gelatinize further and swell. These events take place with partial collapse of the aerated structure of bread and breadlike products. The solution to this problem lies in using lower molecular weight gluten proteins, small and less swollen starch granule like that in rice flour and by largely immobilizing the moisture in the product.

The firmness problem of the bread interior associated with the large diameter, re-swollen starch granules can be solved by simple dilution with small starch granules [74] or chemically modified potato starch [18]. The size of partially swollen, gelatinized granules can be reduced by the use of fats and emulsifiers that inhibit gelatinization. But substitution of wheat flour with 10% rice starch or chemically modified starch and addition of emulsifiers in a higher fat formula solves only the problem of interior firmness. Solutions to the problem of toughening on the exterior involve depolymerizing the gluten protein by methods other than mechanical overworking [18, 74].

The gluten protein can be reduced in size by breaking the disulfide bonds. Reducing agents (L-Cysteine, glutathione, yeast autolysate as source of glutathione, or potassium or sodium bisulfite) are employed in conventional formulations for a similar effect. For microwave reconstitution, the reduction process has to be more extensive at 2 to 3 times higher use level of these agents. Enzymes such as protease and deaminase can serve a similar function. The mid-1980s concept of "new protein systems" for microwaveable bread, in effect, had to do with reduction in bread protein elasticity. Also, pH and ionic strength of a new formulation can be adjusted to induce maximum electrostatic repulsion that should minimize interchain interactions within the gluten protein network.

All microwave-induced effects of polymers have their origin in rearrangements of ionic bonds, hydrophobic interactions, and hydrogen bonds. The availability of free water is very critical for such interactions. Sodium aluminum phosphate or sodium acid phosphate, added as leavening agents, also influence ionic strength and reduce water activity. Water activity is further reduced by the addition of dextrose and corn syrup solids. In addition to reduction of water activity, such additives inhibit gelatinization and water migration within starch granules. Moisture migration and localized reabsorption can be also be prevented by adding texturizing agents of high water-holding capacity. Texturizers interrupt the continuity of protein networks.

The overall formulation strategy for baked products should therefore focus on (1) designing a low-moisture dough whose water activity is further reduced by salts and dextrose, (2) incorporating higher than normal levels of shortening and emulsifiers in order to prevent gelatinization and swelling of starch granules, (3) reducing the size of gluten proteins by reducing agents or enzymes, and (4) including texturizing agents not only for a balanced texture but also for preventing the extent and rate of moisture migration and relocalization. Just in the last 10 years, a lot of these concepts have been translated into proprietary knowledge [11, 19, 47].

Native starches, modified starches, vital gluten, and dietary fibers have definite roles in formulation of microwaveable foods. The soluble carbohydrate such as sucrose, glucose, and other polyhydroxy food ingredients exhibit high dielectric activities in aqueous mixtures [85, 86]. The rheology of reduced gluten proteins, increased firmness due to highly swollen large diameter starch granules, and moisture release and its undesirable redistribution are the key factors to be managed in formulation of microwaveable bread or breadlike doughs.

2. Proteins and Protein Hydrolysates

Solid food proteins are dielectrically inert (Table 3). However, both dielectric constant and dielectric loss increase with increasing amount of water in a range of

Ingredient Interactions and Product Development

2–12%. The dielectric behavior is related to composition, water activity, structure, and their interactions with water and electrolytes [68]. The solvated or hydrated form of protein, protein hydrolysates, and polypeptides are much more microwave reactive. The dielectric property of proteins depends on their side chains that can be nonpolar with a preponderance of alanine, glycine, leucine, isoleucine, methionine, phenyl-alanine, and valine or polar with a preponderance of tyrosine, tryptophan, serine, threonine, proline, lysine, arginine, aspartic acid, aspergine, glutamic acid, glutamine, cysteine, and histadine.

Free amino acids are also very reactive [82]. Free amino acids and polypeptides contribute to increase in dielectric loss. Polypeptides could be considered as strings of peptide bond dipole moments [82]. The dipoles are also associated with polar side chains and hydrogen bonds between $C={O}$ and $N—H$ groups. As suggested earlier, protein dipole moments are a function of their constituent amino acids and vary with pH. A considerable difference in microwave reactivity is expected in cases of cereal, legume, milk, and meat proteins in terms of net charge at a given pH and surface charge distribution. The water adsorbed on proteins affects their dielectric behavior due to interfacial polarization. As dipolar ions, amino acids are super polar molecules surrounded by intense electric fields. Their aqueous solutions produce large permitivity increments (Table 4).

Dielectric activity of proteins may have four origins: (1) high activity due to charge effects of ionization of carboxyls, sulfhydryls, and amines; (2) hydrogen and ion binding as affected by pH; and (3) net changes on dissolved proteins; and (4) relatively low activity due to relaxation and conductive effects. Such activities are important for hydrolyzed proteins and free amino acids that are often part of high sugar food formulations [101, 102]. The microwave activity of proteins in high-moisture foods, however, is largely due to their colloidal nature. They modify the aqueous phase due to their charged surfaces that can bind ions [69]. Similarly, starches and gums in high moisture foods are expected to be dielectrically active mainly because of volume exclusion and hydrogen bonding effects [28]. In general, emulsoid lipids and colloids derive their activity from volume exclusion.

Of critical importance are the properties of hydrated or fully solvated proteins. Even more critical is the knowledge of differential water absorption by charged proteins or proteins like zein that may dissolve only in water-polyol mixtures. Proteins have polar and nonpolar side chains, which act like connected dipole moments. The dielectric contribution of such chains and additive dipole moments of polypeptide chains of commonly used food proteins are not understood very clearly. Peptones, oligosacharides, and monosodium glutamate are very microwave reactive food ingredients.

3. Ionic and Nonionic Gums

Water bound to hydrocolloids may have different critical frequency, and polymer-solute binding may change both dielectric constant and dielectric loss. Hydrocol-

Table 4 Dielectric Increments of Polar Ingredients in Water

Polar ingredients	Dielectric constant	Dipole moment Dubye unit = 10^{-16} esu-cm	Dielectric increment
Acetic Acid	6.15	1.74	
dl-Lactic acid	22.00		
Glycine		13.30	22.60
L-Aspertic acid			27.80
Glutamic acid			26.00
Proline		20.00	21.00
Glycine heptapeptide			290.00
Lysyl glutamic acid			345.00
Glycerol	42.50	2.58	–01.80
Glucose		5.30	–04.30
Ethanol	24.55	1.69	
Sorbitol			–02.75
Oleic acid	2.46	1.18	
Potassium or sodium			–03.00
Chloride			–03.00
Hydroxyl			–13.00
Magnesium			–24.00
Sulfate			–07.00
Magnesium			–24.00
Egg albumin		250.00	00.10
Water	78.60	1.90	

Source: Dean [124]: *Lange's Handbook of Chemistry*, McGraw-Hill, New York, 1985.

loids may influence relaxation of water and other dipoles and may influence ion acceleration irrespective of the continuous or discontinuous nature of phases present in the food. Hydrocolloids exhibit both direct and indirect effects in microwave heating of foods. Direct effects may be conductive in case of ionic colloids such as starch phosphate, starch esters, carboxy starches, carboxymethyl cellulose, alginate, carrageenan, pectinates, xanthan gum, and gum arabic. The nonionic types of gums, such as native starch, dextrins, guar gum, and locust bean gum, have an indirect effect. They can immobilize a significant portion of the water and make it less microwave reactive and delay the relaxation by increasing the solution viscosity.

Changes in density and specific heat can also influence the rate of heating. Although we know that specific heat of foods is dependent on moisture content, our knowledge of specific heats of foods where moisture is immobilized by various colloidal ingredients, albeit to the same degree, is very meager.

Major hydrocolloids used in foods are starches, proteins, and gums of plant or bacterial origin. The food product development scientist should have information on their concentration-dependent water activity, viscosity, and conductivity effects. In the range commonly used (0.1–2.0%), bound water could range from 0.25 to 0.55 gram per gram resulting in between 25 to 60% of water being less mobile. Since bound water of food polymers is additive, foods containing single or multiple polymers of varying bound water are expected to have different degrees of total water mobility and polarizability because bound water doesn't contribute to permittivity at higher frequencies. The rotational mobility of bound water is rather restricted, and it behaves as if it is integral part of the polymer and thus unavailable to solutes.

D. Microparticulated Components

Many food products consist of suspended particulates. A case in point is microwaveable soup. Every piece of particulate material presents a new surface, new geometry, new shape, and in many cases a new composition. Microwave reflection and transmission patterns through these particulates will vary accordingly. Therefore, microwaveability would depend on size and size variation of particulates, compositional differences in case of mixed particulates, and volume fraction of particulates. Marinating or infusion treatments and hydration and ionization of polymers will also impact their microwaveability. While there is some empirical and proprietary work in this area, quantitative information on the dielectric properties of particulate foods is very much lacking.

1. Oils and Emulsions

Although loss factor of fats/oils and their emulsions is low, they heat in the microwave oven largely by internal reflection. Oil-in-water emulsions have higher electrical conductivity than water-in-oil emulsions [1]. Emulsion properties relevant to microwaving are electrical conductivity, dielectric constant, dielectric loss factor, and electrophoretic mobility. Such properties depend on droplet size and nature of the electrical double layer. Every droplet can have its own surface and surface charge. Proteins are known to spread on water-in-oil interfaces; they can substantially change an emulsion's microwave reactivity. Sauce and sausage compositions thus make excellent candidates for microwaveable foods but their dielectric properties are poorly understood [3].

2. Micellular Structures

In micellar emulsions (microemulsions), the ratio of outer to inner phase is often 50:50. Their water activity is rather low [17], and their incorporation will have a significant effect on microwaveability of foods. Similarly, colloidal suspensions, depending on the dielectric property of the dispersed colloidal phase, will impact the dielectric loss and heat dissipation characteristics.

3. Vegetables and Whole Animal and Plant Foods

Multicellular tissues with solid and liquid phases present another unique challenge. Starch inclusions, glycogen in liver cells, cell walls, and cell membranes are a few typical examples. Inclusion complexes may differ in size, shape, charge density, and concentration. These differences will dictate the dielectric behavior of whole organ and tissues of animal origin and whole and diced vegetables and fruits of different sizes. Mere arrangement of various vegetables in terms of size and moisture content makes a big difference in their microwave interaction in the microwave oven. Microwaveability of bacon strips exemplifies this point.

E. Layered and Multiphasic Foods

The concept of layered products denotes differential heating between two or more food products positioned as layers, fillings, coatings, or even embedded deposits. To do so requires proper selection of ingredients with appropriate dielectric properties, thermal mass (specific heat × mass), a media differentiating film, and thickness. The objective is to manage reflection, transmission, and absorption of microwaves in a controlled manner. A few general guidelines for the selection of ingredients are:

- High moisture sauces will heat faster than a low-moisture sauce.
- Thicker layers tend to heat to lower temperatures.
- High fat foods will offer better impedance matching and coupling.
- Salt, sugar, and polyol based compositions will heat to a higher temperature.
- Highly absorptive foods will act as good shield to microwave propagation; they prevent transmission. (For example, a highly absorptive sauce will protect heating of the food that it coats or covers).
- The role of water on microwave heating can be modified with gums and salts.
- Reflection can be designed into a food prodcut. For example, butter reflects 25% of incident power, and a sauce reflects almost 60% of incident power.

Pizza is a good example of layered and multiphasic food, and its microwaveability has been attempted at commercial scale. Microwaveable pizza crust, toppings, and interfaces between crust and topping are covered under proprietary technologies. Low-density, aerated pizza crust compositions with dome-shaped blisters and spongy structure or overlayered crust, where two layers have differential dielectric behavior, have been disclosed [9]. In many cases, physical susceptors are a necessary complement for browning and acceptable crust development. Layers of edible substances can be used as sprays or coatings in order to control transmission and surface heating [4, 78, 116]. With proper control of thickness and dielectric property of single or multiple layers, high final temperatures needed for browning and crisping can be attained. At least qualitatively, 50% high-amylose batter-coated dough [119], prebaked pastry system [79], and cation-adjusted juicy meat [39] are good examples of use of sound theory in predicting microwaveability. Physical arrangement of one dielectrically defined food medium against the other can also be used to control the rate of heating and the degree of shielding. Examples are discussed in Sec. VI.

A coating or a film can serve as a microwave shield for a microwaveable product. Coatings and sprays of different dielectric properties can be applied in succession. Their thickness, thermal mass, and dielectric property can control the amount of energy transmitted and the final temperature of the shielded food component [4]. This approach has been commercially practiced for ice cream layered with topping where the latter absorbs most of the microwaves leaving the ice cream still frozen.

F. Browning and Crisping Agents

At ambient pressure, high-moisture foods cannot be heated above the boiling point of water (212°F) within a microwave oven. As an alternative to metallic susceptors, food compositions containing high-boiling solvents and plasticizers and ingredients with positive temperature coefficient of loss have been developed to induce the rapid rate of heating necessary for browning reactions to take place. These are known as "chemceptors" (chemical susceptors).

Several patents exist on chemical susceptors. U.S. Patent 5,223,289 describes a lipid, reducing sugar, protein, and water composition where lipid is 50% by weight [30]. The pH-controlled blend is preheated and condensed into a paste that can be applied on foods for proper browning in the microwave oven. Another patent discusses the preparation of a spray-dried material from lactase hydrolysed milk as a source of reducing sugar and lysine. The powder having undergone Amadori rearrangement during spray drying is an excellent browning agent [48].

Three ingredients must be present for proper browning: amino acid, reducing sugar, and a pH-adjusting agent. U.S. Patent 5,089,278 devises a method to permit timely reaction of these ingredients [45]. The pH-adjusting agent and amino acid are encapsulated in a liposome, and the reducing sugar is part of the external continuous phase. The liposome is composed of a phospholipid bilayer enclosing an aqueous core containing pH-adjusting agent, flavors, vitamins, and colors. The pH is targeted to be below the pK of the amino acid, and the molar ratio of amino acid to sugar is 0.5 to 1.0. The emulsion contains a film forming agent such as microcrystalline cellulose or crosslinked starch for effective coating. Yet another example is a water-in-oil emulsion, composed of 30 parts aqueous discontinuous phase and 67 parts oil phase, which is stabilized with 3 parts of emulsifier and that contains carbonyl containing browning reactant [109]. The emulsion can be used as a coating for biscuits, pizza, pie crust, and hash brown potatoes for final browning during microwave reconstitution.

Other browning compositions described in the literature include (1) a mixture of precursors of Maillard reaction (proline and reducing sugar) in glycerine [13] which can be applied to dough for microwave baking; (2) a mixture of 60 parts bread crumb, 25 parts oil, 10 parts dextrin, and 3 parts soy protein [20]; (3) a mixture of 60 parts flour, 30 parts salt (potassium acetate, potassium chloride, and potassium bicarbonate) and 10 parts seasoning [46]; (4) a mixture of 10 parts yeast extract, 30 parts flour, 25 parts shortening, 20 parts water, and 15 parts reducing sugar [37]; and (5) a mixture of melted, caramelized, and foamed disacharide alone or in combination with minor amounts of monosacharide subsequently dissolved in water to be applied as coating to meat, poultry, fish, eggs, and pastries [66]. In most of these inventions, a high rate of heating to temperatures above 120°C is brought about by utilizing a combination of salts and high boiling liquids.

Browning and crisping is a problem for high-moisture foods but not for foods with moisture in the close range of the bound water of constituent food solids. Low-moisture food products, e.g., snacks, cold cereals, breadsticks, nuts, and cereal and legume nuts can be rendered crispy because food solids (starch, proteins, dietary fiber) devoid of water can be heated to their combustion temperature within a short time by microwave heating.

G. Edible Susceptors

Browning and crisping compositions to be applied as layers or coatings are in a sense edible susceptors often with a combination of resistive loss and lossy behavior built into the dielectric property of the ingredient blend. Ionic loss increases with temperature. The higher the ionic conductivity of a food, the quicker and, at relatively lower temperature, the ionic loss begins to dominate. Hence, ionizable salts and charged proteins can be used to create an edible susceptor. Egg white–like proteins, protein hydrolysates, salt and leavening, ionic emulsifiers, polyol, and sugar-based compositions applied as film or coating with the aid of food gums and oxidized, high-amylose starches can serve as edible susceptor. To be effective, their compositions should have low-moisture content.

All browning compositions include sugar, salt, glycerol or propylene glycol, and amino acid or peptone blends for an efficient microwave absorption and a fast rate of heating. Precursors of Maillard reaction, high-boiling solvents (glycerol, propylene glycol, sorbitol, and triacetin) along with selected electrolytes are often used in these browning compositions. The solvent can raise the surface temperature of foods to a range of 350–370°F in less than 120 s [14]. These compositions can also include flavor precursors [15]. Special blends have been patented for coatings and marinating proteinaceous foods [51]. The microwaveable flavor ingredients are available in encapsulated form [52]. A very useful formula of a browning composition includes 20% salt, 27% nonfat dry milk, 22% glucose, 20% glycerine,

1% egg white, and 10% sodium carbonate. A 10% solution of the blend in water serves as a coating. Another formula uses 3.5 g of a blend of 40% glycine, 40% glucose, and 20% sodium carbonate with 284 g of pie crust to prepare a coating to which skim milk, yeast extract, and egg white solids could be added.

V. FOOD PRODUCT DESIGN FOR MICROWAVE HEATING

The key to designing microwaveable foods is to preferentially depositing microwaves in foods. The total energy absorption by the food depends on both dielectric constant and loss factor. It also determines the field intensity and intrinsic impedance. The latter in turn determines reflection, transmission, and power coupling. With a good understanding of the dielectric properties of various food components and their interaction during microwve heating, it is possible to develop successful microwaveable foods.

Water, as the dominant dipole, is often the continuous phase in most food systems; the presence of ionic solutes, pH and ionic strength, and viscosity of continuous phase are very critical to conductivity and, therefore, to microwave absorption. Water, polar molecules of flavors and seasonings, ionized salts, peptones and peptides, yeast autolysate, monosodium glutamate, soya sauce, sugars and oligo-saccharides, high DE matodextrins, fats and oils, emulsifiers, and high-boiling plasticizers, and food polymers (by way of viscosity, volume exclusion, and surface charge) are key microwave interactive constituents in foods. Microwave reactive foods include, egg albumin and bacon with a positive temperature coefficient of loss, sauce compositions, browning aids, sugar solutions of different concentration and sugar types, and salts and ions. Product by product dielectric matching with respect to their ingredients per se and the physical state of the ingredients in the food system is important for controlled heating of a foods in the microwave.

As described in Chapter 3, impedance dictates reflection and transmission, and refraction characterizes absorption. All of these depend on the food composition, viscosity, temperature, and frequency. A complete discussion on the electromagnetics and dielectric properties are provided in Chapters 1–4, and the critical factors in the design of microwaveable foods are emphasized here.

A. Impedance Matching

The efficiency of microwave coupling with the food load in a given microwave oven depends on the intrinsic impedance of the food material within the oven. For maximum coupling, intrinsic impedance of the food composite needs to be close to that of free space or air which is 377 ohms. A typical low moisture solid food reflects less and transmits more than water with an intrinsic impedance of 42 ohms. Foods with a low dielectric constant will reflect the least and their intrinsic impedance values will be very high. Such foods couple incident radiation well. On the other hand, a moist potato, water, 0.1% sodium chloride in water have intrinsic impedance values of 46, 42, and 42 ohms, respectively, and reflect much more of the incident energy. A complete impedance match is impossible but a closer match to 377 ohm value is attainable by designing a product based on low moisture, high level of fat or oil, low HLB emulsifiers, and nonionic food gums. The impedance matching concept is especially critical for layered foods, where electric fields (and thus heating rates) can be substantially different in the various layers, depending on the impedence mismatch.

Resonance and quarter wavelength principles can also be used to improve impedance matching and coupling. Internal focusing by combined effects of penetration depth and impedence matching can avoid reflection and increase absorption of microwaves and the rate of heating. For zero reflection from a food product, its thickness should equal

$$(\tfrac{3}{4} \times 10^8) \div (\nu \sqrt{\epsilon'_{food}})$$

where

ν = frequency in Hz

In essence internal focusing by $\tfrac{1}{4}$ wavelength matching avoids reflection at the "air-food" interface.

Low-loss materials have longer penetration depth, and much of the microwave is transmitted back into them. Low-reflecting food materials like oils and wheat flour dough (with 25% moisture) heat by this mechanism. This is also responsible for standing waves within the food and can cause hot spots and focusing of microwaves.

B. Penetration Depth

Penetration depth is defined as the depth from the food surface to a point where electric field strength is $(1/e)$th or 37% of that in the free space. A low value implies high-loss leading to high microwave absorption. A microwave coating can be designed as a microwave shield where attenuation is maximized at very short penetration depth. Microwaves penetrate deeper in oils and dry foods because of longer penetration depth. When penetration depth is close to size of the food load, center heating or focusing is unavoidable in cylindrical and spherical foods. To avoid center heating, size should be increased.

C. State of Water and the Dissolved Constituents

Bound water, nonionizing salts have low dielectric activity. Free water and ionized (dissociated) salts have high dielectric activity. Protein-water complexes shift wavelength intermediate to that for pure protein and water. Sugar and alcohol mixtures in water have a similar but more synergistic effect. Colloids, oil-in-water and water-in-oil emulsions, liposomes, and gums depress both dielectric constant and dielectric loss. Ions, salts, peptides, amino acids, protein hydrolysates depress dielectric constant but increase the dielectric loss factor.

A number of dissolved food constituents change the dielectric constant of water as shown in Table 4. Dipole moment relates to polarizability and dielectric constant. The table shows that while amino acids have positive increment, sugars, oligosaccharides, and polyols have negative increment. The dielectric constant of the continuous aqueous phase containing these solutes equals dielectric constant of water plus dielectric increment of the solute multiplied by its concentration. This phenomena is additive, and therefore the moisture content and concentration of sugars, polyols, aminoacids, peptide/peptones, and other polar additives to the food can be used as design parameters while developing a microwaveable food product.

D. Food Volume and Surface Area

The total power absorbed by a food depends on the oven power level and the size of the food product, as discussed in Chapter 2. The total power absorbed by the food product increases with the size of the load, but the power absorption per unit volume decreases. The total power absorption also increases with the surface area of the food. Thus, a high surface-to-volume ratio is desired for rapid heating.

E. Food Geometry

The geometry of the food product impact coupling of microwave energy. Microwave heating is capacitive in nature, and thus surface discontinuities at edges and corners can account for charge accumulation and thus create nonpropagating fields that are known to overheat and burn edges and corners of low-moisture foods like potato and corn chips. Curved shapes can concentrate or focus microwaves. Focusing depends on the penetration depth and the wavelength in the food (see Chapter 2). Focusing can be deliberately excluded or included by changing the dielectric properties and adjusting the shape and size of the food product.

F. Other Food Characteristics

In formulated and prepared foods, food characteristics such as size and composition of particulates, and multicellular tissues, are important. Often the food material is also anisotropic and nonhomogeneous, i.e., properties vary with direction and position as in the case of muscle foods. These factors will also impact the heating rate.

G. Use of Active Packaging to Modify the Microwave Absorption

The microwaveable food product and the package should be viewed together in terms of the overall microwave interaction. Packaging can significantly impact the heating rates. More importantly, active packaging, such as the use of aluminum foil, can be used to obtain desired heating patterns within a given food product. Extensive discussion on different packaging techniques available to enhance the heating characteristics of microwaveable foods is available in Chapter 12. Advances in packaging and susceptor technology have given rise to a multitude of advanced package designs that can control the amount of incident microwaves and the field distribution within the food product. However, the high costs associated with these technologies has slowed their acceptance by the food industry.

VI. PRODUCT PERFORMANCE TESTING

The microwaveable product should be thoroughly tested at various stages of product development. The tests should be conducted during the initial stages of product development, for product characterization as affected by ingredients and the package selection, and during the final stages of the product development to come up with heating instructions. Additional discussion on microwave cooking at home by consumers is available in Chapter 10.

A. Selection of Ovens and Testing Protocol

The product testing should be conducted in a variety of microwave ovens representing different wattages, sizes, feed system, turntables, shelves, etc. The test kitchen should maintain different ovens of different nominal power outputs ranging from under 500 Watts to over 1000 W. Testing in a conventional oven and in a combination microwave and convection oven is also recommended.

A database should be maintained on overall oven-to-oven variability and how this variability affects the heating characteristics of foods. Then a set of microwave ovens which represent the "typical," "worst case," and "best case" scenario should be identified. Using this information, product testing can be designed to monitor product heating rate and quality in selected microwave ovens. Based on the results from these, heating instructions can be developed for maximum convenience, quality, and safety [129].

The microwave ovens should be characterized prior to testing the products. Each oven should be characterized for its actual power output and field distribution and heating patterns. The line voltage to the microwave oven should be monitored and controlled for better repeatability. This can be done manually by using

Ingredient Interactions and Product Development

a good multimeter and a rheostat, or a line voltage conditioner, can be used. A 2 liter and IEC approved microwave power absorption testing procedure should be used to measure the actual power output of the microwave oven. This power absorption test should be conducted periodically for each oven and recorded. When evaluating products that are smaller in size, power absorbed in a cup of water should also be monitored. Power vs. load volume testing should also be done on each oven in order to determine the power coupling efficiency as a function of the size of the load in the microwave oven. Power distribution within a microwave oven can be determined using a thermal paper or by using a thin layer of egg white or pancake batter. Additional discussion is available in Refs. 126 and 128.

Product temperatures should be continuously monitored during microwave heating and at the end of heating (during standing time). In instances where very viscous products are being heated, it is advisable to also monitor the pressure buildup within the product. This pressure buildup is responsible for "bumping" of the product and should be minimized during the product heating cycle. It is also recommended that an infrared camera be used to do thermal imaging of the heating patterns. The instrumentation needed for these measurements are described in detail in Chapter 8.

B. Heating Instructions

Foods formulation and heating instructions should be developed such that consumer safety is not compromised during routine cooking or reconstitution. Clear heating directions can promote optimum performance of microwaveable food products. Rotating the food during cooking, elevating the food during cooking, covering the food, stirring the food, use of "standing" times or rest periods in between and at the end of cooking, and the use of a lower power settings can all improve the uniformity of heating in a microwave oven [127].

C. Product Abuse Testing

One important aspect of heating instructions for the microwaveable product is to ensure consistent product quality and performance when prepared in different microwave ovens. Tolerance testing of the product should be conducted because of the short heating times and variation between different ovens [125]. If the product is more sensitive to time of heating or power variations, it should be optimized for performance at the bottom of the power range. Product reformulation, including a change in shape or geometry, and appropriate package selection can be used to mitigate poor product performance.

Tests should also be conducted to simulate product abuse by consumers due to misinterpretation of the instructions and to expose safety-related issues due to over heating or due to pressure buildup within the product such as in sauces and soups and other viscous products. Abuse testing of the product should be conducted against worst possible cases of superheating, eruptions, and explosions. These occur due to presence of hot and cold spots within the product during microwave heating and are discussed in Chapter 2. Instructions to use a lower power setting or duty cycle and to stir the product several times are safeguards against these occurrences. Arcing, fires, and burns can also occur under improper use, and these are discussed in Chapter 13.

VII. ADVANCED TECHNOLOGIES FOR MICROWAVEABLE FOOD PRODUCT DEVELOPMENT

Although not every patent has been commercially reduced to practice, proprietary technologies incorporate numerous concepts for layered foods, puffable snacks, batters and breadings, browning and crisping compositions for baked goods, meat, edible shields, cake mixes and brownies, pancakes, soup, pizza crust, multicomponent dinner entrees, casseroles and shelf-stable foods. The approaches utilized, depending on the nature of the product, include (1) selecting the dielectric property of the food, food coating, food layer, and their positioning; (2) selecting foamed verses dense foods; (3) selecting foods of similar dielectric properties; (4) marinating for dielectric adjustments; (5) designing cylindrical or spherical shaped food for center heating with diameters not to exceed 0.8–3 in; (6) using high-boiling liquids as solvents and plasticizers; (7) using salts and proteins for ionic conductance; (8) designing large surface area for high rate of heating; (9) keeping slow-heating foods at the rim with their thick parts to the outside; (10) keeping fast-heating foods in the center; (11) designing a nonabsorbing edible food laminate to be layered with an absorbing adhesivelike food composition; (12) dampening the effect of product thickness by selecting high dielectric constant and dielectric loss values; (13) reducing the effect of water by food gums and other solutes, i.e., reducing water activity; (14) optimizing target dielectric constant and dielectric loss by maintaining a moisture range between 25 to 40%; (15) and creating a low thickness effect by increasing viscosity; (16) inhibiting transmission by a high loss layer; and (17) putting an edible barrier film between food layers so that they remain immiscible and create a sharp interface with respect to their dielectric constant and dielectric loss properties.

Pillsbury Company has been the leader in developing microwaveable dough, bread, cookie, brownie, shielded products, pancakes, self-topping cakes, and pastry compositions. Other companies involved in proprietary technology development are International Flavors and Fragrance on browning compositions, Nabisco on cookies, National Starch and Cerestar USA on starch products applications, Kraft/General Foods on frozen confections, stuffing mix, and browning compositions, and General Mills on microwave puffing of RTE cereals. Proctor

Ingredient Interactions and Product Development

and Gamble's cake mix was one of the more acceptable products in late 1980s. Golden Valley MW Foods, Inc. led work on microwaveable popcorn including flavor and coating compositions. The patents by Atwell et al. [4], Boehm and Fazzolare [11], and Pesheck [76–78] has a lot of empirical information for microwaveable food product development.

A summary of selected proprietary technologies is included below as examples to support arguments and speculations presented in earlier sections. Key food patents are categorized under main product headings in Table 5.

A. Aroma Coating

Aroma producing coatings are designed with 0.5–5.0% salt, 5.0–10.0 milk solids, 5.0–10.0% fat/oil, aroma sauce, and water for topical coating [57]. Such coatings are capable of producing desired aroma upon exposure to 2450 MHz microwave for no more than 30 s. Such compositions heat fast to higher temperature because of the oil phase and thus permit browning and other temperature-dependent flavor-producing reactions. This concept can be extended to reaction flavors in general.

B. Batters and Breadings

Microwaveable batter and breading compositions can be used as coatings on poultry, fish, meat-based fillings, and vegetables or as coatings to a regular wheat flour–based dough. A general dough composition is wheat flour—68%, double acting leavening acids—2.5%, high amylose starch—15%, dextrin—10%, fat—10.0%, salt—1.5%, and sugar—1.5. It may contain up to 0.8% carboxymethyl cellulose. An example calls for addition of high amylose corn flour in place of high-amylose starch [119]. The batter is prepared at 70% solids. The microwave reactivity derives from seration and low-density, high-loss effects of monosac-

Table 5 Patented Microwaveable Food Products

Product category	References	Product category	References
Aroma coating	57	Layered products	4, 77, 78
Bacon	USP 5, 132, 126	Meat product	39
Baked products	18, 19, 24, 40, 52, 74	Pancake	32, 33
Butter	60, 81, 118, 119	Pasta product	38
Bread	2, 24, 34, 65, 77, 78	Pastry	79
Brownie cookie	11, 58, 117	Pizza	9, 16, 75
Browning compositions	29, 30, 45, 48, 50, 109	Popcorn	53
Dough	2, 18, 37, 47, 74	Sauce	3
Cake mix	64, 87	Snack food	61, 112
Cereal (puffed)	12, 47, 57	Stuffing mix	94

charides, sugar, and oligosaccharides, high-temperature gelatinization and molecular ordering of amylose, high loss due to salts, and positive temperature coefficient of different salts [3, 60, 119]. This is specially true for a high-solids batter where solutes are concentrated in only 30% water. U.S. Patent 5, 531, 944 [60] calls for a dry batter mix composed of 4% double-acting baking powder, up to 20% high-amylose starch, 5.0% dextrin whose amount is inversely proportional to that of double-acting baking powder, and 1.0% food gum [60]. This mix can be applied to frozen foods of categories mentioned above up to an extent of 30%. U.S. Patent 4,595,597 discloses an even simpler microwaveable batter mix composed of at least 50% high amylose corn starch or flour (with 70% amylose) which could be (1) unmodified, (2) acid-converted high-amylose corn flour, and (3) 5 parts unmodified high-amylose corn flour + 1 part high-amylose corn dextrin [2, 59]. The batter is used for prefried microwaveable foods and the novelty resides in hard-to-gelatinize linear amylose molecules. A more complex formulation involves a core food, an emulsoid batter with 60% fat and 1% salt, breading which is finally coated with a 5% solution of egg white solids in water [65]. The emulsion-like batter adheres to core comestible, couples microwave, and heats at a high rate. An egg white coating imparts characteristic of a susceptor film with positive-temperature coefficient of loss.

The addition of waxy or resistant starch, use of moisture-resistant breading, and low water content of the breaded composition are most useful. Breading could be made from pregelatinized barley flour, yellow corn meal, modified corn starch, and double toasted Japanese-style breading. Two compositions stand out: (1) 47% modified starch, 52% water, and 1% salt and (2) 44% waxy maize and 46% oil.

Differential dielectric behavior of starch and protein as a function of hydration, particle size, density, and texturization needs to be better characterized. Enhancement of microwaveability by incorporating oil, chemical leavening agents, and salt will be integral to such investigations [65].

C. Bread

Cooling rate, moisture loss, cooling time, and bread texture details have been disclosed in U.S. Patents 4,842,876 and 5,035,904 [2, 47]. A simple formulation for baked bread containing at least 11% shortening and 10 dietary fibers is disclosed in U.S. Patent 5,164,216 [34]. The bread can be reconstituted easily in a microwave oven due to higher temperature heating effects of shortening and water barrier effect of dietary fiber. Yet another microwaveable bread formula utilizes 5% oil or shortening, 1% monodiglyceride, and 20% of reground bread or pregelatinized starch, all expressed as percent of wheat flour [90]. The dough-making procedure calls for first preparing an emulsion of 55% water (based on wheat flour), 1% monodiglyceride or other emulsifiers, 20% pregelatinized starch (or reground bread crumbs), and 5% shortening to which is added wheat flour. The entire composition is finally mixed into a dough for bread making. Reportedly, microwaveability derives from the effects of the emulsion. A higher sugar and shortening content and incorporation of egg solids, gluten, and rice starch helps microwaveability to a great extent.

D. Brownies and Cookies

These products are good examples of high fat dielectrics containing oil-in-water emulsions. Internal standing waves are common to heating such products and so are hot spots. The effects can be different in magnitude for oil-in-water emulsions in comparison to water-in-oil emulsions [106]. A maximum internal temperature of 125–140°C is developed due to combined effects of sugar and polyols in the formula [117], and steam release is facilitated by high-sugar content, low moisture, and the level of emulsifiers. The resultant microwave brownie has a controlled texture. U.S. Patent 4,948,602 discloses a very novel filled cookie composition [11]. A baked graham or chocolate dough shell containing an oil-based and another water-based filler. The water-based filler is deposited on top of the oil-based filler wherein the upper surface of the cookie shell has an upper opening through which the water-based filler such as marshmallow can flow during microwaving and at least partially coat the cookie shell. Both fillers contain flavoring like vanilla or chocolate. The graham dough and oil-based fillers are coextruded, baked for expansion, and finally injected with marshmallow.

E. Cake Mixes

Most cake mixes are designed with double-acting leavening salts. The sugar used is, of necessity, a 50/50 blend of powdered and granular in order to ensure quick dissolution and inhibit starch gelatinization. The requirement for water and leavening is higher than for cake mixes for conventional oven. Xanthan gum, no more than 0.25%, is part of many mixes. A key requirement is high hidrophile-lipophile balance (HLB) emulsifier system [83, 92]. Another cake mix formula [87] utilizes sugar and flour in a ratio of 1.4 to 1.0, leavening up to 4%, emulsifier up to 3%, and shortening up to 16% [87]. The emulsifier system is 2/1 mixture of propylene glycol monoester and lactylated monodiglyceride. A 60 part portion of this mix is added to 15 parts egg solids, 20 parts water, and 5 parts oil in order to prepare the cake batter for microwave baking. A cookie dough, disclosed in U.S. Patent 4,911,939 [58], contains 100 parts flour, 30 parts blend of sugar and high fructose corn syrup, 25 parts shortening, 4 parts leavening, 5 parts pregelatinized starch, and 3 parts vital gluten [58]. The composition is par-baked, leaving part of the leavening unreacted in order to create a protective skin. The par-baked product can be sold to the consumer for reconstitution in the microwave. High-temperature heating in the microwave results from the combined effect of salts, leavening, oil, an almost complete immobilization of water by added gluten and pregelatinized starch. These two ingredients together can immobilize at least 16 parts of water coming from high fructose syrup, wheat flour, any coming from the pregelatinized starch. Similarly, a pastry/strudel dough of reduced surface density is formulated with 100 parts wheat flour, 7% starch, 10% water, and 20% glycerine-like plasticizer [79]. Bread crumbs of 0.5 cm diameter serve as crisping agent; they are surface-embedded in a dough coating. A self-topping cake is made in two layers: (1) an unleavened, fat free, starch- and sugar-based topping of 400 centipoise in viscosity and (2) a leavened cake batter of a lower 300 centipoise viscosity composed of 40% sugar, 30% flour, 3.5% leavening, and 12% shortening [64]. Sugar and starch ratio in the topping is 1/1 for effective gelatinization control. The two layers, cake mix with sugar-starch topping, keeps their phase integrity during microwave baking. The combination brings about a satisfactory cake crumb structure and texture. Chocolate flavor precursors (sugar, phenyl alanine, and leucine added to a glycerol-propylene glycol blend at pH 10.0) can be applied internally or coated on to the dough for the production of a flavored baked product by microwave baking in less than 2 minutes [15].

F. Ready-to-Eat Cereals and Snack Foods

A coextruded cereal that can be puffed in a microwave is described in U.S. Patent 5,558,890 [12] wherein at least three layers of cereal sheets of below 35% in moisture and each layer 0.25 inch in thickness are laminated under mechanical compression [12]. In one example extruded pellets represent a wheat dough layer, a fruit paste layer, and a corn dough layer, one of the layers being 100% cereal based. The pellets are dried to 11% moisture from 30% and then puffed in a high intensity microwave field strength in the range of 150–300 V/cm. Pellets are compression formed in a pressure range of 100–6000 psia. Such puffed cereals can be sugar coated and flavored. The novelty lies in conversion of most of the 11% moisture into superheated steam at a fast rate under a high-strength microwave field. Also, a oat flake or oat bran cereal containing small amount of lecithin can be microwave cooked into hot cereal [57] wherein the composition could be a dry oat-lecithin product in bulk or packaged in single-serve quantities. In either applications lecithin may be encapsulated as a particulate powder. Cold cereals are puffed at around 11% moisture where both loss and attenuation are low. Upon microwave exposure, there establishes an internal standing wave causing quick rise to high temperature. This mechanism alone accounts for approximately 25% of total power dissipation.

A microwaveable filled snack composition consisting of an outer dough layer and an inner one-to-one oil and flavoring core is disclosed in U.S. Patent 5,124,161 [112] wherein the outer dough casing is composed of 10% pregelatinized or chemically modified starch, 2% salt, 4% sugar, 4% polyol, and 0.5%

acetylated monodiglyceride. It is cooked to full gelatinization, dried, tempered, and extruded through a corotating twin screw extruder. The composition of the filling and moisture of outer layer are critical. Of special interest is a disclosure about a high-protein snack food [61]. This invention calls for complete lactose fermentation of whey with *Kluyveromyces fragilis,* heating the broth at 85°C in order to precipitate whey proteins, preparing a protein paste, mixing the protein paste with 1.5% salt, 0.75% egg solids, 2% leavening agent, 20% corn starch or potato flake or mixture thereof, shaping the blend into snack pieces, frying at 204°C for 15 s, and finally completing cooking in the microwave oven for 3 min. The resultant product is a crisp snack food of superior texture.

G. Conventional Dough Formulations

A high-amylose starch-based dry batter mix coated with an adherent wet batter composition can be par-fried and then baked in a microwave oven in order to produce an outer crust of acceptable color and crispness [119]. The combination can also be used for filled dough or dough-enrobed food stuff. A 50% high-amylose starch is used to an extent of 30% in the mix which, in addition, contains 3% double-acting baking powder, up to 10% shortening, and 1% carboxymethyl cellulose. All ingredients are calculated as percent of 70% wheat flour in the mix. Proper adhesion is ensured by maintaining a pH of 6.4. In another invention, crisp crust is produced by microwaving a preformed, yeast or chemically leavened, frozen dough pieces that have been frozen after baking at 75°C and tempering back to a moisture level of 30% [40]. In essence, salts, lower moisture, and already gelatinized starch and denatured gluten are responsible for optimum crust development. In an other patent a combination of protease and deaminase enzymes, emulsifiers (monodiglycerides or ethoxylated monodiglcerides, or DATEM (diacetyl tartaric acid ester of mono- and diglycerides), oxidizing agents such as potassium bromate, reducing agents such as cysteine or glutathione from yeast hydrolysate are used in a dough and the dough is baked to 15% moisture. It is finished in a microwave oven after aging at 45°F. The salable consumer product is the baked and aged dough piece [47]. Partially baked and cooked dough piece of defined surface area to weight ratio containing at least 10% shortening is reconstituted in the microwave after slow cooling [2]. Low moisture of such products is the critical factor in microwave reconstitution. A frozen dough containing chemically modified starches of acetyl, hydroxypropyl, succinyl or a combination thereof is claimed to have good microwaveability [18]. Replacement of chemically modified starches with rice flour is claimed to produce similar effects [74]. A microwaveable reactive dough composition is disclosed in U.S. Patent 4,448,797 which, applied as a coating, undergoes browning upon exposure to microwaves [37]. The formulation utilizes 5% yeast extract as source of amino acids, 40% flour, 25% shortening, 1% salt, 2% reducing sugar, and 37% water.

H. Meat-Filled Products

An aqueous, ionic solution, in a pH range of 5.0–9.0, containing a single and multivalent cation normality ratio of (sodium + potassium)/calcium = 8/3 is intimately contacted with meat in order to allow a water gain of 7–8%. Thereafter a breading consisting of starch, protein, and flavoring is applied as coating. When cooked in a microwave, the coating creates a crosslinked, starch-protein skin which controls heating rate and penetration appropriate for a juicy and tender poultry preparation [39]. A dry mix preparation, suitable for uncooked or partially cooked meat as either a dry mix or liquefied dry mix, which develops as thick sauce during microwave cooking of meat, has been described to contain 0.4% xanthan gum, 1.5% uncooked starch, and other flavor and nonflavor additives [116] in a pH of range of 3.5–4.0.

I. Pancake Compositions

A pancake composition containing 40% wheat flour with 5% corn flour by weight, 30% water, 20% high fructose corn syrup, 3% butter milk solids, 3% leavening agent, 1.5% salt, and 1% egg yolk solids is disclosed in U.S. Patent 5,447,739 [33]. The pancake composition is cooked to 8% moisture to be sold to consumer for reconstitution in the household microwave oven. Another frozen, cooked pancake of 50% moisture is sold to consumers for reconstitution in the microwave but formulated differently as 40% wheat flour, 35% water, 15% high fructose corn syrup, 8% oil, 3% butter milk solids, 1.2% salt, 6% leavening, 1% albumen, and 1% egg yolk solids [32]. Added oil and albumen produce equivalent microwave reactivity at substantially higher moisture content. A foamed (whipable) toping consisting of 30% margarine, 0.5% gelatin, 0.2% carboxymethyl cellulose, 0.5% maple flavor, 12% powdered sucrose, 48% corn syrup, and 9% water works well with microwaveable pancakes. The topping added to pancake accounts for the gelling effect by gelatin.

J. Pasta Products

Dull waxy starch is covered in U.S. Patent 5,397,586 for manufacture of microwaveable pasta product [38]. The composition contains 15% amylose extender dull (aedu) starch, 30% water, 70% wheat flour, and 3–5% egg solids [38] and the dough can be used both for flour tortilla and pasta production.

K. Pizza

A microwaveable dough, based on salt, yeast leavening, and bread crumb, for pizza crust composition is disclosed in U.S. Patent 4,834,995 wherein the crumb is surface embedded [16]. Advantage is taken of a 5% moisture lower cracker crust layer topped with a no more than 30% moisture bread dough, both acting as a microwaveable pizza crust [9]. The novel effects reside in low-moisture and electrolyte composition of the overlayer of the bread dough. A blistered and aerated dough gives an acceptable pizza crust for reconstitution in the microwave embodying the microwave reactive effects low density and surface geometry. Similar effects can be obtained by using an upper prebaked layer atop a raw dough core. A microwave reactive coating can be prepared for pie dough using 38% wheat flour, 18.5% shortening, 27.9% water, 0.85 salt, and 1.3% dextrose [37]. The microwave effect presumably is that of an emulsion. A color coating composition of 5% tea, coffee, and chicory as coloring agent dispersed in a 15% aqueous mixture of margarine, milk solids, egg white solids is applied to an extent 3–5% by weight of pizza crust for acceptable crust color [75].

L. High-Temperature Cooking Sauce

An emulsoid sauce composition of pH 5.00 is prepared with a 70 parts emulsion (of 3% browning mix, 2% caseinate, 35% fat, 2% modified starch, 0.5% lysophospho-lipoprotein type emulsifier, and 57.5% water) and 30 parts cheese [3]. The emulsifier is a phospholipase A_2 converted egg yolk.

M. Stuffing Mix

A 20 lb/ft^3 density dry stuffing mix is based on 85% bread crumb, 10% oil, and 5% seasoning [117]. The novelty lies in low moisture, use of pregelatinized ingredients, and incorporation of oil in the formula.

N. Sweetener Coating for Microwaveable Popcorn

A recent disclosure presents a good example of a sugar-based microwaveable flavoring [53]. A sweetener blend of crystalline sucrose and oil and a glaze composed of an salt, emulsifier, oil, and corn syrup are applied to popcorn, the former at a sweetener-to-popcorn ratio of 1:1 and the latter sufficient to afford complete coating.

REFERENCES

1. Adamson, A. W. 1960. Physical Chemistry of Surface, chapt. VI, pp. 317–343, Interscience Publishers, New York.
2. Anderson, K. H. 1989. Method of microwave heating of starch based products, U.S. Patent 4,842,876.
3. Andreae, C. F., Dazo, P. E., Kuil, G. M., Gerardus, A., and Mulder, J. F. 1998. High temperature cooking sauce, U.S. Patent 5,738,891.

4. Atwell, W. A., Pescheck, P., Krawiecki, M. M., and Anderson G. R. 1990. Microwave food products and method of their manufacture, U.S. Patent 4,926,020.
5. Bengtsson, N. E., and Ohlsoon, T. 1979. Application of microwave and high frequency heating in food processing. Food Process Engineering Second International Congress on Engineering and Foods. 8th European Food Symposium, Finland.
6. Bengtsson, N. E., and Risman, P. O. 1971. Dielectric properties of foods at 3 GHz as determined by a cavity perturbation technique. II. Measurements on food materials. J. Microwave Power 6(2):107–124.
7. Banner, B. A, Richardson, L. V., and Darley, K. S. 1987. Food coating compositions, U.S. Patent 4,675,197.
8. Benzason, A., and Edgar, R. 1976. Microwave tempering in the food processing industry. Electron Progress XVII (1):8–12. Rathyon Co.
9. Bone, D. P., and Manoski, P. M. 1981. Frozen pizza crust and pizza suitable for microwave cooking. US Patent 4,283,424.
10. Bookwalter, G. N., Shukla, T. P., and Kwolek, W. F. 1983. Microwave processing to destroy Salmonellae in corn-soy-milk blends and effect on product quality. J. Food Sci. 47:1683.
11. Boehm, R. G., and Fazzolare, R. D. 1990. Filled cookie, U.S. Patent 4,948,602.
12. Brown, G. E., Schwab, E. C., and Herrington, T. R. 1996. Multi-layered puffed T-T-E cereal and high intensity microwave method of preparation, U.S. Patent 5,558,890.
13. Buckholz, L., Byrne, B., and Sudhol, M. A. 1989. Process for microwave browning and product produced thereby, U.S. Patent 4,882,184.
14. Buckholz Jr., L. L., Byrne, B., and Sudhol, M. A. 1990. Process for microwave browning and product produced thereby, U.S. Patent 4,904,490.
15. Byrne, B., Miller, K. P., Tan, C.-T., Buckholz Jr., L. L., and Sudhol, M. A. 1990. Process for microwave chocolate flavor formation in and/or on foodstuffs and products produced thereby, U.S. Patent 4,940,592.
16. Canzoneri, S. 1989. Method of preparing pizza dough, U.S. Patent 4,834,995.
17. Chmiel, O., et al. 1997. Food microemulsion formulations, U.S. Patent 5,674,549.
18. Cochran, S. A., Cin, D. A., and Veach, S. K. 1989. Microwaveable baked goods, U.S. Patent 4,885,180.
19. Cochran, S. A., Cin, D. A., and Veach, S. K. 1990. Microwaveable baked goods, U.S. Patent 4,957,750.
20. Coleman, E. G., Wanger, D., Ballard, D. J., Stone, C. E., Swallow, N. A., and Carey, N. L. 1987. Coating composition for microwave cooking, U.S. Patent 4,640,837.
21. Condon, A. U., and Odisha, H. 1958. Handbook of Physics, chapt. 7, McGraw Hill, New York.
22. Cone, M., and Snyder, T. 1986. Mastering Microwave Cooking, Simon and Schuster, New York.
23. Copson, D. A. 1975. Microwave Heating, AVI, Westport, CT.
24. Corbin, D., and Corbin, S. 1992. Process of making a microwaveable bakery product, U.S. Patent 5,116,614.
25. D'Amico, L. R. 1988. Process for producing a freeze-thaw stable microwaveable pre-fried foodstuff, U.S. Patent 4,778,684.

Ingredient Interactions and Product Development

26. Decareau, R. V. 1983. Microwave in food processing. Food Technology in Australia 35:18.
27. Decareau, R. V. 1984. Microwave in food processing. Food Technology in Australia 36:81.
28. Decareau, R. V. 1984. Microwave in the Food Processing Industry. Academic Press, New York.
29. Domingues, D. J., Atwell, W. A., Beckman, P. J., Panama, J. R., Conn, R. E., Matson, K. L., Graf, E., Feather, M. S., Fahrenholtz, S. K., and Huang, V. T. 1992. Process for microwave browning, U.S. Patent 5,108,770.
30. Domingues, D. J., Atwell, W. A., Graf, E., and Feather, M. S. 1993. Process for forming a microwave browning composition, U.S. Patent 5,223,289.
31. Edgar, R. 1986. The economics of microwave processing in the food industry. Food Technology, June, p. 106.
32. Emaluenson, R. L., et al. 1994. Process of making misted microwaveable pancake, U.S. Patent 5,277,925.
33. Emaluenson, R. L., et al. 1995. Misted microwave pancake, U.S. Patent 5,447,739.
34. Engelbrecht, D. A., and Spies, R. D. 1992. Microwaveable bread product, U.S. Patent 5,164,216.
35. Federal Register. 1968. Food Additive, Subchapter 6—Radiation and radiation sources intended for use in the production, processing, and handling of foods. Revised paragraph 121.3008, food additives regulation. Part 121, vol. 23(45), March 6.
36. Frohlich, H. 1958. Theory of Dielectrics, Oxford University Press, Ely House, London.
37. Fulde, R. C., and Kwis, S. H. 1984. Brownable dough for microwave cooking, U.S. Patent 4,448,791.
38. Furcsik, S. L., Stankus, C. A., and Friedman, R. B. 1995. Pasta products, U.S. Patent 5,397,586.
39. Gagel, S., Sheen, S. S., and Moyer, J. 1995. Process for preparing tender, juicy microwaveable meat, U.S. Patent 5,384,140.
40. Gantwerker, S., and Walsh, G. E. 1994. Process for preparing a precooked microwaveable frozen baked food product, U.S. Patent 5,281,433.
41. Gerling, J. E. 1987. Microwave oven power. A technical review. J. Microwave Power E^2 22:199.
42. Goeldeken, D. L. 1997. Dielectric properties of pre-gelatinized bread system at 2450 MHZ as a function of temperature, moisture, and specific volume, J. Food Sci. 62(1):145.
43. Hasted, J. B. 1961. The dielectric properties of water. Progress in Dielectrics 3:101–150. Haywood and Co. Ltd., London.
44. Hasted, J. B. 1973. Aqueous Dielectrics, Chapman and Hall, London.
45. Haynes, L. C., et al. 1992. Microwave browning composition, U.S. Patent 5,089,278.
46. Hsia, S. T., and Ogasawara, P. 1985. Food coating composition for foods cooked by microwave, U.S. Patent 4,518,618.
47. Huang, V. T., et al. 1991. Starch based products for microwave cooking or heating, U.S. Patent 5,035,904.

48. Hsu, C. J., and Melachourius, N. 1993. Method for producing a microwave browning composition, U.S. Patent 5,196,219.
49. Jain, R. C., and Voss, W. A. G. 1986. Dielectric measurements of browning agents, additives, and other high loss materials. J. Microwave Power E^2 21(2):120.
50. Kang, Y. C., et al. 1991. Microwave browning composition and process for producing the same, U.S. Patent 5,059,434.
51. Kang, Y. C., et al. 1991. Process for microwave browning proteinaceous fibrous meat products, U.S. Patent 4,985,261.
52. Kang, Y. C., et al. 1992. Process for microwave browning uncooked baked goods foodstuffs, U.S. Patent 5,091,200.
53. Jensen, M. L., and Risch, S. J. 1995. Composition for sweetening microwave popcorn; method and product, U.S. Patent 5,443,858.
54. Kent, M. 1987. Electrical and dielectric properties of food materials, Science and Technology Publishers, England.
55. Kenyon, E. M. 1970. The feasibility of continuous heat sterilization of food products using microwave power. U.S. Army Natick laboratory Tech. Report 71-8-FL (AD-715853).
56. Keyser, W. L., Medrow, R. K., and Milling, T. E. 1991. Method of preparing hot oat cereal in a microwave oven, U.S. Patent 5,069,917.
57. Kunz, G. F. 1991. Microwave food aroma composition, U.S. Patent 4,992,284.
58. Lou, W. C., and Fazzolare, R. D. 1990. Shelf-stable microwaveable cookie dough, U.S. Patent 4,911,939.
59. Lenchin, J. M., and Bell, H. 1986. Batters containing high amylose flour for microwaveable pre-fried foodstuffs, U.S. Patent 4,595,597.
60. Melvej, H. S. 1995. Batter mix for frozen food products and method of making. U.S. Patent 5,431,944.
61. Mickle, J. B., Smith, W. J., and Dieken, L. M. 1980. Method of manufacturing a high protein snack food, U.S. Patent 4,183,966.
62. Meisel, N. 1973. Microwave applications to food processing and food systems in Europe. J. Microwave Power 8(1):143–148.
63. Misra, D. K. 1981. Permeability measurement of modified infinite samples by a directional coupler and sliding load. IEEE Transactions. MTT 29(1):65.
64. Moder, G. J., et al. 1993. Self-topping cake, U.S. Patent 5,215,774.
65. Monagle, C. W., and Smith, J. C. 1985. Process for breading food, U.S. Patent 4,518,620.
66. Moody, R. D. 1981. Microwave cooking browning composition, U.S. Patent 4,252,832.
67. Moore, N. H. 1968. Microwave energy in the food field. 20th Anniversary Meeting of R & D Associates, Misc. Eng. Applications Newsletter 1(1):5–7.
68. Mudgett, R. E., Goldblith, S. A., Wang, D. L. C., and Westphal, W. B. 1971. Prediction of dielectric properties in solid foods of high moisture content at ultrahigh and microwave frequencies. J. Food Process and Preservation 1:119.
69. Mudgett, R. E. 1982. Electrical properties of foods in microwave processing. Food Technology, February, p. 104.
70. Mudgett, R. E., and Schwartzberg, M. G. 1982. Microwave food processing—pasteurization and sterilization: a review. Food Process Engineering 78:1–11.

Ingredient Interactions and Product Development 393

71. Mudgett, R. E. 1986. Microwave properties and heating characteristics of food. Food Technology. June, pp. 84–93, 98.
72. Ohlsoon, T., and Bengtsson, N. E. 1973. Dielectric food data for microwave sterilization processing. J. Microwave Power 10(1):93–108.
73. Osepchuk, J. M. 1984. A history of microwave heating applications. IEEE Transactions on Microwave Theory and Techniques 32(9):1200–1224.
74. Ottenberg, R., and Ottenberg, L. 1984. Yeast raisable wheat based food products that exhibit reduced deterioration in palatability upon exposure to microwave, U.S. Patent 4,463,020.
75. Peleg, Y. 1993. Microwave reconstitution of frozen pizza, U.S. Patent 5,260,070.
76. Pesheck, P. S., et al. 1991. Microwave food products and method of their manufacture, U.S. Patent 4,988,841.
77. Pesheck, P. S., et al. 1991. Microwave food product and methods of their manufacture, U.S. Patent 5,008,507.
78. Pesheck, P. S., et al. 1992. Microwave food product and methods of their manufacture and heating, U.S. Patent 5,140,121.
79. Pesheck, P. S., McIntyre, T., and Levine, L. 1996. Pre-baked microwaveable pastry dough, U.S. Patent 5,576,036.
80. Pei, D. C. T. 1982. Microwave baking—new developments. Baker's Digest. February, pp. 8–10, 32–33.
81. Petcavich, R. J. 1989. Method and container for producing batter based baked good, U.S. Patent 4,865,858.
82. Pething, R. 1979. Dielectric and Electronic Properties of Biological Materials, John Wiley & Sons, New York.
83. Pinegar, R. K. 1986. Process for preparing parfried and frozen potato products, U.S. Patent 4,590,080.
84. Richman, P. O., and Bengtsson, N. E. 1970. Dielectric properties of foods at 3 GHz as determined by cavity perturbation technique—measurement of food materials. J. Microwave Power 6(2)101–124.
85. Rothymayer, W. W. 1975. Food Process Engineering. Proceedings Royal Society, London B191:71–78.
86. Roebuck, B. D., and Goldblith, S. A. 1972. Dielectric properties of carbohydrate-water mixtures at microwave frequencies. J. Food Science 37:199.
87. Roedbuck, R. M., and Palumbo, P. D. 1983. Microwave cake mix, U.S. Patent 4,396,635.
88. Rosenberg, U., and Bogl, W. 1987. Microwave thawing, drying, and baking in food industry. Food Technology, June, pp. 85–91.
89. Rosenberg, U., and Bogl, W. 1987. Microwave pasteurization, sterilization, blanching, and pest control in the food industry, June, pp. 92–99.
90. Saari, A. L., et al. 1991. Method of preparing microwave bread, U.S. Patent 5,049,398.
91. Shanbhag, S. P., 1991. Process for microwave reheatable french fried potatoes and product thereof, U.S. Patent 5,004,616.
92. Shanbhag, S. P., and Cousminer, J. J. 1990. Process for preparing potato granule coated french fried potatoes, U.S. Patent 4,931,296.

93. Shaw, T. N., and Galvin, J. A. 1949. High frequency heating characteristics of vegetable tissues determined from electrical conductivity measurements. Proceeding REE 37:83.
94. Sabhlok, J. P., Horan, W. J., and Carlton, D. K. 1990. Microwaveable stuffing mix, U.S. Patent 4,940,591.
95. Schiffman, R. 1971. Application of microwave energy to doughnut production. Food Technology 25(6)7:718–722.
96. Schiffman, R. 1973. The application of microwave to the food industry in the United States. J. Microwave Power 8(2):137–142.
97. Schiffman, R. 1986. Food product development for microwave processing. Food Technology, June, pp. 54–98.
98. Schmidt, W. L., et al. 1988. Method of preparing a packaged frozen confection, U.S. Patent 4,794,008.
99. Schmidt, W. 1961. The heating of food in a microwave cooker. Phillips Tech. Rev. 22(3):89–101.
100. Shukla, T. P., and Miara, D. K. 1989. Unpublished data.
101. Shukla, T. P. 1988. Microwaveable foods: dielectric properties of ingredients. Intl. Conference on Microwaveable Foods, Scotland Business Research Inc., Princeton, NJ.
102. Shukla, T. P. Microwaveable Food Ingredients, Cereal Foods World, 43(10):770.
103. Shukla, T. P. 1989. Hydrocolloids for microwaveable foods, Intl. Conference on Formulating Foods for the Microwave Oven. The Packaging Group, Inc., Milton, NJ.
104. Shukla, T. P. 1993. Bread and bread-like dough formulations for the microwave, Cereal Foods World, 38(2):95.
105. Shukla, T. P. 1993. Batter and breadings for traditional and microwaveable foods, cereal Foods World 38(9):701.
106. Shukla, T. P. 1995. Microwave technology in cereal Foods Processing, Cereal Foods World, 40(1):24.
107. Singh, S. P., and Miara, D. K. 1981. A magic T and sliding short technique for measuring the permittivity of modified infinite samples. AEDU 35(3):137.
108. Smith, F. J. 1979. Microwave hot air drying of pasta, onions, and bacon. Microwave Eng. Applications Newsletter 2(6):6–12.
109. Steinke, J. A., et al. 1990. Browning agent for food stuffs, U.S. Patent 4,968,522.
110. Suzuki, T., Balwin, R. E., and Korsdigen, B. M. 1984. Sensory properties of dextrose and sucrose cured bacon: microwave and conventionally cooked. J. Microwave Power 19(3):9.
111. Tinga, W. R. 1969. Multiphase dielectric theory—applied to cellulose mixtures, Ph.D. dissertation, University of Alberta, Edmonton, Alberta, Canada.
112. Van Lengerich, B. H. 1992. Filled, microwave expandable snack food product and method and apparatus for its production, U.S. Patent 5,124,161.
113. Vermeulen, F. E., and Chute, F. S. 1987. The classification of processes using electric and magnetic fields to heat materials. J. Microwave Power EE:187.
114. VonHippel, A. R. 1954. Dielectric materials and Applications, MIT Press, Cambridge, MA.

Ingredient Interactions and Product Development 395

115. Voss, W. A. G. 1969. Advances in use of microwave power, HEW Seminar, April 25, paper 008, p.2.
116. Wilson, M. N. 1991. Dry mix for preparation of in-situ sauce for foodstuffs, U.S. Patent 5,008,124.
117. Yasosky, J. J., Hahan, P. W., and Atwell, W. A. 1990. Controlling the texture of microwaveable brownies, U.S. Patent 4,940,595.
118. Yasoski, J. J. 1993. Microwaveable batter-coated, dough-enrobed foodstuff, U.S. Patent 5,194,271.
119. Yasosky, J. J., Pesheck, P. S., and Levine, L. 1996. Microwaveable batter coated dough, U.S. Patent 5,520,937.
120. You, T. S., and Nelson, S. O. 1988. Microwave dilectric properties of rice kernels. J. Microwave Power 23(3):150–159.
121. Nelson, S. O., Forbus, Jr., W. R., and Lawrence, K. C. 1994. Microwave permittivities of fresh fruits and vegetables from 0.2 to 20 GHz. Trans. ASAE 37(1):183–189.
122. Roebuck, B. D., and Goldblith, S. A. 1972. Dielectric proeprties of carbohydrate-water mixtures at microwave frequencies. J. Food Sci. 37:199–204.
123. Nelson, S. O. 1985. A mathematical model for estimating dilectric constant of hard red winter wheat. Trans. ASAE 28(1):234–238.
124. Dean, J. A. 1985. Lange's Handbook of Chemistry. McGraw Hill, New York.
125. Buffer, C. R. 1993. Microwave cooking and processing: Engineering fundamentals for the food scientist. AVI. 169 pp.
126. Schiffmann, R. F. 1987. Performance testing of products in microwave ovens. Microwave World 8(1):7–14.
127. Schiffmann, R. F. 1990. Problems in standardizing microwave oven performance. Microwave World 11(3):20–24.
128. Schiffmann, R. F. 1991. Oven considerations for the food product developer. Paper presented at the 26th Annual Symposium of International Microwave Power Institute, Buffalo, NY.
129. Stanford, M. 1990. Microwave oven characterization and implications for food safety in product development. Microwave World 11(3):7–9.

12
Packaging Techniques for Microwaveable Foods

Timothy H. Bohrer
Ivex Packaging Corporation
Lincolnshire, Illinois

Richard K. Brown
Fort James Corporation
Sarasota, Florida

I. INTRODUCTION

The earliest microwave packages were created shortly after it was discovered that the electromagnetic radiation from a magnetron tube was capable of heating food. As the story goes, Percy Spencer, who was then working for Raytheon after having been involved in magnetron development during World War II, stood in front of a magnetron and noticed that his hand became warm. He also realized that the candy bar in his pocket was melting [1]. Thus, Spencer not only discovered microwave heating of food; he also created the first microwave package. A U.S. patent [2] issued in 1949 with Spencer as the inventor showed a pouch specifically designed for microwave heating of an entire cob of popcorn. This is the most commonly accepted version of the genesis of microwave packaging, although other companies, including General Electric, had been exploring the use of microwave radiation for food thawing and other uses.

Since the initial uses of microwave ovens were principally in food service, packages were initially designed for this generally controlled food distribution and preparation channel [3]. In food service, the consumer rarely sees the package, and is usually oblivious to its use in food preparation. Product protection is the primary functional requirement in this end use. While food service uses for microwaveable foods will be mentioned, the majority of this discussion of packaging for microwaveable foods will focus on packaging for consumer use.

Retail packages for general consumer use were not developed until home ovens began to proliferate and food manufacturers perceived consumer demands for the inclusion of microwave heating instructions on packages. To do this meant satisfying the three key functional requirements for all consumer packages:

1. *Protection:* Maintaining product integrity and quality through consumer use
2. *Utility:* Assisting the consumer in the use of the product
3. *Motivation:* Attracting a consumer to purchase and use the product

Successful consumer packaging for microwaveable foods must fulfill these functions as well. The uniqueness of microwave packaging is the requirement for the package to be part of the microwave heating cycle of the food. In this regard, the package assists the consumer in the use of the product by serving as a food heating container or device.

To cope effectively with this added complexity, a holistic design approach is required, as package components usually must perform multiple functions in the finished package. For example, a high-gloss coating on a paperboard folding carton that is present on the carton for marketing reasons must also be heat resistant if the carton will be put into the microwave oven for the cooking cycle. Complicating the design of microwaveable food packaging is the wide range of materials and approaches available.

To undertake an effective discussion of options, one should adopt the underlying premise that in all cases, performance is only meaningful in relation to the cost and ease of implementation of a specific option. The measure of what a package does functionally against its total cost in use is different for the cooking needs of each food item and the product positioning that a food company's marketing group chooses. Is the product to be positioned as premium or value? Does preparation need to be one step and foolproof, or does the likely consumer of the product expect to take an active role in a complicated preparation? When these and other questions are answered for a wide variety of products, a wide variety of solutions are developed and used in the market. More solutions are always in various stages of development.

For microwave foods, the function of utility poses a special challenge, that of accommodating the different way that food cooks in a microwave field, as has been discussed in other chapters. Food manufacturers have worked closely with material and packaging suppliers to develop a wide range of microwave packaging approaches.

This chapter discusses the evolution of packaging for microwaveable foods, describes approaches available and in use today, highlights key performance differences between these options, and suggests areas for further development. The primary division among package types is between active and passive packages. These two primary package types will be defined, performance requirements will be presented, and alternative materials and configurations will be described and compared.

One final introductory note: The words "cooking" and "heating" will be used interchangeably in this discussion to refer to the process in which microwave energy in the oven cavity is used to bring the food product to a finished thermal state.

II. PASSIVE PACKAGES

A. Definition

Passive packages, as their name implies, are made of materials that do not appreciably react to the microwave field of the oven or appreciably modify the power distribution in the oven. This definition means that we will consider here only those that are microwave transparent, leaving reflective packages, which modify the power distribution, for the later discussion on active packages (see Sec. III). The principal function of a microwave transparent package is to serve as a container during storage, distribution, and heating. In recognition of the overriding importance of the containment function, a patent for an early passive microwave package is simply titled "Receptacle" [4]. If these containers are primary packs, they will be likely to also be a vehicle for printed information. In cooking, however, they only serve to constrain the product in a desired configuration, prevent spilling or leakage of the product, and facilitate removal of the cooked product from the oven. In some cases, the packages are designed to encourage the consumer to serve the cooked product in the package.

B. Performance Requirements

The passive package used with many microwaveable foods must also be able to tolerate conventional oven heating. This requirement existed with many of the earliest introductions of microwaveable foods. Because microwave ovens had not penetrated a majority of households, food manufacturers did not want to limit the use of their product to microwave oven heating, risking the loss of potential sales that might otherwise have been made to consumers not owning microwave ovens. As a result, when food manufacturers began to introduce foods with microwave heating instructions, they continued to include conventional oven heating instructions. This ability to work satisfactorily in microwave and conventional ovens came to be known as "dual oven use" or "dual ovenability." A principal difference between conventional oven heating and microwave heating is the temperature that the package reaches during the heating cycle.

From a practical design standpoint, the temperature-resistance requirements for "microwave-only" passive packages are less rigorous than those for dual ovenable passive packages. As long as dual ovenability is a requirement for a passive package, designers default to a package capable of enduring the conventional oven temperature exposure for the cooking instructions. Given the large temperature variations inherent in home conventional ovens, proper testing must anticipate abusive use and conditions.

Foods are heated in a conventional oven by thermal transfer from the hot air in the oven by convection and conduction, and from the heating elements and hot walls of the oven by radiation. Since the package surrounds the food and contacts the hot air and is in a direct line of incidence for infrared radiation, it can easily reach and sustain temperatures above that of the bulk of the food. This is especially true for areas of the package not in contact with the food, where little or no heat transfer from the package to the food moderates the temperature of the exposed package segment.

The temperature stability required of a passive package for microwave use is dependant on the formulation of the food to be heated, as the temperature that the package reaches can be no higher than that reached by the hottest part of the food. In foods with high moisture contents, temperatures rise to the vicinity of 100°C, and then plateau in that range until substantial dehydration occurs. While dehydration will allow higher temperatures to be reached, continued microwave heating after that point moves from the realm of "cooking" to "ruining," and most microwave oven users are familiar with the extreme toughening that can occur from overheating a food in the microwave.

Foods with significant amounts of free oil can reach much higher temperatures under microwave exposure in the areas where free oil exists. Test cooks with intended packaging materials and the food product are recommended, as these tests are generally sufficient to identify any potential problems of this nature.

To ensure that the package is truly passive, designers must run tests to confirm that no components of a passive package are strongly interactive with microwave radiation. For example, in one early package, a hot melt was used to seal the end flaps of a folded paperboard tray. Components of the hot melt absorbed microwave energy, causing very high temperatures to be reached by the hot melt during the microwave heating cycle. Not only could consumers potentially burn themselves handling what they anticipated was a relatively cool section of the package after heating, the package bonds could release if the hot melt exceeded its melt point. Tests of empty packages under high applied power loads can readily identify these kinds of problems before packages are supplied to the marketplace.

The following discussion of materials that have been used for passive microwave packages focuses on their relative strengths and weaknesses, as well as current applications. The materials and fabrication methods chosen depend on the needed levels of

- Temperature resistance
- Grease resistance

- Moisture and oxygen barrier
- Food release
- Sealability
- Cost

C. Packaging Alternatives and Performance

Table 1 summarizes the performance characteristics of a selection of commonly used options for passive packages designed for microwave heating. These and other options are described in more detail in the following sections.

1. Glass

A comment about a widely used packaging material that predated microwave cooking is appropriate at this point. On first examination, glass is an attractive candidate as a passive package for microwaveable foods, since it is almost totally transparent to microwave energy, and offers excellent product visibility and barrier properties. Drawbacks include high weight for contained volume, relatively high cost, and fragility. In addition, there are several problems associated with the microwave heating of foods in glass, one the result of the products typically packed in glass and the second the result of the predominant geometry of glass packages.

Most foods packaged in glass that have microwave heating instructions are relatively viscous sauces or pureed foods. The viscosity of the food results in inhibition of convection currents, which are useful in transferring heat from hot spots to cold spots. Foods of this type need to be well stirred prior to serving to avoid the potential of mouth burns, such as have been reported for baby foods heated in glass jars in the microwave. This problem is present for all passive packages used to heat viscous foods.

A related potential problem has come to light—the explosive boiling and ejection of viscous foods. Since most jars are in the shape of vertical cylinders, a pressure gradient exists from the top to the bottom of the contents. In the case where a viscous food is heated in such a container, if localized heating causes boiling, the gases released may not easily make their way to the surface, and build pressure rapidly. Much like stovetop heating of spaghetti sauce, small portions of sauce may be ejected and spatter from the top of the jar, potentially burning the cook. There is anecdotal evidence that ejection can be precipitated by the insertion of a spoon to stir the contents. While this result is a function of geometry, not package material, glass jars most commonly possess the geometry prone to the problem.

Products packaged in glass intended to be heated in the microwave should have very conservative heating instructions provided to consumers. Those

Table 1 Passive Microwave Packaging Options

Package style	Material(s)	Maximum service temperature, °C	Oxygen barrier	Moisture barrier	Grease resistance	Sealability	Dual ovenable?	Comments
Tray, sleeve, support board, or overwrap	LDPE coated paper or paperboard	95	Poor	Fair	Poor	Good	No	Light duty reheating only
Overwrap	Heat sealable oriented PET film	220	Poor	Good	Good	Good	No	Need venting during reheat, barrier improved by coatings
Overwrap	Heat sealable oriented PP film	110	Fair	Fair	Good	Good	No	Need venting during reheat, barrier improved by coatings
Tray	PP coated paperboard	125–135	Poor	Fair	Good	Good	No	Pressed or folded trays
Tray	PET coated paperboard	205	Poor	Fair	Good	Fair	Yes	Pressed or folded trays
Thermoformed tray	LDPE	75	Poor	Good	Fair	Good	No	Light duty reheating only
Thermoformed trays, clamshells	PS and foamed PS	80	Good	Fair	Good	Poor	No	Light duty reheating only
Thermoformed tray	PP	110	Poor	Good	Good	Good	No	Brittleness an issue at freezer temperatures
Thermoformed tray	CPET	220	Good	Fair	Good	Fair	Yes	Generally pigmented white or black

instructions should lead the consumer to heat the product in increments, stirring and testing the temperature frequently. While less convenient than just heating in one "shot," the gradual temperature rise that will result is much safer for those preparing and eating the food. At least one pasta sauce product now has microwave heating instructions that specifically instruct the consumer to not heat the contents in the jar, but pour the contents into a microwave safe container for heating with frequent stirring.

More recently, the advantageous barrier and microwave transparency properties of glass have been incorporated into flexible packages that contain thin layers of glass or other inorganic barrier layers deposited by physical or chemical vapor deposition [5]. These packages often take the form of stand-up pouches and are growing in use for liquid food products.

2. Paper, Paperboard, and Film Structures

Uncoated paper and paperboard are at the low end of the performance spectrum, and are typically used in vending applications. The performance required is low, as short shelf-life requirements, relatively short heating cycles, and a low expectation on the part of the consumer for appearance characterize the application. Also important is the fact that there is no consumer expectation that these packages will work in a conventional oven. The whole concept of the vending environment is to provide microwave ovens adjacent to the vending machines to satisfy the purchaser's needs for one-stop buying, heating, and eating. Heating cycles are relatively gentle, as the food is fully precooked and microwave exposure is intended only to bring it to an acceptable eating temperature. Often these packages come in the form of simple plates, trays, sleeves or support boards that are contained in a lightweight film overwrap. They may have a light grease-resistant treatment, but generally are not printed; labels adhered to the overwrap communicate the product identity and required ingredient statement.

Printed overwraps of treated paper are a step up in decoration, although at the sacrifice of product visibility, which is generally desired for vending. Film overwraps are also printable, and are increasingly used for small, hand-held items for quick reheating. An example is single tamales sold in the freezer case at a supermarket or convenience store. Unless these wraps are opened prior to the microwave heating cycle to provide some venting of moisture, unacceptably soggy textures result.

The boil-in-bag is another kind of flexible structure that can be used in microwave ovens. Initially designed for immersing in boiling water, these packages generally contain frozen vegetables, vegetables with sauce, rice, and other side dishes. These packages are made from coextrusions or laminations containing sealant layers of medium density polyethylene (MDPE), whose melt point is above the boiling point of water, and are compatible with microwave heating, provided that the foods do not have large amounts of free oil. Other, higher temperature-resistant versions of pouches have been developed for more demanding cook cycles [6]. A variety of venting approaches can be used to relieve the pressure that is created when moisture in the package is vaporized. These range from package instructions that call for the consumer to cut slits in the package prior to heating, to complex concepts that involve selective portions of the package that soften during heating to open holes [7], to microwave reactive inks that melt holes in the polymer film during heating [8].

Obtaining high levels of gas barrier is difficult with many of these resins, and a patent in the late 1980s proposed an aluminum foil laminate as a barrier pouch stock [9]. The patent describes the removal of the foil layer prior to placing the pouch in the oven prior to heating. This approach has been superseded by structures like those described in Ref. 5.

Plastic-coated paper or paperboard is a common substrate used for passive packages, combining the rigidity of the fiber component with the chemical resistance and sealability of the polymer component. The materials are combined using extrusion coating of a molten polymer resin, adhesive or extrusion lamination of a previously fabricated polymer film, or by roll coating of a polymer solution.

Low-density polyethylene (LDPE) coated paperboard can be used for light-duty microwave applications, limited to microwave heating for which the temperature reached by the polymer does not exceed the boiling point of water. LDPE has poor grease resistance during cooking and it is not suitable for conventional oven exposure. A number of carton structures have been proposed and used for PE coated paperboard for microwave heating [10]; frozen vegetables are packaged in this type of structure.

Polypropylene (PP) extrusion coatings may be used for more rigorous microwave applications as a result of the polymer's higher softening and melt points. It has better grease and temperature resistance than LDPE, but is still not suitable for conventional oven usage. While high-density polyethylene (HDPE) offers better temperature and grease resistance, its higher cost and greater difficulty in sealing have kept it from becoming a common coating for paper or paperboard.

Flexible composite structures of paper and polymer films were used for early passive packages for popping of corn, sometimes in combination with rigid elements [11]. Functional challenges included the need to preserve the desired moisture content of the corn, withstand oil penetration and survive the steam generated during the popping process. These passive packages are no longer in use, as susceptor packages (see Sec. III.C.10) for microwave popcorn have captured the entire market.

Paper and paperboard are suitable for dual ovenability when extrusion coated with polyester (PET) resins [12, 13] or adhesive laminated with PET film. This substrate can be formed into pressed trays, heat-sealed folded trays, or used as support disks for food products. It has also been the subject of designs for folding cartons [14]. Grease resistance is good, although PET is only a modest gas barrier during storage. It is capable of withstanding the full range of microwave conditions to be expected for food cooking, and can be used for conventional oven recipes with temperatures up to 205°C, assuming that product fill is reasonable. Areas of the package not in good contact with the food will begin to brown as cook time increases at this temperature. While not harmful, the embrittling that accompanies severe browning can make the package weaker and more difficult to handle. The dual ovenability of these packages was arguably more important at their time of introduction, when household penetration of microwave ovens was low and growing. Today, the vast majority of these packages are used in microwave ovens.

Folded trays have heat sealed lids of the same material as the bases, and cooking instructions for microwave heating instruct the user to peel back the lid at one corner for venting or to make a knife cut. After cooking, the lid is removed to provide access to the food [15]. The decoration capabilities of these packages are high, with folded trays and lids capable of accepting complex process printing and being used extensively as primary packs. Press-formed trays are generally contained in an outer carton or overwrap, but can accept simple decoration.

Some work has been done to find alternates to PET extrusion coatings or film laminations. This work has focused on the use of water-based solutions of polymers designed to provide grease and moisture resistance as well as to tolerate the required temperature conditions [16]. This approach has been successful on only a limited basis, as the relatively thin coatings are susceptible to pinholing and are more difficult to heat seal.

3. Polymer Trays

Thermoformed polymer trays are useful heating containers for microwaveable foods. Depending on the required temperature resistance, the designer can choose from a good selection of materials, several of which are dual ovenable [17]. Complex shapes are easily produced, including raised bottom structures and multiple compartments. Decoration is a primary limitation for these packages, which are typically contained in printed paperboard cartons or sleeves that provide product depictions, ingredient listings and cooking instructions.

LDPE trays are suitable only for extremely light-duty applications, as distortion will begin at temperatures as low as 75°C. This is more stringent than the temperature requirement for LDPE coated paper or paperboard packages. In the case of coated fiber substrates, heat distortion of the polymer is less important than for standalone tray structures, since the fiber component provides rigidity and shape retention at temperatures above the heat distortion temperature but below the melting point of the polymer.

PP trays are suitable for applications in which a service temperature slightly above the boiling point of water is required, as these trays have distortion temperatures of about 110°C. The drawback of homopolymer PP is low-temperature brittleness, which can cause tray damage or breakage during frozen distribution and handling; copolymer versions have somewhat improved low-temperature durability. For controlled cooking applications, however, trays made from these materials continue to be used due to their optical clarity and low resin cost.

The desire to incorporate insulating properties in a polymer tray has led to the consideration of expanded or foamed polystyrene (PS) as a tray material. It is widely used in carryout packaging, and refrigerated meat and poultry trays. However, at temperatures as low as 80°C it will suffer significant distortion, even to the point of causing package leaking. Any consumer who has attempted to reheat restaurant leftovers in a foamed PS clamshell has firsthand knowledge of the limited applicability of this material for microwave heating. High-impact polystyrene (HIPS) is capable of microwave reheating for many foods, but is severely limited by low-temperature brittleness. As a result, while widely used for carryout packaging, HIPS is not generally used for frozen foods.

Crystallized polyester, or CPET, trays have become the most widely used polymer trays for microwave heating. Capable of conventional oven exposure up to 205°C, these trays are fully dual ovenable, providing very high use versatility for food manufacturers. In most applications, CPET trays are pigmented to enhance their appearance; in recent years, black has become the most popular color for these trays. Heat seal films, typically PET, are used for lidding these trays. Cook instructions generally call for several small punctures to be made in the lid film for venting. Variations on the CPET theme abound, including multilayer structures combining PET with other polymers, even including foamed layers [18].

One high-cost approach to polymeric structures that was in commercial use in the 1980s, but has faded from the scene, is thermoset-filled polyester. Trays of this material were used initially in food service, and then were used for plated frozen meals sold in supermarkets. Extremely rigid and durable, packages made from this material eventually were seen by consumers as too costly for disposable packaging. In the end, consumers are not interested in purchasing permanent dishware along with their food; instead, consumers want cost-effective packages that meet their needs for cooking the food they contain.

Almost every new high-temperature polymer is tested for application in trays for microwave packaging. New candidates are most often disqualified for reasons of high cost, difficulty in forming trays at high production rates, interaction with microwave energy and transfer of trace components to the food.

D. Passive Package Outlook

Passive packages will continue to be used for products whose cooking characteristics are sufficient to meet the consumers' expectations of quality vs. cost. In general, this means small portions of food, foods reheated from refrigerated conditions, brief reheating of prepared foods purchased warm and brought home for consumption, and foods not requiring surface browning or crisping. Cost reduction will be a primary focus of activity for passive packaging, and will be accomplished through light weighting for all structures and through reduction in coating costs for paper/polymer composites.

As new high-temperature resistant polymers capable of satisfying FDA requirements for direct food contact at use temperatures are developed, they will be considered as alternates for these uses, either in the form of coatings or as sheets to be thermoformed into trays. Extending the temperature range of dual ovenability for passive packages will be an area of continuing effort.

Representative patents describing passive packages suitable for microwaveable foods are included in the Appendix at the end of this chapter.

III. ACTIVE PACKAGES

A. Definition

In the context of packages for microwaveable foods, "active package" means that the package performs functions beyond that of just containment and has been designed to positively change the way the food cooks. An active package augments the cooking process by either adding capability not normally present in microwave heating or by overcoming inherent deficiencies in the way microwaves interact with the food being heated.

B. Performance Opportunity

The underlying assumption that accompanies the use of an active package is that the food in a passive package does not respond to microwave radiation to yield a satisfactory cooked result. Types of problems that can easily occur with passive packaging approaches for some food items include:

- Sogginess, or lack of crisping development
- Lack of browning or color development
- Nonuniformity of moisture loss, leading to localized toughening
- Nonuniform temperature distribution
- Boil-out or run-off of sauces and toppings
- Inappropriate heating rates for proper cooking of multiple component food products

Many of these problems are present with popular food items when cooked in a microwave oven. As food companies wished to take advantage of growing microwave oven ownership by consumers, they needed to develop products and packages to overcome these problems and provide high-quality, convenient foods for microwave preparation. Two principal avenues of package development have been pursued in the search for improved microwave cooking, application of surface heating and power distribution modification.

Materials used to achieve these ends must balance three interactions with incident microwaves: reflection, absorption, and transmission (RAT characteristics). These properties are easily measured by standard nondestructive techniques, appearing to the casual observer be constant and suggesting to some that it is a simple matter of choosing the right material, incorporating it into a package, and inserting the product. Unfortunately, the challenge is far from that easy, as the RAT characteristics of a material can change dramatically with the applied power level, as is demonstrated in Fig. 1. If microwave exposure changes the properties of a structure, its response will be dynamic even during constant microwave power exposure.

Chapter 3 demonstrated that properties of foods in microwave ovens are highly variable. Even different starches behave differently in microwave cooking. The use of active packaging to cook foods is therefore subject to complex, dynamic phenomenon. This is the challenge facing the microwave scientist and packaging designer in the search for predictable cooking enhancement through the use of active packages. Close collaboration between all involved parties is required, as it is necessary to view the oven, food, and package as an integrated,

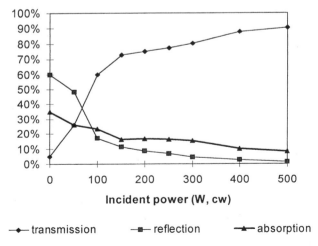

Figure 1 Changes in observed reflection, absorption, and transmission characteristics of metallized film susceptors at increasing incident power.

codependent system. Without this level of collaboration, failure is more probable, and success, if achieved, takes longer and costs more.

A significant criterion for passive package design is whether dual ovenability is required (see Sec. II.B). This characteristic as it applies to active packaging will be discussed for surface heating and field modification packages in Secs. III.C.11 and Sec. III.G below.

Table 2 summarizes the performance characteristics of the most useful options for active microwave packages. These and other options are described in more detail in the following sections.

C. Surface Heating

1. Rationale—Unsatisfied Consumer Needs

Overcoming the tendency of food surfaces in the microwave oven to be cooler than the bulk of the food is the goal of surface heating. In normal microwave cooking, these surfaces do not reach the temperatures required for meaningful browning and crisping until excessive dehydration of the bulk of the product has taken place. The air surrounding the product remains near room temperature throughout the cook cycle, and the problem is exacerbated by the continual movement of moisture through the food surface from the bulk of the product, which receives microwave energy directly from penetrating waves. Development of the surface temperatures needed for Maillard browning is inhibited until after substantial dehydration of the bulk of the product. By that point, the food has been rendered unpalatable due to dehydrative embrittling and toughening, and loss of flavor.

The concept behind surface-heating elements is to provide a means to generate thermal energy in close proximity to the food surface. When surface heating is supplied, the surface of the food begins to dehydrate early in the cooking cycle, and rises to the temperatures where Maillard browning can be effective. This happens simultaneously with the movement of moisture from the bulk of the food to the air in the oven that surrounds the food, approximating the mechanisms by which foods brown, crisp, and cook in a conventional hot-air or radiant heat oven or in a frying pan. A useful way to think about surface-heating structures is to consider them to be microwave energy powered hot-air ovens or frying surfaces, which can be brought to close proximity to the food and cause "normal" cooking to accompany microwave radiation cooking.

Two primary approaches have evolved to create the surface-heating function. The first is to use a "lossy" material, a material with the characteristic of quickly converting microwave energy to sensible heat, which is then conducted and radiated to the food surface. These materials are resistive in the electrical field of the oven and the resulting resistive losses generate heat. Susceptors in their various forms are examples of lossy elements.

Table 2 Active Microwave Packaging Options

Package style	Material	Function	Examples of food applications	Comments
Susceptor pad	Susceptor metalized film laminated to paperboard (board susceptor)	Bottom browning and crisping	Prebaked crust small pizza (<125 mm)	Patterned susceptor more efficient for directing heating to crust shapes
Susceptor sleeve	Paper or board susceptor	Surface browning and crisping	Precooked pocket sandwich, enrobed dough snacks, waffles	May need side vents to assist in moisture transport; paper-based sleeves can be fully sealed and used as primary pack
Raised susceptor tray	Board susceptor	Bottom browning and crisping	Prebaked crust larger pizza (100–225 mm)	Minimizes heat loss to oven floor
Susceptor trays	Press formed susceptor board	Surface browning and crisping	Small (<500 gram) meat pies	Top crust can be browned using paper susceptor patched onto inside top panel of carton
Even heating tray	Patterned aluminum foil sandwiched between PET film and paperboard	Uniform temperature distribution after heating	Symmetrical food items of at least 350 gram weight	Even heating patterns can work with a broad range of products, but may need to be modified for specific food properties
Differential heating tray	Same as even heating tray	Desired temperature distribution after heating nonsymmetrical products	Multicomponent food products	A custom pattern will be required for each food combination
High-energy crisping	Patterned aluminum foil sandwiched between susceptor metalized PET firm and paper or paperboard	Surface browning and crisping of difficult products and partial shielding	Raw dough and large meat and fruit pies, raw dough snacks, egg rolls	For pies, used in pressed tray and top panel carton patch. Can be used as sleeve for dough-enrobed products and rolls.

The second approach is to use materials that are primarily electrical conductors in the microwave field, and take advantage of edge or fringe effects, localized field concentrations at the edges of the conductive materials. Slot line heaters use the fringe heating effects of conductors for surface heating. Both approaches will be described in more detail below.

2. Permanent Appliances—A Reference Point

The design of permanent cooking devices designed to provide surface heating functionality and sold as "browning dishes" is a helpful introduction to disposable packaging. One type of browning dish is made by bonding a thin metal layer, often tin oxide [19], to the food contact surface of a piece of ceramic cookware and overcoating the layer to prevent abrasion. Instructions call for preheating the empty dish on the highest oven power setting for several minutes. The oscillating electric field in the oven induces localized electric currents in the metal layer. Because the layer is electrically thin, it acts as a resister, and the resistive losses from the currents generate heat, which raises the temperature of the entire tray. The food is placed on the preheated dish, which is returned to the oven for additional microwave exposure. The high temperature of the browning dish acts like a preheated skillet on a range, searing the contact surfaces, and mimicking frying while the microwave field heats the bulk of the food.

These devices are capable of providing browning and crisping when used properly, but have the following drawbacks:

- Preheat time extends the duration of food preparation.
- The consumer must handle a very hot utensil, especially for long preheat cycles.
- The dish's shape often does not match the food's.
- Only the contact surface of the food is browned. Two piece, hinged browning dishes styled like a waffle iron can overcome this problem, although at added cost and complexity.
- The dish needs a high thermal mass to brown items like raw flesh foods, and maintains a high temperature long after the microwave cook cycle is complete, creating a burn hazard for consumers.
- Splatters require cleaning of the oven cavity.
- The need to wash the dish after use is not consistent with convenience.

The safety concern of overheating is addressed by ferrite technology [20]. Ferrite-containing materials interact principally with the magnetic component of the microwave field to create heating. In addition, they have the useful property of a Curie temperature, a temperature at which the magnetic interaction ceases and microwave energy conversion to sensible heat also stops. Because this is a reversible transition, heating restarts when the ferrite cools below the Curie point. Dishes incorporating this technology have been commercialized.

3. Disposable Surface-Heating Packages

The drawbacks associated with permanent appliances for surface heating resulted in limited ownership and use, and drove a quest for low thermal mass, disposable surface-heating elements that could easily be customized for the shape and size of the food product to be heated and which could heat multiple surfaces of the food simultaneously. Terminology for lossy surface-heating elements varies, with "susceptor" the most commonly used term, after the characteristic of the element to be susceptible to microwave heating. Other early terms included "receptors" and, simply, "heaters." No matter what the name, these elements directly affect the surfaces of the food by accelerating surface heating and texture/color development. Susceptor elements have enabled the cost-effective introduction and commercial success of a wide variety of foods designed specifically for microwave heating.

While the microwave energy field has both electric and magnetic components, most packages exploit interactions with the electrical component of the field. Early attempts included mixtures of inorganic salts [21], whose polar nature when compounded with water makes them lossy. These structures are temperature limiting, because their heating ability subsides beyond a temperature determined by composition. These mixtures must be relatively thick, compared to most packaging components, to be effective in absorbing and converting energy, which complicates package fabrication and automated filling operations and increases packaging cost.

Other investigators searched for alternative lossy materials and chose to focus on thin resistive layers of electrically conductive materials, like metals [22–24]. This avenue of exploration generated the most commonly used susceptor structure.

4. Metallized Film Susceptors

A comprehensive theoretical treatment of microwave interactive thin films has been published elsewhere [25]. The following discussion provides a general understanding of the function of microwave susceptors.

As described above, very thin layers of conductive material have significant electrical resistance and generate localized resistance heating where they are exposed to a rapidly oscillating electric field, as shown in Fig. 2. Thin metal susceptors interact primarily with the electrical component of the energy field and provide sensible heat to the food. For aluminum, metal thickness on the order of 30–60 Å with surface conductivity of roughly 0.01 mhos generates the most heat. Because metal films that thick are not self-supporting or abuse resistant, a carrier is required. Oriented polymer films are good carriers for the metal layers, and are commonly used in vacuum deposition, the most cost-effective way to uniformly apply the layers.

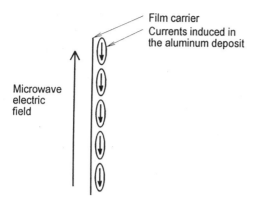

Figure 2 Localized energy dissipating currents induced in a metallized film susceptor.

In this process, a thin film is passed over molten metal in a vacuum chamber. Evaporated metal molecules move in line of sight travel from the source and impinge and freeze on the film surface (see Fig. 3). Vacuum-metallizing machines used for lightweight industrial metallizing are generally capable of depositing metal layers with the uniformity required for susceptor applications. Metallizing machines designed to produce barrier metallized films typically operate at deposition thickness of 300–400 Å and are generally not capable of susceptor uniformity.

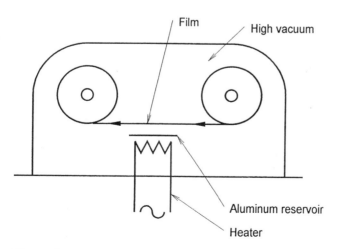

Figure 3 Schematic of vacuum metallization used to deposit resistive metal layer on polymer film for susceptor applications.

Typically, oriented, heat-set PET film is used as the metal deposition surface and carrier film. This film is stable during the metallizing process, is widely available at 12.2 μm thickness, and meets high-temperature USFDA food contact requirements. The film is unsuitable on its own as a carrier for providing extended heating to food surfaces, due to residual orientation stresses. Indeed, a piece of unsupported PET metallized to the proper metal level for efficient heat conversion will shrink rapidly and catastrophically immediately upon exposure to the power found in microwave ovens, losing its the capability to heat.

Mounting the metallized film on a substrate dimensionally stable at the temperatures reached by the metal layer during cooking was an important discovery in the course of the development of microwave susceptor technology. In most cases, this is somewhat in excess of 200°C. Paper and paperboard serve this purpose well and are the most commonly used stabilizing materials used in susceptors today. This combination opened the door to the production and use of low-cost, high-volume susceptors.

Aluminum is the most commonly used vacuum deposited metal. Stainless steel and some nickel alloys have been vacuum deposited using sputtering, a slower and somewhat more expensive process, rather than heat evaporation as used for aluminum. While these alternate materials were tested in a search for more robust heating performance, the general conclusion reached by food processors and package producers is that aluminum provides the best cost/performance relationship. The other metals have only seen limited commercial use, implying that few reproducible performance advantages exist.

Heating levels of metallized film susceptors are a function of the conductivity of the metal layer, the orientation and annealing characteristics of the carrier film, the nature of the bond of the metallized carrier to the dimensionally stable base stock, and the weight of the base stock. These variables dictate the amount of thermal energy that can be transferred to the food and, thus, the amount of cooking augmentation that can be achieved.

5. Self-Limiting Metallized Film Susceptors

If the film were perfectly bonded to the base stock, and neither the film nor the stock experienced any dimensional changes during heating, the susceptor would theoretically continue to rise in temperature until it reached an equilibrium state in which the amount of thermal energy it produced was in equilibrium with the amount of thermal energy conducted and radiated from the structure. This is analogous to the potentially dangerous nature of the browning dish discussed above. Temperatures reached in this runaway heating scenario could cause burns to consumers and thermal damage to the oven cavity, as well as to the package and the food.

Paper and paperboard spontaneously combust at 234°C (remember Ray Bradbury's book *Fahrenheit 454?*). Below that temperature, browning and thermal degradation occur, the degree dependant on both the temperature and the duration of exposure. The food products to be heated also do not have an unlimited ability to withstand elevated temperatures, as evidenced by personal experience as well as the frequent comedic treatment of burnt food. Susceptors that do not exhibit runaway heating are thus a requirement if practical application is to be made of them. Susceptors that overcome this tendency are said to be "self-limiting."

Self-limiting susceptor behavior requires a mechanism by which the microwave absorption of the susceptor is reduced as the temperature is elevated. This reduction can either be permanent, in which case the heating ability of the susceptor is forever limited, or reversible, in which case the susceptor has the ability to cycle in heat generation capability. Metallized film susceptors utilize permanent heating capability reduction to self-limit.

Metallized film susceptor self-limiting results from a weakening of the bond between the metallized film and the base support stock. As the bond weakens during heating, the residual shrink energy in the metallized film causes it to shrink and distort, even to the extent of evidencing cracking, or crazing. This distortion changes the conductivity characteristics of the metal layer and reduces its ability to generate heat energy. As the temperature of the susceptor stops rising as a result of these changes, the susceptor continues to heat, although at a reduced rate, enabling some surface heating to continue. Variables affecting the rate and extent of this bond weakening are the original bond strengths and the sensitivity of the bond to degradation as temperature increases. Film properties, metal surface characteristics, adhesive chemistry, application method, and amount and degree of cure all affect the preheating bond and rate and extent of bond weakening during heating.

Two mechanisms minimize the potential negative impact on food cooking of this heating reduction phenomenon. First, where the susceptor is in extremely close proximity or good contact with the food, the rate of heat transfer from the susceptor to the food is high. Heat generated by the susceptor transfers effectively to the food, preventing the susceptor from reaching temperatures where crazing and significant heating reduction will occur. Second, quickly dehydrating the food surface and raising its temperature is the most effective way to improve browning and crisping. Because the energy is preferentially absorbed by the susceptor compared to frozen components, its performance early in the heating cycle is a key factor influencing browning and crisping. Thus, well-designed susceptors in good heat transfer arrangement with the food create surface browning and crisping, while avoiding overheating of the susceptor. Precooking of the surface components, which are usually flour based, facilitates good browning results with susceptor structures.

Designers have attempted to develop metallized film susceptor structures with predictable self-limiting heating behavior. Their functional basis is described here; manufacturing methods are discussed immediately following in Sec. 6 below. One approach creates concentrating paths for current flow. For packages in which food contact is highly variable, this approach "turns off" the susceptor where it is not in close proximity to the food load. The susceptor breaks down early in the cook cycle in those areas, limiting heating interaction and the temperature reached in that immediate area.

A regular pattern of X's that do not have heating capability will generate current concentrating paths when imposed on a susceptor film [26], as shown in Fig. 4. The pattern forces a concentration of currents between the aligned points of the X's and generates a series of sites at which the breakdown occurs if heat transfer is poor. A grid pattern of lines of active microwave heating function serves a similar role, and both are in commercial use.

Matching the active heating area to the shape of the product and varying the heating rate applied to different areas of the product provides better browning and crisping performance than overall susceptor functionality of one heating level. Several approaches have been developed to achieve this goal, including patterning of metallized film susceptors and development of printable susceptor materials.

6. Patterned Metallized Film Susceptors

Metallized film susceptors can be patterned in a variety of ways, most of which have been used commercially. Die cutting the laminated susceptor structure to fit

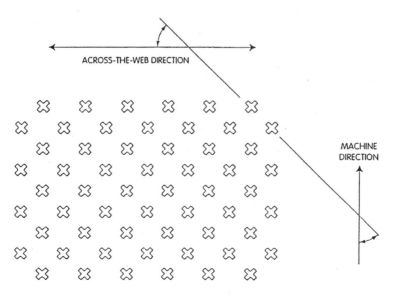

Figure 4 Pattern of Xs creating current concentrating paths. (From U.S. Patent 5,530,231.)

the shape of the product is the simplest approach, and it works well to heat items such as pizza, for which the shape of the product is well defined and uniform. However, it is difficult to pick up a hot pizza supported solely on a die cut susceptor pad, which also does not prevent hot topping from running off or spilling and creating a mess that must be cleaned up.

Consumers want to minimize both cooking and cleanup requirements for a convenience food product and want confidence that children can safely use the packages without close supervision. A package whose perimeter extends beyond the boundaries of the food product, but for which the heating function corresponds closely to the food shape, would provide better cooking and convenience to the consumer. Fabricating this type of package is accomplished in several ways.

The metallizer is the earliest stage in the production process to pattern the susceptor. Machine direction stripes of excluded metal are easy to create using a mask in the vacuum chamber to prevent metal deposition in segments of the moving web of film. Stripe metallized film used for the susceptor lamination provides a reasonable match between product configuration and heating function for rectangular products, such as French bread pizza.

For more complex patterns, several options were developed. One metallizer prints a pattern of oil on the film in the vacuum chamber prior to exposure of the moving film to evaporating metal. The oil prevents adhesion of the metal, creating a pattern of metallization on the film [24]. This step complicates the metallization process, but is workable for low-resolution patterning.

Demetallization is used to make patterns after metallization in several ways, including printing a chemical resist coat and subsequent immersion in an etching bath followed by washing [27] or direct etchant printing followed by washing [28]. (This is similar to the process shown later in Fig. 16.) Either of these approaches can be combined with in-line lamination of the film to paper or paperboard. A simpler process prints an etchantlike solution on the film and then laminates the film to paper or paperboard [29]. This process avoids the washing step by converting the contacted metal to a metallic salt, which remains in the laminated structure. "Deactivated" areas in which the metal salt is created do not interact with microwave energy to create sensible heat. These areas appear to be the same color as the supporting paper structure, normally white, in contrast to the gray appearance of active susceptor areas. Packages employing demetallization or deactivation approaches to patterning are visually distinctive, providing consumers cues that more advanced technology has been used.

Either the demetallization approaches or the chemical deactivation approach to patterning heating functionality offer the highest resolution and control. They create intricate patterns that match the shapes of the surface heating functionality and the food or vary the heating effect across the profile of the food [30]. The later is used when different areas of the product need different heating rates. Grid patterns with varying line thickness and separation of the active metallization create extremely effective gradations in heating response for metallized film susceptors.

Mechanical patterning approaches deactivate the heating function by disrupting the metallized layer locally [31]. Mechanical approaches have not been used commercially, due to their inherently lower resolution capability compared to etching approaches, which possess the same resolution as the printing process used.

7. Printable Susceptor Coatings

Researchers have pursued the elusive goal of developing an easily printable susceptor material as an alternative to patterning metallized film susceptors. They attempted to formulate a material that could be positioned on the package with the accuracy of printing. Varying the application rate in the pattern would create varying heating rate, and no waste would be involved, since heating material would only be placed where heating was wanted, as opposed to etching processes, where material is removed and discarded.

Early attempts focused on suspending particles of conductive materials in an inklike polymer matrix and laying them down in a thin printed layer such that the proper electrical resistance was achieved [32]. A controlled amount of interparticle contact must be maintained for proper conductivity and current flow. Balancing the consistent contact with control of heating rate has proven to be more difficult than initially envisioned.

Carbon black, which can be obtained with predictable particle size distributions, is prone to runaway heating. Metal flakes are interesting due to their conductivity properties, but it is difficult to obtain the proper contact between the individual flakes, as contact density varies greatly with flake orientation. The greatest functional success has come with blends of carbon black, metal flakes, other additives to control particle separation, and a carrier polymer which has a high coefficient of thermal expansivity [33].

This structure balances the different heating characteristics of carbon black and metal flakes, and the polymer provides a moderating mechanism. As the carrier expands with temperature, particle separation reduces the microwave heating interaction. Cooling of the structure is accompanied by thermal contraction, which brings the particles back into a degree of contact that recreates the heating interaction. The temperature range over which this reversible transformation occurs can theoretically be adjusted by material selection, yielding a high degree of flexibility in design.

Unfortunately, these formulations have not succeeded in the marketplace. Runaway heating plagued the simplest formulations, and the more complex approaches have proven difficult to handle commercially and have not offered significant performance advantages over the metallized film susceptors. Work continues in this area, as the potential benefits are compelling.

8. Slot Line Heaters

Fringe effect, or slot line, structures possess the potential for higher energy conversion than susceptors [34], but require careful design to avoid overheating. Designers of slot line structures must work within inherent design limitations, such as a need to create structures with radial symmetry to avoid directional effects. In addition, surface browning using slot line structures is limited to the area of the food directly adjacent to the edges of the conductive material, which creates a pattern that is generally at odds with the food manufacturer's desire for uniform browning. Surface heating with fringe effect heaters has been the subject of patent activity (see Fig. 5), but has seen only limited commercial usage.

Fringe effect heaters combined with metallized film susceptor structures are an effective avenue to higher-powered susceptors, and will be discussed in Sec. III.E. Other methods of using conductors in conjunction with resistive elements for surface browning, including resonant elements and antennas, will also be discussed there.

9. Regulatory Compliance of Susceptor Elements

Susceptors reach high temperatures in close proximity to food items intended for human consumption, and their safety needs continual confirmation. High temperatures increase the mobility of trace materials that may be contained in the package structure. Other components may undergo chemical changes at elevated temperatures, and chemical reaction products could migrate to the food in either gaseous or liquid form. Susceptor manufacturers, aware of potential concerns in this area, continually test susceptors and the materials used to produce them to ensure that public safety is not jeopardized by the broad use of these packages. Test-

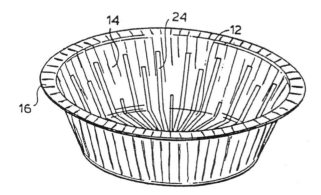

Figure 5 Slot line heater structure incorporated into paperboard pressed tray. (From U.S. Patent 5,117,078.)

ing began when packages were first introduced and producers and users of susceptor packages have collaborated on the selection and testing of materials incorporated in susceptors.

In the mid-1980s, an industry group worked with the USFDA to develop a set of test protocols and to agree to voluntary migration standards for microwave susceptor packages. These tests quantify volatile and nonvolatile materials available for transfer to food from susceptor structures exposed to high temperatures. The working group evaluated existing susceptor packages using these standards and confirmed that no adverse risks are presented to consumers who consume food heated in properly made susceptor packages. Industry participants continue to use these voluntary protocols and standards to ensure safety for consumers. The F02.30 Subcommittee of the American Society for Testing and Materials (ASTM) published and keeps current the following procedures for evaluating susceptors [35].

> F1308-98 "Standard Test Method for Quantitating Volatile Extractables in Microwave Susceptors Used for Food Products"
> F1519-94 "Standard Test Method for Qualitative Analysis of Volatile Extractables in Microwave Susceptors Used to Heat Food Products"
> F1349-98 "Standard Test Method for Nonvolatile Ultraviolet (UV) Absorbing Extractables from Microwave Susceptors"
> F1500-98 "Standard Test Method for Quantitating Non-UV-Absorbing Nonvolatile Extractables from Microwave Susceptors Utilizing Solvents as Food Simulants"

To date, the FDA has chosen not to issue regulations regarding microwave susceptors, indicating confidence on their part that the voluntary guidelines provide assurance of consumer safety.

10. Successful Applications

Commercial use of metallized film susceptor packages began with a 10 cm. square frozen pizza. This product was sold in convenience stores and was introduced in a complicated package [36, 37] that included

- a folded paperboard tray, containing
- a single-face corrugate pad, on which rested
- a metallized film paperboard susceptor, on which rested
- a release treated parchment baking cup, on which rested
- the pizza, which was covered by
- a foil laminated paperboard vented lid, all contained in
- a printed film overwrap

This structure, while complicated, worked well in conjunction with the specially formulated pizza and yielded a crisp crust with correctly done topping. This breakthrough product accelerated development work with products and susceptor package technology and was followed by paperboard packages for supermarket-distributed 18 cm frozen pizzas as well as French fries and popcorn [38].

Susceptor bags for microwave popcorn followed and were the first susceptor packages that incorporated flexible materials instead of paperboard. Susceptor packaging is the standard for popcorn, not for browning or crisping, but for improvements in both the percentage of kernels that pop and the volume of each popped kernel. The susceptor is located directly under the kernel/oil mixture (see Fig. 6) and provides a boost to popping performance by quickly heating the oil and kernels to popping temperature, speeding completion of popping and popping those kernels that would not pop under normal microwave exposure.

Producers quantify popping performance by measuring the popped volume for a given weight of kernels and the percentage of unpopped kernels at a given cook time. The size of the "popping window," the time between practical completion of popping and onset of scorching of kernels from overheating, is another measure of performance. Consumers are generally instructed to turn off the oven when pops are about one second apart, and prefer brands that give them greater latitude in their determination of the heating end point.

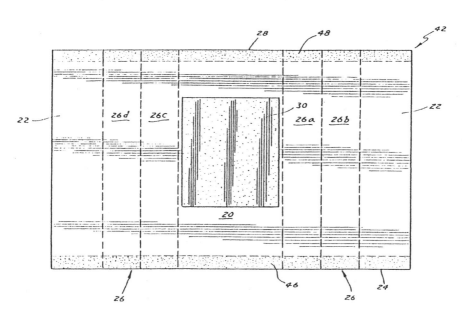

Figure 6 Unfolded microwave popcorn bag showing susceptor patch location. (From U.S. Patent 5,498,080.)

Microwave popcorn is the single largest use for microwave susceptor packages. Recent refinements have been directed to increasing manufacturing efficiencies and reducing material and fabrication costs. Opportunities for quality improvement include increasing the size of the popping window and minimizing the toughening that occurs in already popped kernels while the remainder of the popping cycle proceeds.

Food processors and package producers continue to extend the application of this technology. Researchers from both industries screen popular food items to determine the potential for metallized susceptor technology to provide acceptable microwave cooking enhancement. Collaboration between companies generates the best results, when food and package are modified in an iterative manner, taking full advantage of the degrees of design freedom available from package and food technology.

Structures that raised susceptors and the pizza off the oven floor (see Fig. 7) [39, 40] and that patterned the susceptor to match the shape of the product facilitated the introduction of larger pizzas. Some of these packages also enhanced cooking by containing the hot air under the susceptor layer. This approach reduces heat losses from the side of the susceptor away from the food and channels more energy into the food side, speeding browning and crisping.

Sleeves (see Fig. 8) [41–43] and trays add the ability to provide surface heating to multiple surfaces of products such as French bread pizzas, waffles, enrobed sandwiches, and other items. These packages gave birth to a new category of microwave-heated hand-held snack foods, and gave further impetus to market expansion of surface-heating packages. Complex packages for fabricated French fries and food sticks surrounded a multiplicity of these elongated

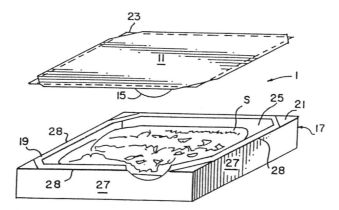

Figure 7 Susceptor removed from carton prior to inversion of carton to provide elevation. (From U.S. Patent 4,661,671.)

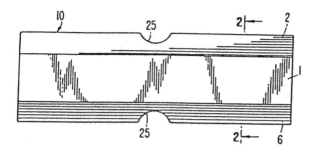

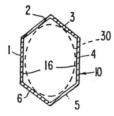

Figure 8 Hexagonal cross-section apertured susceptor sleeve. (From U.S. Patent 4,948,932.)

food items with susceptor material (see Fig. 9) [44], better mimicking oven heating or deep-frying.

Flexible structures that contain metallized film susceptors are suitable for direct filling of products on horizontal form-fill-seal equipment [45]. These structures are similar in some respects to popcorn bags, but are used for heavier dough-

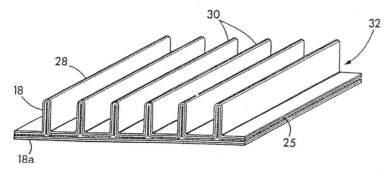

Figure 9 Folded susceptor structure for multiple elongated food items. (From U.S. Patent 4,943,439.)

based products that require browning and crisping on multiple surfaces. These susceptor packages act as primary packs, and the PET film provides barrier protection for the product during shipping and distribution. Multicolor printing and heat seal or cold seal material makes them fully functioning primary packages. Instructions direct the consumer to open one end of the pack to provide some venting of moisture, although controlled release heat seals can be incorporated to provide automatic venting.

Frozen pot pies commonly require cook times of up to 60 mins in a conventional oven. Pressed trays made from laminations of paperboard and metallized susceptor film can cook pot pies in the microwave in 5–12 min, depending on weight. While crust formulations specifically designed to brown and develop good texture during microwave cooking have been pursued [46], supplemental heating is crucial to achieving desired browning and baking levels. In addition to susceptor trays, some microwave pot pie packages include a susceptor patch suspended from the inside top panel of a folding carton that provides browning assistance for the top crust [47].

To summarize package choices, flat susceptor pads work well for products like pizza, where only one surface needs browning and crisping. U-boards, sleeves, and pouches become the preferred package styles for products that require browning on multiple surfaces. Trays are preferred for products that are not self-supporting, or where leakage of filling is possible.

11. Dual Ovenability of Susceptor Packages

To determine whether it is necessary or desirable for a susceptor package to be dual ovenable, three questions need to be answered:

1. Is the microwave food formulation compatible with conventional oven heating?
2. Is the food contained in the package in a way that is compatible with conventional heating?
3. Is it likely that consumers would find conventional heating instructions to be of value for the specific product?

For the pot pies described in the previous paragraph, both microwave and conventional oven instructions are included on the package. This is the result of the way that these three questions can be answered for this food/package combination.

First, the pot pie food formulation is compatible with conventional oven heating as well as heating in a susceptor package. This is not the case for many of the products packaged in susceptor packages; frozen pocket sandwiches are good examples of products for which the formulation required to create good cooking performance in a microwave with a susceptor package does not work well for conventional oven heating. In these cases, the package only includes microwave heating instructions.

Second, for the microwave pot pies, the conventional oven instructions call for placing the trays on baking sheets and then in the oven, the same instructions that have existed for pot pies packaged for conventional oven-only heating. The susceptor tray structure described above is essentially the dual ovenable passive package structure described in Sec. II.C.2. The only difference is the metallization on the PET film; in the conventional oven, this package performs the same as the nonmetallized version. Because the pot pie is in good contact with the entire package, there is little danger of excessive browning of the package in the conventional oven. (In contrast, flexible susceptor structures tend to not be as useful for conventional oven heating, as they tend to have large areas not in good contact with the food and would be expected to become browned and brittle in conventional oven heating). Other products, like frozen pizza, would not likely use the microwave susceptor element to advantage in conventional oven heating. If conventional heating instructions were included, they would instruct the consumer to discard the susceptor element and place the pizza directly on a baking sheet; in this case dual ovenability would not be required.

Finally, since some consumers may prefer to heat frozen pot pies in their conventional oven, the food manufacturer can capture additional sales from a dual ovenable pot pie. This would not be the case for popcorn packaged in susceptor bags for microwave popping. This product is bought strictly for its convenience aspects; those who wish to pop popcorn on their stovetop will purchase bulk popcorn and oil.

In summary, dual ovenability of susceptor packages is sometimes useful, but mostly is not. Increasingly, the foods involved are formulated to optimize microwave browning and crisping, often at the expense of conventional heating performance. This requirement must be evaluated on a case-by-case basis, as each food/package situation is different.

D. Field Modification

1. Rationale—Unsatisfied Consumer Needs

Field modification packages for microwaveable foods redirect the energy of the microwave field in a predictable fashion to optimize the microwave heating performance of foods, overcoming two major drawbacks of microwave cooking with respect to heating rates. The first is the tendency for overdone edges and underdone centers of food items. The second is providing different heating rates to different components of multicomponent meals.

The first problem is observed when heating a 1 kg frozen lasagna. Conventional oven instructions call for at least a 50 min cook time, sufficient for thermal conduction from the edges of the food to heat the center of the food to desired temperatures. Microwave heating is a desirable approach for reducing the long cook time associated with conventional oven heating and for minimizing freezer to table time. However, the consumer who attempts to realize this savings of time using a passive package is in for a disappointing result. After 20 min on high power in an 850 W home microwave oven, the cheese and topping at the edges of the lasagna are dehydrated and overdone, sauce has boiled out at the perimeter, and spots in the center may still be frozen, as shown in the temperature profile in Fig. 10 [48].

Overexposure of the perimeter of the food and poor exposure of the center, especially the bottom center, creates this problem. Incident energy illuminates the edges of the food from the top, side, and bottom, while the center is only exposed well from the top. The problem of poor center bottom exposure results from an effective microwave field "shadow" that the product casts under itself, and which prevents significant amounts of energy from being reflected to the center bottom of the food. This shadow, shown schematically in Fig. 11 with a pizza as the modeled food product, results from absorption of energy as a propagating wave repeatedly hits the bottom of the food. The nonabsorbed portion reflects off the oven floor back to the food, but each successive reflection toward the center of the food contains exponentially less energy.

The shortened cook time that is a convenience advantage of microwave cooking significantly reduces thermal conduction, further compounding the nonuniform exposure of the food. Heat energy in hot areas of the food is not effectively conducted to cool areas, preserving the temperature differences resulting from uneven microwave exposure when the food is removed from the oven for consumption. Until recently, food companies have been forced by this deficiency to make conventional oven heating of products of this type and size the preferred approach.

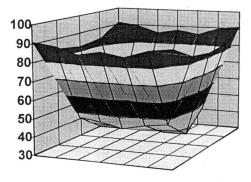

Figure 10 Temperature (°C) profile of lasagna microwave heated from frozen in transparent tray.

Figure 11 Schematic of exposure "shadow" resulting from reduced power of successive reflections.

Consumers often perceive that microwave foods cool faster than foods cooked conventionally. This common complaint is a manifestation of uneven heating [49]. A thermodynamic analysis of the situation concludes that the consumer may mistakenly believe that all of a microwave-heated food is hot, when only areas of the food really are. Consumers often stop heating food as soon as one area of the food appears to have reached the proper temperature, expecting it to be "done." When significant temperature gradients are present, the total heat energy contained in this food is substantially less than the total heat energy of a conventional oven-heated product, which has reached proper center temperatures. The cause of the consumer confusion is clear; food items with less total thermal energy *do* cool faster. The mistake made by the consumer is assuming that the food had been heated uniformly in the microwave. Fortunately, the perception of faster cooling for microwave heated foods can be overcome when uniform temperatures are achieved in microwave heated foods.

Given the allocation of the electromagnetic spectrum, nonuniform bulk heating is a given. The wavelength of 2450 MHz radiation is the same order of magnitude as the dimensions of the food we want to cook. This guarantees the presence of standing waves, hot and cold spots, and other interactions that are unfriendly to even heating. Even heating field modification packages create uniformity of cooking by overcoming this inherent problem.

The second problem is seen when heating a multicomponent meal, such as a two-component meal with adjacent vegetable and meat portions. When this meal is heated in a passive package in a microwave oven, the vegetables are overcooked and mushy by the time the meat barely reaches an acceptable temperature. Figure 12 illustrates a typical temperature profile for such a product. The food manufacturer does not deliver to the consumer the quality hoped for, because the cooking of each component has been suboptimized in the interest of selecting a cook time that turns out to be an unhappy medium. Consumers have developed reduced expectations for these products, and seem to value them primarily for convenience, not for quality [50].

Food companies typically deal with the difficulties in obtaining proper heating by building elaborate cooking instructions that include wait times, stirring, multiple heating steps, and for ovens without turntables, package rotation between

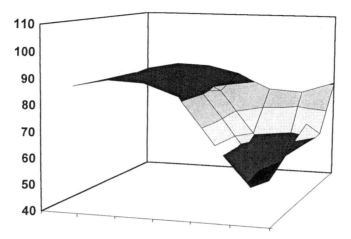

Figure 12 Temperature profile (°C) of two component meal cooked in transparent tray. Vegetables are overcooked while meat is underdone.

heating steps. None of these requirements are consistent with the promise of convenience surrounding microwave cooking. Consumers want to be able to put the food into the microwave, set the power and time, and come back when the product is done. Research reports a drop in confidence in the provided heating instructions [51], indicating that consumers are disappointed with the results they achieve.

Consumers have learned that even brief overcooking of microwave popcorn can lead to badly overdone and scorched kernels, and this is one product for which they appear to accept close monitoring of the cook cycle. With a maximum cook time of 5 min in most ovens, and elimination of cleanup, microwave popcorn has overcome resistance to monitoring the pop cycle. For large food items with cook times approaching 20 min, consumers have been clear. They want to be freed from the requirement to monitor and manipulate.

Field modification technologies attempt to solve these problems through deliberate control of energy in the oven by including elements in the package that are good conductors of electricity. Conductive elements can act as shields to the passage of microwave energy, can intensify energy locally, or can transmit energy to a point in the package that otherwise would receive relatively little direct microwave exposure.

2. Shielding Package Structures

Simple shielding approaches for use in microwave cooking are not new [52]. Microwave cookbooks have suggested the use of aluminum foil around the ends of

Packaging Techniques for Microwaveable Foods 429

large portions or around small, protruding areas of the food, like the ends of turkey drumsticks and wings [53], often on the same page that includes a warning against the use of metal objects in the oven. Shielding packages seek to expand upon that approach and achieve more sophisticated results without burdening the consumer with excessive manipulation of the package or the food. The earliest examples of shielding were incorporated into permanent appliances, similar to the early development of browning technology as discussed above.

In one early use, metal sleeves with different size holes (see Fig. 13) were placed around plates or trays containing multicomponent food items that were heated in microwaves in food service [54]. The holes were located over different food components to permit controlled amounts of microwave energy to impinge on the components and, in theory, create optimum cooking conditions for each component. In an early example of shielding packages for retail products packaged in compartmented aluminum trays, apertured foil lids were included to accomplish the same end. While each of these is effective in highly controlled en uses, neither is sufficiently robust or flexible for broad consumer usage.

Solid metal elements have high current carrying capability and create intense field strengths at edges and points of the sleeve or tray. When they are placed in close proximity to oven cavity walls, arcing between the metal element and the oven wall is likely. Their fixed geometry makes them inflexible for different food products, because different menu items are not easily accommodated by a single design, and many sleeves may be needed for food service outlets with large menu variety. Cost and storage issues make the sleeves impractical for consumer use.

Aluminum trays with apertured or removable foil lids are in a cost range suitable for consumer packages. However, cooking in such a package results in no

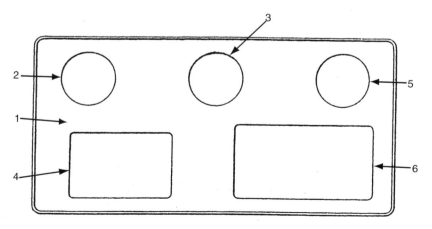

Figure 13 Example of apertured metal element for microwave shielding. (From U.S. Patent 3,615,713.)

microwave exposure of the food from the bottom or sides; all the energy must impinge upon the food from the top. While an apertured foil lid can control the amount of energy incident upon the top surface of each of the multiple food components, the lack of side or bottom heating means that food profiles must be thin or cook time long for effective conductive heat transfer, a constraint for removable foil lids as well. Cook instructions generally require that the consumer leave the aluminum tray in the over-carton during cooking to minimize arcing potential. Unsupported foil lids are fragile and damage poses an arcing risk, and certainly would affect the cooking result.

Aluminum foil tray manufacturers have attempted to overcome arcing concerns by coating the trays with opaque materials that mask the foil appearance and reduce the potential for arcing if the tray contacts the walls of the oven. Domed plastic lids can provide sufficient physical separation of the tray from the oven wall, but add cost and complexity. Today, cooking instructions for most frozen foods packaged in aluminum trays either have no microwave instructions, or require removing the food from the aluminum tray and placing it in a "microwave-safe" dish for heating.

Metal can manufacturers have attempted to develop approaches that could make their packaging compatible with microwave reheating. These include using a low height-to-diameter geometry, tapering the sidewalls to provide for a smaller cross section at the base that at the top opening and coating the can with an electric insulator to avoid arcing to the oven floor or walls [55, 56]. Both patents also describe the use of a microwave transparent lid that is applied to the top of the can after the metal end has been removed. Despite such efforts, consumers are loath to place metal cans into their ovens and this remains an elusive target for metal can makers.

Shielding packages are designed to limit the amount of energy reaching a specific spot on the food. Relatively large elements of conductive material are typically used and are placed near areas of the food that would otherwise overheat. Close proximity of the shield to the food is helpful in most cases, and primary packs are the most effective place to incorporate shielding elements. Patterned shields offer the best cooking performance, because they can be tailored to the specific food. Pressed paperboard trays are especially useful, as continuous patterned shields can be included in stable arrangements. These packages will be discussed in more detail below.

3. Field Intensification Package Structures

Field intensification attempts to focus electromagnetic energy in areas where the package designer desires a higher heating effect than would otherwise occur. While similar in structure to shields, which are designed to exclude energy from an area, and used in conjunction with them, field intensification elements are used to boost localized heating performance, and are capable of providing surface and bulk heating.

Field intensification approaches include elements that work as lenses [57, 58], using controlled-size apertures to focus energy. Other elements act as one-way mirrors, allowing energy to penetrate to the food from oven cavity, while trapping and re-reflecting power reflected off the surface of the food [59]. Still other concepts propose elements that absorb energy and reradiate it to the food at higher energy modes [60]. The theory behind the higher modes is to create or enhance multiple, smaller-energy modes within the food; conduction within these smaller regions can occur during the timeframe of cooking and a more uniform temperature distribution is claimed to result.

Conductive loops of electrical lengths equal to integral multiples of energy wavelength resonate vigorously by creating constructive interference [61]. These elements can be used in package designs to increase the energy absorbed at a spot, and can be used in conjunction with the energy transmission devices described in the next section to create more even heating patterns. Examples of the use of loops in a commercial package are shown later in Fig. 17.

4. Energy Transmission Package Structures

Energy transmission approaches transfer energy from areas of high impingement to areas of limited energy impingement. The concept is similar to that of an antenna, and, indeed, the word "antenna" is used to describe these elements in some of the patent literature [62, 63]. The structures themselves (see Fig. 14) are quite reminiscent of the "rabbit ear" antennas on older radios and televisions and work

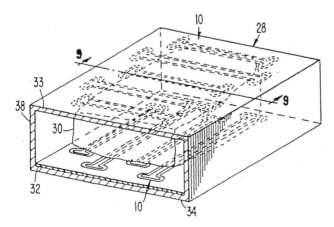

Figure 14 Sleeve containing multiple antenna elements. (From U.S. Patent 5,322,984.)

on the same principle. For the antenna approach to work, three actions must be achieved. First, the energy must be collected, it then must be transmitted to the desired area, and finally, the energy must be dissipated usefully.

As discussed below in Sec. III.E, antennas can be used in conjunction with susceptors to deliver energy to areas of the susceptor that are normally not well illuminated, facilitating higher energy browning and crisping. They can also be easily used in combination with shielding elements when both types of elements can be fashioned simultaneously during conversion. This opportunity leads one to the following discussion of conductive element patterning.

5. Patterning of Conductive Elements

Successful implementation of a field modification approach often requires the ability to design and position complex patterns of conductive material, with small, discrete elements often playing a key role in developing functionality. The challenge to package designers is thus threefold: to create conductive material patterns that achieve the desired heating functionality, to ensure that those patterns are safe and robust in consumer use, and to develop high-volume, cost-effective methods to produce useful packages incorporating these patterns. The various design objectives noted above lend themselves to execution in aluminum foil of 6–7 μm thickness that is combined with other higher-strength carrier materials.

It is useful at this point to stress that true conductive elements are required for full shielding and effective energy redistribution performance at the power levels encountered during both normal cooking and abuse conditions. Current carrying capacity is the key performance criterion for material selection. The following equation calculates skin depth, the dimension of a skin layer from the surface of a material in which a large portion of microwave induced electric currents are contained:

$$\delta_s = \sqrt{\frac{2}{\sigma \omega \mu_a}}$$

where δ_s = skin depth
σ = conductivity
ω = angular frequency
μ_a = absolute permeability

The skin depth of pure aluminum at 2450 MHz is 1.7 μm. Any thicker layer of aluminum effectively acts as a full shield to microwave energy of this frequency, and is capable of carrying the full power load expected to be encountered in commercial and consumer ovens. Because, in practice, both sides of the aluminum layer are exposed to incident microwave energy, the layer should be at least 3.4 μm thick to avoid interference and stress amplification effects. Conductors thinner than two skin depths cannot achieve full shield performance in high-energy situations, because they are unable to carry the full current imposed on them. In this instance, the metal layer breaks down, causing small islands of metal to form that are capable of carrying the divided current load, but at a significant loss of shielding and energy transmission performance.

While aluminum foil has received the most attention as a shielding component of packaging, metallized films with metal thickness of 200–400 Å have been proposed as shielding elements. This proposal stems from similarity in appearance to aluminum foil, and low-energy RAT measurements that indicate the material possess near 100% microwave reflection properties. However, when these materials are exposed to levels of electrical stress typically encountered during cooking in microwave ovens, they break down, reflection drops off substantially, and the materials behave very much like standard metallized film susceptors. Figure 15 shows the results of reflection measurements at increasing power exposure for 7 μm foil and 3.5 optical density metallized film structures. This data shows that high-optical-density metallized films, whose metal thickness is an order of magnitude greater than metallized film susceptors, do not possess sufficient aluminum thickness to carry the imposed currents, and thus, have only limited usefulness in field modification packages.

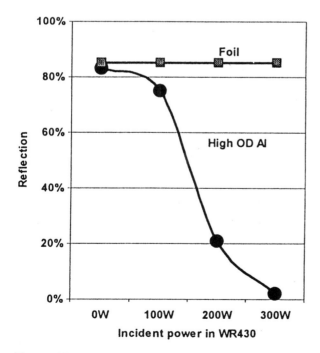

Figure 15 Reflection performance at various power exposures for aluminum foil and high optical density metallization.

As a practical matter, defect-free unsupported aluminum foil is currently only commercially available in thickness of 6 μm and higher, well in excess of the theoretical minimum of 3.4 μm described above. With electrical current carrying capability not an issue for foil, package converters choose foil thickness and grade best suited for the particular patterning processes that they employ.

The problem of taking a continuous web of (electrically thick, but mechanically thin) aluminum foil and transforming it into the desired patterns for field modification has been approached in several ways. In most cases, the foil patterns are laminated to a rigidity-supplying carrier and a food contact surface. Combinations of PET film, aluminum foil, and paper or paperboard have been commonly used, although combinations of foil and semirigid plastics can be envisioned as well.

Overall foil laminations to paperboard with die cut holes for selective energy transmission and steam venting were used in conjunction with early susceptor packages, as described above [28, 29]. This approach is limited, as it does not easily pattern the perimeter of the foil area, removing only sections from the bulk and leaving gaps in the package surface.

A relatively straightforward way to create a foil pattern covering only a portion of the package is to laminate narrow strips of foil to wider paperboard and then, if desired, laminate PET film over the foil [64]. While simple in execution, this approach can only create the simplest patterns and is not capable of making packages with sophisticated functionality.

Chemical etching is the current commercially used approach for creating sophisticated foil patterns [65]. In this approach, illustrated in Fig. 16, chemicals are used to dissolve the foil not wanted in the final design. The commercial process begins with laminating aluminum foil to PET film, which provides a carrier for the foil through the etching step. This roll of composite is taken to the next process operation in which a chemically resistant material is printed in a pattern corresponding to the areas where foil is desired. Following cure of the resist coat, the

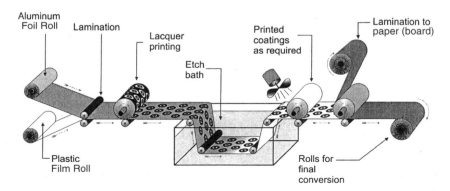

Figure 16 Chemical etching process.

composite is immersed in a bath of etching chemical, most commonly concentrated caustic soda. Following a suitable dwell time in the etching bath, the web exits, is rinsed and dried, and is ready for subsequent processing. Generally, lamination to paper or paperboard in-line with etching protects the foil and simplifies further conversion steps. The resulting lamination can be converted into a variety of structures, including cards, sleeves, cartons, folded trays, and pressed trays. All can deliver field modification functionality.

In another approach to creating patterns, foil is laminated to paperboard using a wet bond adhesive applied in a pattern corresponding to the desired final foil pattern [66]. While the adhesive is still uncured, a rotary cutting die is used to penetrate the foil, part way into the paperboard in register with the adhesive pattern. Vacuum or other means strips away the foil not in contact with the adhesive, creating the pattern. While this process lends itself to higher production speeds than chemical etching, it is unlikely to be suitable for complex patterns and small elements. It has not yet been used commercially.

Laser cutting of the foil has also been proposed and may evolve as a next generation approach [67]. For this to work in a commercial setting, a relatively high-powered laser controlled by a high-speed aiming device is required. This is difficult to achieve in practice, because complex patterns require many centimeters of outline to be cut per meter of lineal substrate. On the positive side, one would expect smooth edges to result from this cutting technique.

6. Effectiveness of Field Modification Packages

Early in the field modification package discussion, several problems associated with microwave cooking were described as the motivation for the development of field modification approaches. Cooking results are now presented for field modification package solutions for those problems.

The first problem presented was the highly nonuniform temperature distribution resulting from cooking a 1 kg lasagna from frozen in a passive tray. Figure 10 illustrated the typical result of overheated perimeter and cold center. The proposed solution to this problem is a paperboard pressed tray incorporating shielding elements on the vertical sidewalls and resonant loops in the bottom of the tray. Figure 17 shows a round version of this type of tray.

The graph in Fig. 18 plots the temperature profile achieved using a rectangular tray incorporating elements similar to those shown in the round version above. In this case, the solution was arrived at after studying the heating profile using the passive package. Since overheating was observed at the edges and corners of the product, the shielding elements on the vertical sidewalls were added to direct energy away from the edges and corners. The resonant loops were added to the bottom of the tray to overcome the tendency for low heating rates by increasing energy coupling and transfer at the bottom center of the product.

Symmetrical patterns lend themselves well to the heating of largely homogeneous foods that have good symmetry. For the case of nonsymmetrical, multi-

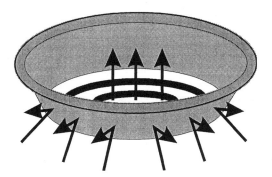

Figure 17 Tray with patterned conductive elements providing edge shielding and bottom resonant loops.

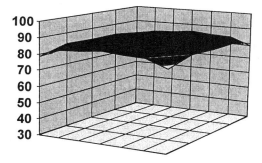

Figure 18 Temperature (°C) profile of lasagna microwave heated from frozen in patterned foil even-heating tray.

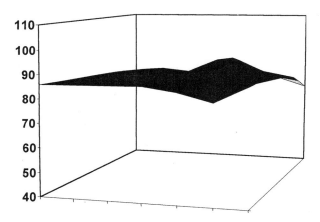

Figure 19 Temperature (°C) profile of two-component meal heated from frozen in patterned foil tray. Meat is cooked thoroughly and vegetables retain texture and flavor.

component foods, nonsymmetrical patterns of conductive elements may be used to tailor the heating of each component. Remembering the two component meal of Fig. 12 above, a tray combining essentially complete shielding on the sides and bottom where the vegetable portions rests with large apertures and resonant loops under the meat can create the temperature profile illustrated in Fig. 19.

An example of the effect of an energy transmitting package for the pizza used in Fig. 11 to illustrate the microwave "shadow" problem is shown in Fig. 20. In this case, antennas made of conductive elements are used to capture incident energy at the edge of the food and transmit that energy to the center of the product.

Energy transmission elements used in conjunction with an overall susceptor layer can achieve uniform browning of the entire bottom crust of a food. Consumer taste and appearance tests were conducted with raw dough top and bottom crust 20.5 cm frozen fruit pies cooked:

1. 50 min per package instructions in conventional oven in the pressed aluminum tray in which the product was supplied
2. 18 min in microwave oven in passive tray in the carton
3. 18 min in microwave oven in a combination antenna/susceptor tray under a combination resonant element/susceptor suspended from the inside lid of the carton

Consumers were asked to taste and rate the pies prior to being told how each were cooked. They unanimously preferred the pies cooked in the field modification package (item 3), over the other two pies on the bases of taste, texture, and appearance [47].

These three examples demonstrate the results that can be obtained with field modification approaches. Field modification packages can create uniform temperature distributions, optimize heating rates for nonuniform foods, and develop high-energy browning and crisping for raw dough and other to brown products. They are powerful tools when used in conjunction with food formulation customization, and can provide consumers with oven quality foods in a fraction of the conventional oven time.

Figure 20 Use of conductive elements to transmit incident energy to bottom center of food.

7. Conductive Element Design Considerations

Care must be taken in designing conductive elements to avoid the inadvertent creation of field concentration points that can lead to arcing. It is also necessary to understand the potential of the perimeter of conductive elements to heat and to design them so that this heating is not detrimental to the cooking result or to package stability.

Overlapping conductive areas must be held in close proximity and effectively capacitively coupled to prevent arcing between them. Laminations of patterns of aluminum foil and paperboard that are formed into pressed trays possess sufficient integrity in this regard, when corner pleats are well formed and bonded. Overlapping conductive layers in folded trays made from similar laminations are inherently less robust, and elimination of overlapping foil areas may be required to prevent potential for arcing to occur in damaged or stressed packs, although at the expense of heating uniformity.

Mechanical patterning approaches are subject to creation of defects at cut perimeters and care must be taken to ensure that knives are well maintained and that smooth, clean cuts are achieved. This edge should be maintained, and the pattern should be designed in such a way that cutting does not disturb this edge. When the foil is patterned using chemical etching, a very smooth, even edge is left on the foil, which is beneficial for stability and avoidance of arcing.

Perimeters of conductive elements are subject to the "fringe effect" heating described earlier and this phenomenon can be minimized by ensuring that perimeter lengths are not close in dimension to whole multiples of the wavelength of the microwave energy that is imposed in the oven. For the case of 2450 MHz radiation, the wavelength is 12.25 cm.

Rigorous stress testing of all field modification structures is required to ensure that packages are safe both for intended use conditions as well as for abusive conditions. The empty package must be capable of surviving full power application for an extended period of time without arcing or otherwise misbehaving. This provides a good foundation for having confidence in the package's suitability for use, but is not sufficient for responsible package introduction into the marketplace. High electrical stress application is key here, as the current carrying capability of the conductive elements must be challenged under the conditions that can be realistically be encountered in a wide range of ovens and use conditions.

Two packages cooked simultaneously, packages touching the oven wall, upside-down packages, packages with a substantial amount of the food removed, packages cooked for twice the label instructions, and packages damaged in shipment are some of the situations that need to be tested. These and other seemingly "dumb consumer" uses will occur in the real world; they must be anticipated and the package designed to accommodate them. Both the food manufacturer and the package supplier have an interest in providing a very robust package to the consumer and must work together in designing and conducting abuse tests that reflect everything they know about use conditions. Safe package designs can provide exceptional cooking performance; achieving this takes a sound understanding of the phenomenon involved and the strengths and weaknesses of the materials used.

E. Hybrid Packages

As described above, it is possible to combine resistive and conductive elements in a single package, either as separate elements, or in combination, to create a true hybrid package structure. In these hybrids, multiple functionality is delivered simultaneously and very unique results can be obtained. Several examples discussed earlier are hybrid packages.

The complex pizza package that represented the first commercial use of a metallized film susceptor also incorporated a shielding lid. While the microwave active elements were present on separate package components, the cooking function achieved was dependent on both functions being provided.

The raw dough fruit pie example given above is illustrative of the results that can be achieved when shielding, energy transmission and susceptors are combined. The tray containing the pie combines solid conductive elements for edge shielding with antennas for energy transmission to the center of the pie and susceptor for browning and crisping of the crust. The lid combines shielding to force more energy to the bottom of the tray with resonant elements to ensure sufficient energy absorption in the susceptor.

Combining elements in an effective fashion is limited by converter process capability and the ability of the food formulator to modify the product to respond well to the active packaging elements. Cost considerations for both package designs and food ingredients and fabrication methods need to be built in to evaluations of whether these more complex packages are justified by higher performance.

F. Commercial Usage of Field Modification Packages

A limited number of field modification packages have been introduced. A simple two-dimensional card is used in the United States and Canada to protect the ends of a prebaked pocket sandwich from overheating and toughening, as shown in Fig. 21. A pot pie in a susceptor tray with a base reflector and resonant element/susceptor lid was introduced in the United Kingdom and had good repeat purchase results. The package used is shown schematically in Fig. 22. Several even heating applications in rectangular packages similar in concept to that of Fig. 17 are in supermarkets, also in the United Kingdom for 400 and 1000 g lasagnas. A North American large meat pie was scheduled for market introduction in 2000.

Figure 21 Patterned foil card to prevent overheating of ends and boost heating in center of prebaked pocket sandwich.

Work between food manufacturers and package suppliers to develop food/package combinations in field modification packages continues, and more introductions are anticipated as more difficult food challenges are undertaken with more sophisticated food and package technology.

Similar to the package geometry choices available for surface heating packages, food geometry is the primary determinant of whether a pad, sleeve, tray, or other package style is best for a given food product. As more field modification packages are introduced, preferred styles for certain types of food should emerge;

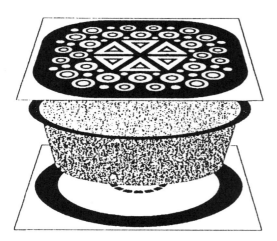

Figure 22 Pot pie active package components. Susceptor tray, base reflector, and shielding/browning lid.

in the early stages, consumers should expect to see experiments with different styles for similar products.

G. Dual Ovenability of Field Modification Packages

The three questions posed for whether a specific browning and crisping package could advantageously be dual ovenable (see Sec. III.C.11) are applicable for determining whether field modification packages should be dual ovenable.

In general, if the objective of using a field modification package is to gain temperature uniformity in microwave heating of large food items, package dual ovenability can be useful. A good example would be the lasagna products described above. Since consumers are accustomed to heating frozen lasagna in the conventional oven, a microwave offering that heats satisfactorily in a conventional oven could simplify a product offering and give consumers flexibility in heating the food.

Fortunately, dual ovenability is generally assured for field modification trays, since the current commercially available packages are constructed of laminations of temperature-resistant oriented films, patterned aluminum foil and paperboard. Similar to susceptor trays, this structure is dual ovenable capable.

For nontray structures, like sleeves or cards, dual ovenability may not be as practical, for the same reasons as described in Sec. III.C.11.

IV. ENVIRONMENTAL CONSIDERATIONS

Energy use comparisons of microwave heating and alternative cook methods should head the list of considerations when evaluating the environmental performance of package alternatives for a given food item. The microwave oven's instant-on/instant-off operation, coupled with its ability to apply energy directly to the items to be heated, gives it an energy usage advantage over conventional cooking methods. In the case of the frozen fruit pie discussed above, microwave cooking entails 18 min of energy application. Conventional oven cooking requires a preheat time of at least 10 min followed by 50 min of cooking. After the oven is turned off, it continues to transfer heat to the kitchen as it cools. Microwave oven manufacturers have consistently touted the energy-saving advantage of microwave cooking, and consumer comments suggest that consumers believe these claims [47].

Packages for microwaveable foods are subject to the same environmental assessments as packages for other products. Observers and commentators on the subject of "environmental friendliness" of packaging make pronouncements that tend to reflect their biases and personal agendas. Not surprisingly, suppliers of a particular material present data "proving" that their material is superior to competitive materials. The most radical environmental activists label anything less natural than wrapping food in plant leaves as wasteful and earth destructive.

The challenge for a user of microwaveable packaging is to construct realistic criteria for determining if a particular package for a particular food is acceptable from an environmental standpoint, given the environmental values of the company and its likely customers. While this may approach may strike some as amoral or cynical; the reality is that there are no widely agreed to standards from which to evaluate packages from an environmental standpoint.

Companies and individuals make hundreds of decisions each day on which materials and products to use, where to get them, how to use them, and how to dispose of any waste associated with them. Each of those decisions represents a cost/benefit decision: Are the perceived benefits from a particular action sufficient to more than offset the perceived costs of taking that action? The decision maker incorporates some kind of assessment of those costs and benefits in the decision process. The reason that there is no widely agreed to environmental packaging standard is that the perceptions of decision makers are highly varied and depend on the factors taken into consideration and the weighting given to each factor.

When comparing microwave packaging alternatives, the rigor applied in making the assessments can range from asking simple questions such as "Is the package recyclable?" or "Which package alternative is lighter?" to undertaking complex life cycle assessments that attempt to quantify the environmental impact of all aspects of making, using, and disposing the package.

Single material packages rate well in assessments that value recycling; composites that contain recycled materials rate well in assessments that value reuse. While both are widely viewed as desirable environmental actions, other considerations may outweigh them. For example, there is no universality in the availability of recycling centers; some easily recycled materials are not collected in some regions. Consumers also tend not to recycle packages with substantial food residues to avoid vermin and odor problems in their recyclable storage areas. Food safety regulations enforced by USFDA or other governmental agencies may preclude the use of some recycled materials in food packages due to concerns over trace materials. The point is that single criterion assessments are limited, as they often ignore realities that must be part of an informed decision process.

It is not the purpose of this chapter to present judgments to the reader on the environmental desirability of specific packages. Rather, we suggest that each maker and user of packages for microwaveable foods consider the criteria they believe are appropriate to make environmental assessments and comparisons of packages. The following lists a number of considerations that we are aware have been considered at one point or another, but is not represented to be complete. We would expect that other criteria may be used and that decision makers will include those criteria specific to their situation. By "packaging" in this list we mean all elements of packaging used to make, transport, and consume the food:

- Energy consumed in cooking/heating
- Energy consumed in producing the package
- Weight and size of the packaging
  - As fabricated
  - As shipped to packer
  - As shipped from packer to end user
- Ease of recycling packaging components
- Recycled material content of packaging components
- Compatibility of packaging components with waste-to-energy systems
- Use of materials or sources thought to be endangered
- Compatibility with sanitary landfill operations

Decision makers will need to stay current on the issues affecting packaging in their markets and incorporate changes in regulations, use and disposal taxes, and public opinion so they can choose appropriate package types.

V. SOME REMAINING CHALLENGES

Despite the progress made in the technologies for passive and active packages, a number of challenges remain. For passive packages, the challenges are largely those of cost efficiency and increased temperature resistance for full dual ovenability, as stated earlier in Sec. II.D.

For active packages, a number of foods continue to be difficult to prepare well in the microwave. Frozen rising crust pizzas have become popular items, but currently only recommend conventional oven preparation. Crust development in the reduced cooking time associated with microwave heating is the technical challenge to be overcome. Dough formulation in combination with package design will be required to solve this need.

Breaded portions of poultry, seafood, and vegetables are popular items that have been difficult to consistently brown and crisp using active packaging. Breading formulations that can stand up to multiple freeze-thaw cycles in distribution and storage are needed, along with packages that can apply high levels of surface heating and not allow moisture to recondense on the breading.

Approximating the grilling or broiling of single servings of raw meat, poultry and seafood has not been achieved to date. This represents the ultimate challenge in "primary" cooking in the consumer microwave, and may not be solved with package technology alone. One can speculate about ovens that modulate frequency during the cook cycle to initially apply energy with low penetration for surface development and then move to frequencies that allow deeper penetration to achieve the proper doneness of the bulk of the food. If feasible, ovens operating in this manner would likely find first applications in food service environments, with later expansion to consumer use.

VI. SUMMARY

The challenge and opportunity of packaging for microwaveable foods is to create a food/package combination that delivers the quality consumers expect from their favorite foods, at a fraction of the cook time, and with no cleanup. Matching food and package is critical to achieving this goal.

Food companies can capture an increased share of the consumers food dollar by meeting the needs of an ever more quality-discriminating and time-stressed audience. Microwave ovens can be a key part of this strategy and packaging can help deliver compelling food experiences.

APPENDIX: PATENTS

This appendix is a listing of U.S. patents that will assist the reader in obtaining a deeper understanding of the technologies employed in packages for microwaveable foods. The listing is broad, but is not represented to be exhaustive. Accordingly, it should not be used as a sole reference for patentability studies. Additional sources, including other U.S. patents, foreign filings and issued patents, published articles, and public use are important elements of a prior art search.

Assignees for patents may change after issuance and such changes are beyond the scope of this listing. Also, patents are abandoned by owners or invalidated in legal proceedings; it is the responsibility of the reader to determine the status of an individual patent at a specific time. No judgment is made, either express or implied, regarding the validity of claims associated with listed patents or whether patents shown in this listing infringe claims of other patents, whether or not included in this listing.

The format of this listing is:

> U.S. Patent number, Title, Last name of first named inventor, Assignee at time of issuance

Issued 1949–1975

> 2480679, Prepared food article and method of preparing, Spencer, Raytheon Mfg.
> 2495435, Method of treating foodstuffs, Welch, Raytheon Mfg.
> 2528251, Receptacle, Spencer, Raytheon Mfg.
> 2564707, Electrically conductive coating on glass and other ceramic bodies, Mochel, Corning Glass Works
> 2564709, Electrically conductive coating on glass and other ceramic bodies, Mochel, Corning Glass Works
> 2600566, Method of heating frozen food packages, Moffett, none

2714070, Microwave heating apparatus and method of heating a food package, Welch, Raytheon Mfg.
2830162, Heating method and apparatus, Copson, Raytheon Mfg.
2889299, Grease resistant cellulosic webs coated with a linear anionic thermoplastic ethyl acrylate-acrylonitrile-methacrylic acid polymer containing a hydrophilic inorganic pigment as extender and composition for manufacture thereof, Daniel D. Ritson, American Cyanamid Company
3179780, Device for heating deep-frozen eatables with the aid of microwaves, Verstraten, North American Phillips Co.
3219460, Frozen food package and method of producing same, Brown, Lever Bros. 3243753, Resistance element, Kohler, none
3243753, Resistance element, Kohler, none
3271169, Food package for microwave heating, Baker, Litton Precision Products
3271552, Microwave heating apparatus, Krajewski, Litton Precision Products
3293048, Food and beverage cooking container and method of using same, Kitterman, none
3302632, Microwave cooking utensil, Fichnter, Wells Mfg.
3353968, Food package for use in microwave heating apparatus, Krajewski, Litton Precision Products
3547661, Container and food heating method, Stevenson, Teckton, Inc.
3573923, Dielectric heating of food, Meiser, Plastics Mfg. Co.
3615713, Selective cooking apparatus, Stevenson, Teckton, Inc.
3783220, Method of apparatus for browning exterior surfaces of food stuff in an electronic range, Tanizaki, Yamizu Shoji Kabushiki Kaisha
3835280, Composite microwave energy perturbating device, Gades, The Pillsbury Company
3835281, Differential microwave heating container, Mannix, none
3857009, Microwave browning means, MacMaster, Raytheon Corporation
3861029, Method of making heater cable, Smith-Johannsen, Raychem Corporation
3865301, Partially shielded food package for dielectric heating, Pothier, Trans World Services, Inc.
3865302, Container for cooking food therein, Kane, E.I. DuPont de Nemours & Co.
3876131, Wedge shaped carton, Tolaas, Hoerner Waldorf Corp.
3924013, Method of cooking food in a polythylene terephthalate/paperboard laminated container, Kane, E.I. DuPont de Nemours & Co.

Issued 1976

3934106, Microwave browning means, MacMaster, Raytheon Company

3936626, Method of heating comestibles, Moore, Chemetron Corp.

3938266, Adhesive system, Cook, Holobeam, Inc.

3941967, Microwave cooking apparatus, Sumi, Asahi Kasei Kogyo Kabushiki

3946187, Microwave browning utensil, MacMaster, Raytheon Company

3946188. Microwave heating apparatus with browning feature, Derby, Raytheon Company

3949184, Folding microwave searing and browning means, Freedman, Raytheon Company

3965323, Method and apparatus for providing uniform surface browning of food stuff through microwave energy, Forker, Jr., Corning Glass Works

3974354, Microwave utensil with reflective surface handle, Long, General Motors Corporation

3973045, Popcorn package for microwave popping, Brandberg, The Pillsbury Company

3985990, Microwave oven baking utensil, Levinson, none

3985991, Methods of microwave heating in metal containers, Levinson, none

3985992, Microwave heating tray, Goltsos, Teckton, Inc.

3987267, Arrangement for simultaneously heating a plurality of comestible items, Moore, Chemetron Corp.

3997677, High temperature resistant hermetically sealed plastic tray packages, Hirsch, Standard Pkg.

Issued 1977

4003840, Microwave Absorber, Ishino, TDK Electronics Company LTD

4013798, Selectively ventable food package and microwave shielding device, Goltsos, Teckton, Inc.

4015085, Container for the microwave heating of frozen sandwiches, Woods, Larry Lakey

4027132, Microwave pie baking, Levinson, none

4027384, Microwave absorbers, Connolly, The United States of America as represented by the Secretary of the Army

4036423, Expandable package, Gordon, International Paper Co.

4038425, Combined popping and shipping package for popcorn, Brandberg, The Pillsbury Company

Issued 1978

4081646, Device for microwave cooking, Goltsos, Teckton

4096948, Cook-in carton with integral removable section and blank therefor, Kuchenbecker, American Can Co.

4122324, Shielding device for microwave cooking, Falk, Teckton, Inc.

Issued 1979

4132811, Food package for assuring uniform distribution of microwave energy and process for heating food, Standing, The Pillsbury Company

4133896, Food package including condiment container for heating food, Standing, The Pillsbury Company

4144438, Microwave energy moderating bag, Gelman, The Procter & Gamble Company

4147836, Polyester coated paperboard for forming food containers and process for producing the same, Middleton, American Can Company

4156806, Concentrated energy microwave appliance, Teich, Raytheon Company

4158760, Seed heating microwave appliance, Bowen, Raytheon Company

Issued 1980

4183435, Polymeric multiple-layer sheet material, Thompson, Champion International Corp.

4190757, Microwave heating package and method, Turpin, The Pillsbury Company

4196331, Microwave energy cooking bag, Leveckis, The Procter & Gamble Company

4204105, Microwave energy moderating bag, Leveckis, The Procter & Gamble Company

4210674, Automatically ventable sealed food package for use in microwave ovens, Mitchell, American Can Co.

4214515, Disposable structure for use in microwave cooking, Kubiatowicz, none

4217476, Individual prepared meal tray, Bellavoine, DePruines ISLCO

4219573, Microwave popcorn package, Borek, The Pillsbury Company

4228334, Dynamic microwave energy moderator, Clark, The Procter & Gamble Company

4228945, Food carton for microwave heating, Wysocki, Champion International Corp.

4230924, Method and material for prepackaging food to achieve microwave browning, Brastad, General Mills, Inc.

4233325, Ice cream package including compartment for heating syrup, Slangan, International Flavors & Fragrances

Issued 1981

4248901, Combination package and sleeve support means, Austin, Champion International Corp.

4258086, Method of reproduction metallized patterns with microwave energy, Beall, General Mills, Inc.

4260060, Food carton for microwave heating, Faller, Champion International. Corp.

4260101, Expandable container and blank therefor, Webinger, Champion International Corporation

4261504, Heat-sealable ovenable containers, Cowan, Maryland Cup

4266108, Microwave heating device and method, Anderson, The Pillsbury Company

4267420, Packaged food item and method for achieving microwave browning thereof, Brastad, General Mills, Inc.

4268738, Microwave energy moderator, Flautt, Jr., The Procter & Gamble Company

4277506, Supportive sidewall container for expandable food packages, Austin, Champion International Corp.

4279933, Expandable food package container, Austin, Champion International Corp.

4280032, Egg cooking in a microwave oven, Levinson, none

4281083, Blends of poly (p-methylstryrene) with polyfunctional monomer, Arbit, Mobil Oil Corp.

4283427, Microwave heating package method and susceptor composition, Winters, The Pillsbury Company

4286136, Cooking container for more efficient cooking in a microwave oven, Mason, none

4292332, Container for prepackaging, popping and serving popcorn, McHam, none

4304327, Container unitizer sleeve and support, Austin, Champion International Corp.

4306133, Microwave pie baking, Levinson, none

4307277, Microwave heating oven, Maeda, Mitsubishi Denki Kabushiki Kaisha

Issued 1982

4312451, Self standing flanged tray with integral lid, Forbes, Jr., Westvaco Corporation

4316070, Cookware with liquid microwave energy, Prosise, None

4320274, Cooking Utensil for uniform heating in microwave oven, Dehn, RTE Corporation

4327136, Polymeric multi-layer sheet material and tray, Thompson, Champion International Corp.

4335181, Microwavable heat and grease resistant containers, Marano, Mobil Oil

4337116, Contoured molded pulp container with polyester liner, Foster, Keyes Fibre Co.

4345133, Partially shielded microwave carton, Cherney, American Can Co.

4351884, Shaped articles from poly (*p*-methylstryrene) blends, Robertson, Mobil Oil Corp.

4351997, Food package, Mattisson, Societe d'Assistance Technique pour Prodiuts Nestle S.A.

4355757, Venting carton and blank therefor, Roccaforte, Champion International Corp.

4358466, Freezer to microwave oven bag, Stevenson, Dow Chemical

4360107, Carton blank and carton for pizza, Roccaforte, Champion International Corp.

4362917, Ferrite heating apparatus, Freedman, Raytheon Company

Issued 1983

4387012, Blends of poly (*p*-methylstryrene) with polyfunctional monomer and shaped article, Arbit, Mobil Oil Corp.

4390554, Microwave heating of certain frozen foods, Levinson, none

4390555, Microwave oven cooking method, Levinson, none

4391833, Method of making and using heat resistant resin coated paperboard product and product thereof, Self, International Paper Co.

4398077, Microwave cooking utensil, Freedman, Raytheon Company

4398994, Formation of packaging material, Beckett, Beckett Industries Inc.

4401702, Film laminate and product pouch or tube therefrom, Hungerford, Mobil Oil Corp.

4404241, Microwave package with vent, Mueller, James River-Dixie/Northern Inc.

4416906, Microwave food heating container, Watkins, Golden Valley Foods Inc.

4416907, Process for preparing food packages for microwave heating, Watkins, Golden Valley Foods Inc.

4419373, Method of heating contents in a self-venting container, Opperman, American Can Co.

Issued 1984

4425368, Food heating container, Watkins, Golden Valley Foods Inc.

4427706, Method for heating par-fried, batter-coated frozen foods, El-Hag, General Foods Corp.

4434197, Non-stick energy-modifying cooking liner and method of making same, Petriello, N. F. Industries, Inc.

4448309, Container for expandable food pouch, Roccaforte, Champion International Corp.
4448791, Brownable dough for microwave cooking, Fulde, Campbell Soup Company
4449633, Ovenable paperboard carton, Johnson, Manville Service Corp.
4450180, Package for increasing the volumetric yield of microwave cooked popcorn, Watkins, Golden Valley Foods Inc.
4453665, Container for expandable food pouch, Roccaforte, Champion International Corp.
4454403, Microwave heating method and apparatus, Teich, Raytheon Company
4456164, Deliddable ovenable container, Foster, Keyes Fibre Co.
4460814, Oven antenna for distributing energy in microwave, Diesch, Amana Refrigeration, Inc.
4461031, Tubular bag and method of making the same, Blamer, Bagcraft Corp. of America
4467052, Tray for packaging food products, Barnwell, Plasti-Pak
4469258, Tray with compound sealed lid, Wright, Champion International Corp.

Issued 1985

4495392, Microwave simmer pot, Derby, Raytheon Company
4497431, Container structure, Fay, James River Corp.
4505391, Cook-in carton with improved integral support structure, Kuchnbecker, James River-Norwalk, Inc.
4505961, Microwavable heat and grease resistant containers and method for their preparation, Lu, Mobil Oil Corp.
4505962, Microwavable plastic containers with heat and grease resistant layer comprising impact polymer, Lu, Mobil Oil Corp.
4507338, Container comprising a polysulfone resin layer with a cellular resin core, Freundlich, Polymer Projections, Inc.
4517045, Apparatus for formation of packaging material, Beckett, Beckett Industries Inc.
4518651, Microwave absorber, Wolfe, E.I. DuPont de Nemours and Company
4529464, Process for manufacturing a food container by extrusion and vacuum forming, Jones, Champion International Corp.
4535889, Frozen food package and cover lid, Terauds, Stouffer Corp.
4542271, Microwave browning wares and method for the manufacture thereof, Tanonis, Rubbermaid Incorporated

4548826, Method for increasing the volumetric yield of microwave cooked popcorn, Watkins, Golden Valley Foods, Inc.
4552614, Demetallizing method and apparatus, Beckett, Beckett Packaging Limited
4553010, Packaging container for microwave popcorn popping and method for using, Bohrer, James River-Norwalk, Inc.
4555605, Package assembly and method for storing and microwave heating of food, Brown, James River-Norwalk, Inc.
4558198, Metal container system for use in microwave ovens, Levendusky, Alcoa
4560850, Container with steam port for use in microwave ovens, Levendusky, Alcoa

Issued 1986

4567341, Side vented and shielded microwave pizza carton, Brown, James River-Norwalk, Inc.
4571337, Container and popcorn ingredient for microwave use, Cage, Hunt-Wesson Foods, Inc.
4574174, Convenience dinner container and method, McGonigle, none
4584202, Microwave popcorn package, Roccaforte, Waldorf Corp.
4585915, Containers with attached lid for storage and microwave cooking, Moore, Dow Chemical
4586649, Food Package, Webinger, Waldorf Corp.
4590349, Microwave cooking carton for browning and crisping food on two sides, Brown, James River-Dixie/Northern, Inc.
4592914, Two-blank disposable container for microwave food cooking, Kuchenbecker, James River-Dixie/Northern, Inc.
4594492, Microwave package including a resiliently biased browning layer, Maroszek, James River Corporation, Norwalk
4595611, Ink-printed ovenable food containers, Quick, International Paper Co.
4596713, Microwave food packets capable of dispersing a food additive during heating, Burdette, None
4604276, Intercalation of small graphite flakes with a metal halide, Oblas, GTE Laboratories Incorporated
4604854, Machine for forming, filling and sealing bags, Andreas, Golden Valley Foods Inc.
4610755, Demetallizing method, Beckett, none
4612221, Multilayer food wrap with cling, Biel, Union Carbide
4612431, Package assembly and method for storing and microwave heating of food, Brown, James River-Norwalk, Inc.

4626641, Fruit and meat pie microwave container and method, Brown, James River Corporation

4631046, Bake in tray and method of forming a blank for the same, Kennedy, Waldorf Company

Issued 1987

4640838, Self-venting vapor-tight microwave oven package, Isakson, Minnesota Mining & Mfg. Co.

4641005, Food receptacle for microwave cooking, Seiferth, James River Corporation

4642434, Microwave reflective energy concentrating spacer, Cox, Golden Valley Microwave Foods

4656325, Microwave heating package and method, Keefer, none

4661671, Package assembly with heater panel and method for storing and microwave heating of food utilizing same, Maroszek, James River Corp.

4676857, Method of making microwave heating material, Scharr, Scharr Industries, Inc.

4678882, Packaging container for microwave popcorn popping, Bohrer, James River-Norwalk

4681996, Analytical process in which materials to be analyzed are directly and indirectly heated and dried by microwave radiation, Collins, CEM Corp.

4685997, Production of demetallized packaging material, Beckett, none

4687104, Microwave carton, Ielmini, Patterson Frozen Foods Ind.

4687117, Frozen food package and cover lid, Terauds, The Stouffer Corp.

4689458, Container systems for microwave cooking, Levendusky, Alcoa

4691374, Cooking bag with diagonal gusset seals, Watkins, Golden Valley Microwave Foods

4698472, Microwave heating stand with electrically isolated reflector, Cox, Golden Valley Microwave Foods Inc.

4703148, Package for frozen foods for microwave heating, Mikulski, General Mills, Inc.

4703149, Container heated by microwave oven, Sugisawa, House Food Industrial Company Limited

4704510, Containers for food service, Matsui, Fukuyama Pearl Shiko Kabushiki

4705707, Polyethylene/Polyester nonoriented heat sealable moisture barrier film and bag, Winter, Presto Products, Inc.

4705929, Microwave Trays, Atkinson, Somerville Belkin Industries

4710522, Foamable composition and process for forming same, Huggard, Mobil Oil Corp.

4713510, Package for microwave cooking with controlled thermal effects, Quick, International Paper Company

4716061, Polypropylene/Polyester nonoriented heat sealable moisture barrier film and bag, Winter, Presto Products, Inc.

4720410, Heat-activated blotter, Lundquist, ConAgra, Inc.

Issued 1988

4721738, Polymeric compositions including microwave energy sensitizing additives, Ellis, Occidental Research Corporation

4724290, Microwave popcorn popper, Campbell, none

4734288, Package for expandable food product, Engstrom, E.A. Sween Co.

4735513, Flexible packaging sheets, Watkins, Golden Valley Microwave Foods Inc.

4738365, Frozen food container, Prater, Ridgway Pkg. Corp.

4740377, Method for Microwave cooking of foods, Dawes, DuPont Canada Inc.

4742203, Package assembly and method for storing and microwave heating of food, Brown, James River-Norwalk, Inc.

4745249, Package and method for microwave heating of a food product, Daniels, Mrs. Pauls's Kitchens, Inc.

4746019, End fill microwavable and/or ovenable container, Prater, Ridgeway Pkg. Corp.

4749008, Method of preparing a packaged frozen confection, Schmidt, General Foods

4756917, Packaging sheet and containers and pouches using the sheet, Kamada, Toyo Aluminum Kabushiki Kaisha

4757940, Ovenable paperboard food tray, Quick, International Paper Company

4763790, Heat treatable containers, McGeehins, Waddingtons Cartons Ltd. (England)

4764399, Method of printing thermoplastic articles with thermoplastic ink, Harrison, Plastona (John Waddington LTD)

4765999, Polyester/Copolyester conextruded packaging film, Winter, Presto Products Inc.

4775560, Heat-resistant paper container and process for preparation thereof, Katsura, Toyo Seikan Kaisha, Ltd.

4775771, Sleeve for crisping and browning of foods in a microwave oven and package and method utilizing same, Pawlowski, James River Corporation

4777053, Microwave heating package, Tobelmann, General Mills, Inc.

4780587, Overlap seam for microwave interactive package insert, Brown, James River Corporation

4785160, Sleeve type carton for microwave cooking, Hart, Container Corporation of America

4785937, Retortable pouch and packaging material for the retortable pouch, Tamezawa, Kabushiki Kaisha Hosokawa

4786513, Package for sliced bacon adapted for microwave cooking, Monforton, ConAgra, Inc.

4786773, Systems and methods for determining doneness of microwave-heated bodies, Keefer, Alcan International Inc.

4794005, Package assembly including a multi-surface, microwave interactive tray, Swiontek, James River Corp.

Issued 1989

4797010, Reheatable, resealable package for fried food, Coelho, Nabisco Brands, Inc.

4797523, Cover for an open container whose content is to be heated in a microwave oven, Kohnen, none

4800247, Microwave heating utensil, Schneider, Commercial Decal, Inc.

4801017, Container, particularly for receiving foods, Artusi, none

4801774, Center-supported microwave tray, Hart, Container Corp. of America

4803088, Container packed with instant food for use in microwave oven, Yamamoto, House Food Industrial Co.

4806371, Microwavable package for packaging combination of products and ingredients, Mendenhall, Packaging Concepts, Inc.

4806718, Ceramic gels with salt for microwave hating susceptor, Seaborne, General Mills, Inc.

4808721, Formed polymer film package for microwave cooking, Mendenhall, Packaging Concepts, Inc.

4808780, Amphoteric ceramic microwave heating susceptor utilizing compositions with metal salt moderators, Seaborne, General Mills, Inc.

4810844, Microwave popcorn package, Anderson, none

4810845, Solid state ceramic microwave heating susceptor, Seaborne, General Mills, Inc.

4813594, Microwavable package, Brown, Federal Paper Board Co.

4814568, Container for microwave heating including means for modifying microwave heating distribution, and method of using same, Keefer, Alcan International Ltd.

4818545, Food material/container combination, Kunimoto, House Food Industrial Co.

Packaging Techniques for Microwaveable Foods 455

4818831, Amphoteric ceramic microwave heating susceptor, Seaborne, General Mills, Inc.
4820893, Two-celled expandable microwave cooking sling, Mode, Waldorf Corporation
4821492, Method of making end fill microwavable and/or ovenable container, Prater, Ridgway Packaging Corp.
4821884, Secondary packaging, Griffin, General Foods Limited
4822966, Methods of producing heat with microwaves, Matsubara, none
4825024, Solid state ceramic microwave heating susceptor utilizing compositions with metal salt moderators, Seaborne, General Mills, Inc.
4825025, Food receptacle for microwave cooking, Seiferth, James River Corporation
4826072, Microwave carton, Hart, Container Corporation of America
4831224, Package of material for microwave heating including container with stepped structure, Keefer, Alcan International LTD
4833007, Microwave susceptor packaging material, Huang, E. I. DuPont de Nemours and Company
4835352, Package material for microwave cooking, Sasaki, Toppan Printing Co., Ltd.
4836383, Microwave food carton with divider panel, Bernstein, International Paper Co.
4836438, Ovenable carton with handles, Rigby, Westvaco Corporation
4836439, Microwave carton, Hart, Container Corporation of America
4841112, Method and appliance for cooking a frozen pot pie with microwave energy, Peleg, The Stouffer Corporation
4848931, Packaging sheet and containers and pouches using the sheet, Kamada, Toyo Aluminum Kabushiki Kaisha
4851246, Dual compartment food package, Esse, General Mills, Inc.
4851631, Food container for microwave heating and method of substantially eliminating arcing in a microwave food container, Wendt, The Pillsbury Company
4851632, Insulated frame package for microwave cooking, Kaliski, E. I. DuPont de Nemours and Company
4857342, Ovenable package for bacon and the like, Kappes, Milprint Inc.
4861958, Packaging container for microwave popcorn popping, Bohrer, James River—Norwalk, Inc.
4864089, Localized microwave radiation heating, Tighe, Dennison Manufacturing Company
4865854, Microwave food package, Larson, Minnesota Mining and Manufacturing Company
4865921, Microwave interactive laminate, Hollenberg, James River Corporation of Virginia

4866232, Food package for use in a microwave oven, Stone, Packaging Corporation of America

4866234, Microwave container and method of making same, Keefer, Alcan International Limited

4871892, Cooking utensil for assuring destruction of harmful bacteria during microwave cooking of poultry and other foods, Samford, General Housewares, Corporation

4873101, Microwave food package and grease absorbent pad therefor, Larson, Minnesota Mining and Manufacturing Company

4874620, Microwavable package incorporating controlled venting, Mendenhall, Packaging Concepts, Inc.

4875597, Convenience packaging, Saunders, Weirton Steel Corporation

4876423, Localized microwave radiation heating, Tighe, Dennison Manufacturing Company

4877932, Microwave container assembly, Bernstein, International Paper

4878765, Flexible packaging sheets and packages formed therefrom, Watkins, Golden Valley Microwave Foods, Inc.

4880951, Food preparation kit for use in cooking food in microwave oven or in thermal oven, Levinson, none

4883936, Control of microwave interactive heating by patterned deactivation, Maynard, James River Corporation of Virginia

4888459, Microwave container with dielectric structure of varying properties and method of using same, Keefer, Alcan International Limited

4890439, Flexible disposable material for forming a food container for microwave cooking, Smart, James River Corporation of Virginia

4891482, Disposable microwave heating receptacle and method of using same, Jaeger, The Stouffer Corporation

4894503, Package materials for shielded food, Wendt, The Pillsbury Company

4896009, Gas permeable microwave reactive package, Pawlowski, James River Corporation

4900594, Pressure formed paperboard tray with oriented polyester film interior, Quick, International Paper Company

Issued 1990

4904836, Microwave heater and method of manufacture, Turpin, The Pillsbury Company

4906806, Cooking kit with heat generating method for microwave oven and methods for microwave cooking, Levinson, none

4911938, Conformable wrap susceptor with releasable seal for microwave cooking, Fisher, E. I. DuPont de Nemours and Company

4914266, Press applied susceptor for controlled microwave heating, Parks, Westvaco Corporation
4915780, Process for making an element for microwave heating, Beckett, Beckett Industries Inc.
4916279, Apparatus for surface heating an object by microwave energy, Brown, James River Corporation of Virginia
4917748, Method of making microwave heatable materials, Harrison, Waddington Cartons Limited
4923704, Methods for microwave cooking in a steam-chamber kit, Levinson, none
4926020, Microwave food products and method of their manufacture, Atwell, The Pillsbury Company
4927991, Susceptor in combination with grid for microwave oven package, Wendt, The Pillsbury Company
4930639, Ovenable food container with removable lid, Rigby, Westvaco Corporation
4933193, Microwave cooking package, Fisher, E. I. DuPont de Nemours and Company
4936935, Microwave heating material, Beckett, Beckett Industries Inc.
4938990, Pattern metallizing, Beckett, Beckett Packaging Limited
4940867, Microwave composite sheet stock, Peleg, The Stouffer Corporation
4943439, Microwave receptive heating sheets and packages containing them, Andreas, Golden Valley Microwave Foods, Inc.
4943456, Microwave reactive heater, Pollart, James River Corporation of Virginia
4948932, Apertured microwave reactive package, Clough, James River Corporation
4955530, Easy opening lid for ovenable carton, Rigby, Westvaco Corporation
4959120, Demetallization of metal films, Wilson, Golden Valley Microwave Foods, Inc.
4959231, Microwave food packaging, Lakey, Marque Foods, Inc.
4960598, Package assembly including a multi-surface, microwave interactive tray, Swiontek, James River Corporation
4962000, Microwave absorbing composite, Emslander, Minnesota Mining and Manufacturing Company
4963424, Microwave heating material, Beckett, Beckett Industries Inc.
4970358, Microwave susceptor with attenuator for heat control, Brandberg, Golden Valley Microwave Foods, Inc.
4970360, Susceptor for heating foods in a microwave oven having metallized layer deposited on paper, Pesheck, The Pillsbury Company

4973810, Microwave method of popping popcorn and package thereof, Brauner, General Mills, Inc.

4977302, Browning utensil for microwave ovens, Merigaud, Degussa Aktiengesellschaft

4982064, Microwave double-bag food container, Hartman, James River Corporation of Virginia

Issued 1991

4985300, Shrinkable, conformable microwave wrap, Huang, E. I. DuPont de Nemours and Company

4990735, Improved uniformity of microwave heating by control of the depth of a load in a container, Lorenson, Alcan International Limited

4992636, Sealed container for microwave oven cooking, Namiki, Toyo Seikan Kaisha Ltd.

4992638, Microwave heating device with microwave distribution modifying means, Hewitt, Alcan International Limited

5002826, Heaters for use in microwave ovens, Pollart, James River Corporation of Virginia

5006684, Apparatus for heating a food item in a microwave oven having heater regions in combination with a reflective lattice structure, Wendt, The Pillsbury Company

5008024, Microwave corn popping package, Watkins, Golden Valley Microwave Foods, Inc.

5011299, Bag construction, Black, Jr., American Packaging Corporation

5019681, Reflective temperature compensating microwave susceptors, Lorence, The Pillsbury Company

5021293, Composite material containing microwave susceptor material, Huang, E. I. DuPont de Nemours and Company

5034234, Microwave heating and serving package, Andreas, Golden Valley Microwave Foods Inc.

5038009, Printed microwave susceptor and package containing the susceptor, Babbitt, Union Camp Corporation

5039364, Method of making selective microwave heating material, Beckett, Beckett Industries Inc.

5039833, Microwave heatable materials, Woods, Waddingtons Cartons Limited

5041295, Package for crisping the surface of food products in a microwave oven, Perry, The Pillsbury Company

5044777, Flat-faced package for improving the microwave popping of corn, Watkins, Golden Valley Microwave Foods, Inc.

Packaging Techniques for Microwaveable Foods 459

5045330, Biased food contact container and container insert, Pawlowski, James River Corporation

5049710, Microwave food carton having two integral layer-divider panels and blank therefor, Prosise, The Procter & Gamble Company

5057331, Cooking food in a food preparation kit in a microwave and in a thermal oven, Levinson, none

5057659, Microwave heating utensil with particulate susceptor layer, Schneider, Commercial Decal, Inc.

5059279, Susceptor for microwave heating, Wilson, Golden Valley Microwave Foods, Inc.

Issued 1992

5078273, Microwave carton and blank for forming the same, Kuchenbecker, James River Corporation of Virginia

5079083, Coated microwave heating sheet, Watkins, Golden Valley Microwave Foods, Inc.

5079397, Susceptors for microwave heating and systems and methods of use, Keefer, Alcan International Limited

5081330, Package with microwave induced insulation chambers, Brandberg, Golden Valley Microwave Foods, Inc.

5084601, Microwave receptive heating sheets and packages containing them, Andreas, Golden Valley Microwave Foods, Inc.

5095186, Method for making selectively metallized microwave heating packages, Russell, Waldorf Corporation

5096723, Microwave food heating package with serving tray, Turpin, Golden Valley Microwave Foods, Inc.

5097107, Microwave corn popping package having flexible and expandable cover, Watkins, Golden Valley Microwave Foods, Inc.

5117078, Controlled heating of foodstuffs by microwave energy, Beckett, Beckett Industries Inc.

5118747, Microwave heater compositions for use in microwave ovens, Pollart, James River Corporation of Virginia

5124519, Absorbent microwave susceptor composite and related method of manufacture, Roy, International Paper Company

5126518, Microwave cooking container cover, Beckett, Beckett Industries Inc.

5126519, Method and apparatus for producing microwave susceptor sheet material, Peleg, The Stouffer Corporation

5132144, Microwave oven susceptor, Parks, Westvaco Corporation

5140119, Package assembly and method for storing and microwave heating of food, Brown, James River Paper Company, Inc.

5149396, Susceptor for microwave heating and method, Wilson, Golden Valley Microwave Foods, Inc.
5153036, Temperature sensing element, Sugisawa, House Food Industrial Company Limited
5153402, Paperboard container for microwave cooking, Quick, International Paper Company
5164562, Composite susceptor packaging material, Huffman, Westvaco Corporation
5171594, Microwave food package with printed-on susceptor, Babbitt, Union Camp Corporation
5171950, Flexible pouch and paper bag combination for use in the microwave popping of popcorn, Brauner, General Mills, Inc.
5175404, Microwave receptive heating sheets and packages containing them, Andreas, Golden Valley Microwave Food, Inc.

Issued 1993

5180894, Tube from microwave susceptor package, Quick, International Paper Company
5182425, Thick microwave susceptor, Pesheck, The Pillsbury Company
5185506, Selectively microwave-permeable membrane susceptor systems, Walters, Advanced Dielectric Technologies, Inc.
5190777, Package for microwaving popcorn, Anderson, American Home Foods, Inc.
5195829, Flat bottomed stand-up microwave corn popping bag, Watkins, Golden Valley Microwave Foods, Inc.
5211810, Electrically conductive polymeric materials and related method of manufacture, Bartholomew, International Paper Company
5213902, Microwave oven package, Beckett, Beckett Industries Inc.
5214257, Tub-shaped packaging container for microwave popcorn, Riskey, Recot, Inc.
5220141, Treatment of paperboard with polar organic compounds to provide microwave interactive stock, Quick, International Paper Company
5221419, Method for forming laminate for microwave oven package, Beckett, Beckett Industries Inc.
5223291, Microwave-core-heating and cooking pasta, pulses, grains and cereals, Levinson, none
5223685, Elevated microwave cooking platform, DeRienzo, Jr., none
5230914, Metal foil food package for microwave cooking, Avervik, Luigino's, Inc.
5231268, Printed microwave susceptor, Hall, Westvaco Corporation
5239153, Differential thermal heating in microwave oven packages, Beckett, Beckett Industries Inc.

Packaging Techniques for Microwaveable Foods 461

5252793, Microwave container assembly, Woods, Waddington Cartons Limited

5258596, Microwave absorber designs for metal foils and containers, Fabish, Aluminum Company of America

5260537, Microwave heating structure, Beckett, Beckett Industries Inc.

5267686, Food package containing separate trays connected together by a single lid structure, Gulliver, Gulf States Paper Corporation

5270066, Double-center wall microwave food package, Pawlowski, James River Corporation of Virginia

5270502, Package assembly and method for storing and microwave heating of food, Brown, James River—Norwalk, Inc.

Issued 1994

5278378, Microwave heating element with antenna structure, Beckett, Beckett Industries Inc.

5280150, Heat generating container for microwave oven, Arai, Sharp Kobushiki Kaisha

5285040, Microwave susceptor with separate attenuator for heat control, Brandberg, Golden Valley Microwave Foods, Inc.

5298708, Microwave-active tape having a cured polyolefin pressure-sensitive adhesive layer, Babu, Minnesota Mining and Manufacturing Company

5300746, Metallized microwave diffuser films, Walters, Advanced Deposition Technologies, Inc.

5302790, Microwave popcorn popping bag, Turpin, Golden Valley Microwave Foods, Inc.

5308945, Microwave Interactive printable coatings, VanHandel, James River Corporation

5310976, Microwave heating intensifier, Beckett, Beckett Industries Inc.

5310980, Control of microwave energy in cooking foodstuffs, Beckett, Beckett Industries, Inc.

5317118, Package with microwave induced insulation chambers, Brandberg, Golden Valley Microwave Foods, Inc.

5322984, Antenna for microwave enhanced cooking, Habeger, James River Corporation of Virginia

Re34683, Control of microwave interactive heating by patterned deactivation, Maynard, James River Corporation of Virginia

5334820, Microwave food heating package with accordion pleats, Risch, Golden Valley Microwave Foods, Inc.

5338911, Microwave susceptor with attenuator for heat control, Brandberg, Golden Valley Microwave Foods, Inc.

5338921, Method of distributing heat in food containers adapted for microwave cooking and novel container structure, Mabeux, Universal Packaging Corporation

5340436, Demetallizing procedure, Beckett, Beckett Industries Inc.

5354973, Microwave heating structure comprising an array of shaped elements, Beckett, Beckett Industries Inc.

5357086, Microwave corn popping package, Turpin, Golden Valley Microwave Foods, Inc.

5391430, Thermostating foil-based laminate microwave absorbers, Fabish, Aluminum Company of America

Issued 1995

5391864, Patterned susceptor for microwavable cookie dough, Bodor, Van den Berg Foods Company

5396052, Ceramic utensil for microwave cooking, Petcavich, The Rubbright Group, Inc.

5402931, Carton with lid sealed to tray end flanges and lid flaps sealed to tray sides, Gulliver, Gulf States Paper Company

5405663, Microwave packaging laminate with extrusion bonded susceptor, Archibald, Hunt Wesson Foods

5410135, Self limiting microwave heaters, Pollart, James River Paper Company, Inc.

5412187, Fused microwave conductive structure, Walters, Advanced Deposition Technologies, Inc.

5414248, Grease and moisture absorbing inserts for microwave cooking, Phillips, Eastman Chemical Company

5416305, Microwave heating package and method for achieving oven baked quality for sandwiches, Tambellini, none

5421510, Container/lid assembly for paperboard food packages which utilized press-applied coatings as a sealing medium, Calvert, Westvaco Corporation

5423449, Multi-compartment ovenable food container, Gordon, International Paper Company

5424517, Microwave impedance matching film for microwave cooking, Habeger, James River Paper Company, Inc.

5425972, Heat sealed, ovenable food carton lids, Calvert, Westvaco Corporation

5433374, Venting/opening for paperboard carton, Forbes, Westvaco Corporation

5464969, Self-venting microwaveable package and method of manufacture, Miller, Curwood, Inc.

5473142, Microwave popcorn container for recreational use and method of using the same, Mass, none

Issued 1996

5489766, Food bag for microwave cooking with fused susceptor, Walters, Advanced Deposition Technologies, Inc.
5493103, Baking utensil to convert microwave into thermal energy, Kuhn, none
5494716, Dual ovenable food trays, Seung, International Paper Company
5498080, Easily expandable, flexible paper popcorn package, Dalea, General Mills, Inc.
5510132, Method for cooking a food item in microwave heating package having end flaps for elevating and venting the package, Gallo, ConAgra, Inc.
5519195, Methods and devices used in the microwave heating of foods and other materials, Keefer, Beckett Technologies Corp.
55302321, Multilayer fused microwave conductive structure, Walters, Advanced Deposition Technologies, Inc.
5552112, Method and system for sterilizing medical instruments, Schiffmann, Quiclave, LLC
5565125, Printed microwave susceptor with improved thermal and migration protection, Parks, Westvaco Corporation
5571627, Temperature controlled susceptor structure, Perry, The Pillsbury Company
5573693, Food trays and the like having press-applied coatings, Lorence, ConAgra, Inc.
5582854, Cooking with the use of microwave, Nosaka, Ajinomoto Co., Inc.
5585027, Microwave susceptive reheating support with perforations enabling change of size and/or shape of the substrate, Young, none
5588587, Dual ovenable food package, Stier, International Paper Company

Issued 1997

5593610, Container for active microwave heating, Minerich, Hormel Foods Corporation
5599499, Method of microwave sterilizing a metallic surgical instrument while preventing arcing, Held, Quiclave, LLC
5601744, Double-walled microwave cup with microwave receptive material, Baldwin, Vesture Corp.
5607612, Container for microwave treatment of surgical instrument with arcing prevention, Held, Quiclave, LLC

5614259, Microwave interactive susceptors and methods of producing the same, Yang, Deposition Technologies, Inc.
5628921, Demetallizing procedure, Beckett, Beckett Technologies Corp.
5645748, System for simultaneous microwave sterilization of multiple medical instruments, Schiffmann, Quiclave, LLC
5650084, Microwaveable bag with releasable seal arrangement to inhibit settling of bag contents; and method, Bley, Golden Valley Microwave Foods, Inc.
5660898, Heat sealed, ovenable food cartons, Calvert, Westvaco
5672407, Structure with etchable metal, Beckett, Beckett Technologies Corp.
5688427, Microwave heating package having end flaps for elevating and venting the package, Gallo, Jr., ConAgra, Inc.
5690853, Treatments for microwave popcorn packages and products, Jackson, Golden Valley Microwave Foods, Inc.
5695673, Microwave cooking device including susceptor retainer and method, Geissler, National Presto Industries, Inc.
5698127, Microwaveable container with heating element having energy collecting loops, Lai, none

Issued 1998

5718370, Partially shielded microwave heating container, Lafferty, Fort James Corporation
5723223, Ultrasonically bonded microwave susceptor material and method for its manufacture, Quick, International Paper Company
5726426, Microwaveable food container with perforated lid, Davis, Ranks Hovis McDougall Limited
5753895, Microwave popcorn package with adhesive pattern, Olson, Golden Valley Microwave Foods, Inc.
5759422, Patterned metal foil laminate and method for making same, Schmelzer, Fort James Corporation
5770839, Microwaveable bag for cooking and serving food, Ruebush, Union Camp Corporation
5770840, Microwave cooking container for food items, Lorence, ConAgra Frozen Foods
5773801, Microwave cooking construction for popping corn, Blamer, Golden Valley Microwave Foods, Inc.
5780824, Expandable and self-venting novelty container for cooking microwaveable popcorn, Matos, Lulirama International, Inc.
5800724, Patterned metal foil laminate and method for making same, Habeger, Fort James Corporation

5834046, Construction including internal closure for use in microwave cooking, Turpin, Golden Valley Microwave Foods, Inc.

5837977, Microwave heating container with microwave reflective dummy load, Schiffmann, Quiclave, L.L.C.

5853632, Process for making improved microwave susceptor comprising a dielectric silicate foam substance coated with a microwave active coating, Bunke, The Procter & Gamble Company

Issued 1999

5858487, Nonstick microwaveable food wrap, Boehler, none

5858551, Water dispersible/redispersible hydrophobic polyester resins and their application in coatings, Salsman, Seydel Research, Inc.

5861184, Packed and frozen sushi product and process for thawing the same, Ishino, Polarstar Company Limited

5863576, Vacuum packed microwaveable lobster package and process, Guarino, Carnival Brand Seafood Company

5863578, Microwaveable vacuum packed seafood package and process, Guarino, Carnival Brand Seafood Company

5863585, Package for food product and method for emptying the package, Sjoberg, Nestec

5864123, Smart microwave Packaging Structures, Keefer, none

5865335, Easy-open closure, Farrell, American National Can Company

5868307, Carton opening feature, Calvert, Westvaco Corporation

5871116, Food service and storage foodstuff holding container assembly, Picchietti, none

5871431, Plant for manufacturing cardboard containers and the manufacturing method for said containers, Vangeli, Vara S.r.L.

5871527, Microwaveable mixture and heating pad, Gubernick, none

5871790, Laminated bag wall construction, Monier, Union Camp Corporation

5873220, Method for producing a self-locking, paperboard pail-like container and product thereof. Haraldsson, Westvaco Corporation

5874715, Heating apparatus with a form of an antenna array plate for a microwave oven, Choi, LG Electronics Inc.

5876811, Microwavable single-serving meal container, Blackwell, none

Re. 36158, Venting/opening for paperboard carton, Forbes, Jr., Westvaco Corporation

5900264, Food package including a tray and a sleeve surrounding the tray, Gics, Gics & Vermee, L.P.

5900293, Collapsible, monolayer microwaveable container, Zettle, S.C. Johnson Home Storage Inc.

5910268, Smart microwave packaging structures, Keefer, none

5916470, Microwaveable heat retentive receptacle, Besser, Aladdin Industries, LLC

5919390, Method and package for microwave roasting of unshelled peanuts/nuts/seeds, Childress, none

5925281, For use in a freezer and in a microwave oven, a microwave-reflective vessel with a cold-keeping agent and methods for its use, Levinson, none

5928553, Sealed bag for microwave heating, Toshima, Kabushiki Kaisha Hoseki Planning

5928554, Microwave popcorn package with adhesive pattern, Olson, ConAgra, Inc.

5928555, Microwave food scorching shielding, Kim, General Mills, Inc.

5935477, Continuous microwave cooking grill having a plurality of spaced segments, Koochaki, Kontract Product Supply Inc.

5939205, Gas barrier resin film, Yokoyame, Toyo Boseki Kabushiki Kaisha

5942320, Barrier composite films and a method for producing the same, Miyake, Daicel Chemical Industries, Ltd.

5945984, Heat retentive food servingware with temperature self-regulating phase change core, Ablah, Thermal Solutions Inc.

5948308, Food product tray with expandable side panels, Wischusen, III, Rock-Tenn Company

5951905, Thawing-heating tray and thawing-heating method, Iwai, Kiyari Co., Ltd.

5952025, Bag and method of making the same, Yannuzzi, Jr., American Packaging Corporation

5958482, Easily expandable nontrapping flexible paper microwavable popcorn package, Monforton, General Mills, Inc.

5961872, Metal container and use thereof in a microwave oven, Simon, Campbell Soup Company

5976651, Propylene packaging laminate comprising methylpentene resin blend layer, Tatsumi, Sumitomo Bakelite Company Limited

5977531, Microwave induced thermal inversion packaging, Pfister, none

5981011, Coated sheet material, Overcash, A*Ware Technologies, L.C.

5985343, Microwave popcorn package, Hasse, Jr., Ryt-Way Industries, Inc.

5986248, Food container for microwave heating or cooking, Matsumo, Snow Brand Milk Products Co., Ltd.

5989608, Food container for cooking with microwave oven, Mizuno, none

5994685, Treatments for microwave packaging and products, Jackson, Golden Valley Microwave Foods, Inc.

5997916, Microwave popcorn fortified with calcium and method of preparation, Dickerson, General Mills, Inc.

6005234, Microwave popcorn bag with cross mitre arrangement, Moseley, Weaver Popcorn Company

6006984, Paperboard package, Chung, none

REFERENCES

1. R. Buderi. The Invention That Changed the World. Simon & Shuster, Inc, New York, 1996, p 256.
2. P. L. Spencer. Prepared food article and method of preparing. U.S. Patent 2,480,679, 1949.
3. D. V. Decareau. Microwave Foods: New Product Development. Food & Nutrition Press, Inc., Trumbull, CN, 1992, p 3.
4. P. L. Spencer. Receptacle. U.S. Patent 2,528,251, 1950.
5. S. Yokoyamo. Gas barrier resin film. U.S. Patent 5,939,205, 1999.
6. J. P. Winter. Polyethylene/polyester nonoriented heat sealable moisture barrier film and bag. U.S. Patent 4,705,707, 1987.
7. R. L. Mueller. Microwave package with vent. U.S. Patent 4,404,241, 1983.
8. R. P. Mitchell. Automatically ventable sealed food package for use in microwave ovens. U.S. Patent 4,210,674, 1980.
9. M. Kamada. Packaging sheet and containers and pouches using the sheet. U.S. Patent 4,848,931, 1989.
10. M. W. Kuchenbecker. "Microwave carton and blank for forming the same." U.S. Patent 5,07,273, 1992.
11. L. C. Brandberg. Combined popping and shipping package for popcorn. U.S. Patent 4,038,425, 1977.
12. W. P. Kane. Method of cooking food in a polyethylene terephthalate/paperboard laminated container. U.S. Patent 3,924,013, 1975.
13. S. W. Middleton. Polyester coated paperboard for forming food containers and process for producing the same. U.S. Patent 4,147,836, 1979.
14. D. R. Prater. Frozen food container. U.S. Patent 4,738,365, 1988.
15. W. R. Rigby. Easy opening lid for ovenable carton. U.S. Patent 4,955,530, 1990.
16. N. Seung. Dual-ovenable food trays. U.S. Patent 5,494,716, 1996.
17. A. Hirsch. High temperature resistance hermetically sealed plastic tray packages. U.S. Patent 3,997,677, 1976.
18. K. P. Thompson. Polymeric multiple-layer sheet material. U.S. Patent 4,183,435, 1980.
19. J. M. Mochel. Electrically conductive coating on glass and other ceramic bodies. U.S. Patent 2,564,707, 1951.
20. D. A. Copson. Heating method and apparatus. U.S. Patent 2,830,162, 1958.
21. W. C. Winters. Microwave heating package, method and susceptor composition. U.S. Patent 4,283,427, 1981.
22. W. A. Brastad. Method and material for prepackaging foods to achieve microwave browning. U.S. Patent 4,230,924, 1980.
23. W. A. Brastad. Packaged food item and a method for achieving microwave browning thereof. U.S. Patent 4,267,420, 1981.

24. O. E. Seiferth. Food receptacle for microwave cooking. U.S. Patent 4,641,005, 1987.
25. C. C. Habeger, Jr. Microwave Interactive Thin Films. Microwave World 18(1):8–22, 1997.
26. G. J. Walters. Fused microwave conductive structure. U.S. Patent 5,412,187, 1995.
27. D. E. Beckett. Formation of packaging material. U.S. Patent 4,398,994, 1983.
28. D. E. Beckett. Demetallizing method. U.S. Patent 4,610,755, 1986.
29. D. H. Hollenberg. Microwave interactive laminate. U.S. Patent 4,865,921, 1989.
30. P. L. Maynard. Control of microwave interactive heating by patterned deactivation. U.S. Patent Re. 34,683, 1994.
31. K. D. Woods. Microwave heatable materials. U.S. Patent 5,039,833, 1991.
32. W. R. Wolfe, Jr. Microwave absorber. U.S. Patent 4,518,651, 1985.
33. K. A. Pollart. Self limiting microwave heaters. U.S. Patent 5,410,135, 1995.
34. D. G. Beckett. Controlled heating of foodstuffs by microwave energy. U.S. Patent 5,117,078, 1992.
35. American Society for Testing and Materials, West Conshohocken, PA.
36. C. H. Turpin. Microwave heating package and method. U.S. Patent 4,190,757, 1980.
37. J. A. Cherney. Partially shielded microwave carton. U.S. Patent 4,345,113, 1982.
38. T. H. Bohrer. Packaging container for microwave popcorn popping and method for using. U.S. Patent 4,553,010, 1985.
39. R. K. Brown. Package assembly and method for storing and microwave heating of food. U.S. Patent 4,555,605, 1985.
40. R. V. Maroszek. Package assembly with heater panel and method for storing and microwave heating of food utilizing same. U.S. Patent 4,661,671, 1987.
41. T. D. Pawlowski. Sleeve for crisping and browning of foods in a microwave oven and package and method using same. U.S. Patent 4,775,771, 1988.
42. R. K. Brown. Overlap seam for microwave interactive package insert. U.S. Patent 4,780,587, 1988.
43. J. L. Clough. Apertured microwave reactive package. U.S. Patent 4,948,932, 1990.
44. D. W. Andreas. Microwave receptive heating sheets and packages containing them. U.S. Patent 4,943,439, 1990.
45. A. A. Smart. Flexible disposable material for forming a food container for microwave cooking. U.S. Patent 4,890,439, 1990.
46. R. C. Fulde. Brownable dough for microwave cooking. U.S. Patent 4,448,791, 1984.
47. R. K. Brown. Fruit and meal pie microwave container and method. U.S. Patent 4,626,641, 1986.
48. All temperature profile data presented in this chapter were generated in the test laboratory of Fort James Corporation's Mississauga, Ontario facility and are based on cook tests conducted there.
49. C. R. Buffler. Perceived Rapid Cooling of Microwaved Foods. Microwave World 12(2):16–18, 1991.
50. Fort James Corporation microwave packaging consumer research, 1999.
51. 1996 Microwave Industry Report: A Survey of Consumer Usage and Attitudes Regarding Microwave Products. International Microwave Power Institute. Manassas, VA, 1996.
52. E. Brown. Frozen food package and method for producing same. U.S. Patent 3,219,460, 1965.

53. M. C. Agnew (ed.). The Southern Living Microwave Cookbook. Oxmoor House, Birmingham, AL, 1988, p. 5.
54. P. N. Stevenson. Selective cooking apparatus. U.S. Patent 3,615,713, 1970.
55. W. T. Saunders. Convenience package. U.S. Patent 4,875,597, 1989.
56. F. E. Simon. Metal container and use thereof in a microwave oven. U.S. Patent 5,961,872, 1999.
57. R. M. Keefer. Microwave container and method of making same. U.S. Patent 4,866,234, 1989.
58. D. G. Beckett. Microwave heating intensifier. U.S. Patent 5,310,976, 1994.
59. R. M. Keefer. Microwave heating package and method. U.S. Patent 4,656,325, 1987.
60. R. M. Keefer. Susceptors for microwave heating and systems and methods of use. U.S. Patent 5,079,397, 1992.
61. L. Lai. Microwaveable container with heating element having energy collecting loops. U.S. Patent 5,698127, 1997.
62. D. G. Beckett. Microwave heating element with antenna structure. U.S. Patent 5,278,378, 1994.
63. C. C. Habeger, Jr. Antenna for microwave enhanced cooking. U.S. Patent 5,322,984, 1994.
64. T. P. Lafferty. Partially shielded microwave heating container. U.S. Patent 5,718,370, 1998.
65. D. G. Beckett. Demetallizing procedure, U.S. Patent 5,340,436, 1994.
66. M. A. Schmelzer. Patterned metal foil laminate and method for making same. U.S. Patent 5,759,422, 1998.
67. C. C. Habeger, Jr. Patterned foil laminate and method for making same. U.S. Patent 5,800,724, 1998.

13
Safety in Microwave Processing

Gregory J. Fleischman
U.S. Food and Drug Administration
Summit-Argo, Illinois

I. INTRODUCTION

The safe use of a technology can be undermined by an inadequate assessment of its associated risks. The possibility of inadequate assessment increases as the technology becomes increasingly sophisticated. This sophistication leads to indirect effects that can go unnoticed by those directly involved with technology development and implementation. Safe use requires an ongoing assessment and awareness of risk and the knowledge of the means by which this risk is reduced or eliminated.

The heating capability of microwaves has been known since 1945 [1]. However, compared to the vast history of the heating of food, it can still be considered an emerging technology. Although microwave ovens are now commonplace in the home, the technology still suffers from misunderstandings that could undermine its safe use. Even when used in a commercial (e.g., hospital food service) or industrial (sterilization/pasteurization for distribution) setting, misunderstandings can obscure steps that must be taken to ensure safety in the food being produced.

This chapter examines the safe use of microwave energy for the heating of food from two perspectives: safety of the food that is microwave processed and safety of the operator. Operator safety covers various phenomena that microwave energy influences. Microbiological aspects dominate the food safety perspective and are examined from the premise that the lethal effect of microwave energy on microorganisms is purely thermal in nature. The food safety perspective also includes a discussion of chemical migration through contact with food packaging during microwave heating.

II. UNIFORMITY OF THERMAL TREATMENT IN CONVENTIONAL AND MICROWAVE HEATING

One of the goals of thermal food processing, regardless of heat source, is the destruction of microorganisms that may exist within the food. At the most basic level, the effectiveness of microbial destruction by heat depends on thermal treatment, defined by the intensity of heat and the duration of exposure to it (cf. Chapter 6). However, in any practical heating situation, not all parts of the food receive the same thermal treatment. Thus adequate thermal treatment must be established on the basis of the part of the food that receives the least thermal treatment in comparison with other parts of the food being processed. In commercial and industrial processing such treatment must be provided consistently, application after application. Therefore, the evaluation of the safety of any thermal processing of food from a microbiological point of view must start with an examination of the variability of the thermal treatment throughout the food being processed.

The considerable experience with conventional heating throughout history has led to processes that produce safe food. Established methods of assessment allow each process to be analyzed, tested, and monitored to ensure safety. Although microwave heating is considered to be a distinctive method of thermal processing, fundamentally different from conventional heating, both methods do share certain characteristics. Through these shared characteristics and understanding the differences between the heating methods, it becomes possible to take what is known about conventional processes and determine the degree of applicability of this knowledge to microwave heating. Therefore, a starting point in the assessment of the safety of microwave heating is an examination of conventional heating. For the purpose of comparison, conventional heating processes can be grouped by the type of heat transfer they utilize—direct contact with a hot medium or absorption of radiation.

A. Direct Contact Heating

The most intuitive of conventional methods by which heating of food occurs is by direct contact between the food and a heating medium. Deep-frying is one example. Another is contact with condensing steam, as in canning (retorting) operations. Heat is transferred by virtue of the temperature gradient existing between the hot medium and the cooler food. This type of heat transfer is termed conduction. The temperature of the food surface rises as the heat is transferred from the heating medium to the surface. Because of the temperature gradient between the surface and the interior of the food, heat is transferred to the interior. Also, temperatures throughout the food are not uniform. Although they approach uniformity as the food is held in contact with the heating medium, many practical heating applications do not heat the food to a point where all of the food is at the same temperature. Thus, for the bulk of the heating time, internal temperatures are nonuniform. For microbiological safety, therefore, it is important that the thermal treatment at the coldest point in the food is sufficient for the destruction of microorganisms. By this reasoning all of the other parts of the food, being warmer for longer periods of time, will have at least a sufficient thermal treatment.

At the heart of any transfer of *heat* energy is conduction. It is enhanced, however, by convection. Convection is the bulk movement of fluids, being either liquids or gases, and thus does not occur in solids. It brings about better contact between warm and cool regions. When convection occurs due to bouyancy differences between warmer and cooler regions in a liquid or a gas, it is called natural convection. When bulk movement occurs by the action of an impeller, for example, it is called forced convection. Either brings better contact between parts of the fluids having different temperatures, thus enhancing conduction.

B. Radiant Heating

A more complex form of conventional heating, yet one taken much for granted, is radiant heating. This type of heating occurs in the common oven. It may be surprising to learn that the transmission of energy from the energy source to the food occurs in the same manner in radiant heating as it does in microwave heating.

In radiant heating and microwave heating, energy is transferred to the food via electromagnetic (EM) radiation. This may cause some confusion since it may be believed that food is heated in conventional ovens by contact with heated air. Indeed, this confusion is aided by the fact that common room radiators transfer heat to the room by heating the air. Although contact with hot air does occur and contributes to the heating process in common ovens, the primary mechanism is radiant heat transfer. Radiant heat transfer is a significant form of heat transfer best exemplified by the warmth felt from sitting near a campfire on an autumn evening. The intervening air is cool but the "warm glow" of the campfire is indeed tangible.

The speed at which the heat energy is transferred between a radiating element and the food is practically instantaneous. All EM radiation, such as light waves, radio waves, and microwaves, travels in free space at the speed of light. Thus, a light bulb appears to cast light instantaneously throughout a room when it is switched on. Likewise a sudden significant flare-up of the aforementioned campfire is felt instantly. Although EM waves all travel at the speed of light, their individual frequencies cause them to interact differently with the various constituent molecules of matter. The frequency of an EM wave is directly proportional to its quantum energy. At lower frequencies, this energy, if absorbed by a molecule, increases its energy by exciting its various translational (pushing), rotational (spinning), and vibrational (shaking) modes. In the order listed, the various modes require increasing energy absorption to be excited. Microwaves, for example, excite rotational modes of water. The higher frequency of infrared waves excite vibrational modes of food molecules. On a macroscopic level, movement is impeded by friction, which converts mechanical energy to heat. In a similar way, this conversion occurs on the molecular level, resulting in the heating of matter.

At higher frequencies in the spectrum, EM waves tend to excite atoms themselves by promoting the electrons of an atom to increasing levels of energy. In the case of gamma radiation, for example, the amount of energy absorbed by an atom is of such magnitude that electrons are ejected from the atom, creating an ion out of the once neutral atom. This radiation is thus termed ionizing. At this point, atomic bonds in molecules can be broken resulting in a change in composition of matter. This does not occur at the level of microwave or infrared waves, where the energy imparted is only sufficient to push, spin, or shake the responding molecules.

One advantage of radiant heating is that infrared waves are relatively easy to create. Unlike the complicated microwave source, the magnetron, radiant heat transfer via infrared waves is a phenomenon of nature and any body having a temperature above absolute zero emits energy in the infrared range. Thus a fire or an electrical heating element emits waves that transfer energy. The higher the temperature, the better heat energy is transmitted.

Infrared waves interact well with food components. Thus all of the energy from an infrared wave is absorbed and turned into heat energy at the food surface, increasing its temperature. The temperature gradient that is created between the food surface and the food interior promotes conductive heat transfer. Whereas the means of heat transmission in common ovens is similar to microwave ovens, conduction is relied upon to deliver the heat imparted at the food surface to the food interior. As with direct contact heating, different parts of the food are at different temperatures.

C. Microwave Heating

Microwaves interact differently with food than the infrared waves used in conventional conduction-based radiant heating. Longer in wavelength than infrared waves (on the order of 0.1 m compared with 0.00001 m for infrared), microwaves interact primarily with water molecules and some solutes such as sugar and salt. Weaker interactions occur with other food components, such as fat. Even with water, however, not all of the microwave energy is absorbed on first contact with the water at the food surface. This allows microwaves to penetrate the food, albeit with continually attenuating energy. Absorption, as discussed in the previous section, converts microwave energy to heat energy. Thus heat will be generated not only at the food surface, but within the food as well, creating the effect of internal heating. A subtle but substantial advantage of this EM penetration is that the microwaves deliver energy inside the food at a rate orders of magnitude faster than conduction owing to the velocity of EM waves. If this were the only practical difference between heating conventionally and with microwaves, microwaves would hold an enormous advantage. However, there are two distinguishing points concerning microwaves that create unique obstacles to ensuring microbiologically safe food.

First, different food components alter the absorption of microwaves. In conventional heating, all of the impinging energy, supplied either by a warm medium or infrared waves, appears as heat at the surface and is slowly carried into the food because of temperature gradients. In microwave heating, different foods absorb microwaves differently owing to their component makeup. It is entirely possible for most of the microwave energy to be transformed into heat energy at the surface of the food if it contains a high concentration of ionic particles. When this happens, microwaves cannot penetrate the food. On the other hand, food may be so transparent to microwaves that they do not attenuate completely before reaching another food/air interface. When this occurs, the wave nature of microwave energy transport becomes important through the phenomenon of interference. This is the second distinguishing point of microwave heating.

Constructive and destructive interference is a natural phenomenon of any type of wave. Dropping a pebble into a still pond one sees circular waves emanating from the point of impact. Dropping two pebbles at different points, one sees a complex pattern that, if carefully observed, will reveal points where the water has a vertical displacement greater than that of the individual waves (constructive interference) or no displacement at all (destructive interference). Although both IR and microwaves are EM waves, the wavelength of an IR wave is so small that interference does not play a role. In fact, its radiation may be regarded as emanating continuously instead of in waves. This is not so for microwaves. The large wavelength of a microwave, about 12 cm, allows wave interference. In the pebble example above, interference was caused by two disturbances. A wave, however, can interfere with itself if it encounters an interface in which some of the wave is reflected. The interference comes from the impinging wave encountering its reflected component. This may establish standing waves. The term describes exactly what happens—the wave no longer appears to be traveling, but standing still. There will be points, called nodes, that experience no fluctuation of energy. All other points experience various intensities of fluctuations up to the maximum fluctuation of the standing wave. This leads to points that experience no microwave heating and those that experience significant heating.

Interference can occur within food, but is more common within the oven itself. Microwaves, once introduced into the oven cavity, reflect off of the cavity walls, creating areas in the oven having high and low intensities of microwave radiation. The existence of such energy patterns creates a complex distribution of temperatures throughout the food. Because temperatures are not the same throughout the food, temperature gradients exist that promote heat conduction. Thus conduction is relied upon in microwave heating to help distribute heat throughout the food.

It is commonly said that microwaves produce an uneven or nonuniform heating pattern. It has been seen in this chapter that all heating methods produce nonuniform heating patterns. Therefore, a more accurate description of the difference between the two heating methods is necessary: Conventional heating produces a more intuitive and predictable pattern of heating, whereas microwave heating produces a complicated, nonintuitive pattern of heating.

A predictable heating pattern leads to the knowledge of the coldest, and therefore least thermally treated, point in the food (for more on this, see Chapter 4, Sec. VI). The monitoring of this point for adequate thermal treatment ensures that all other points have an adequate thermal treatment as well. The challenge for any microwave-based heating is to determine the location of this point or to establish some indicator that ensures this point is receiving adequate thermal treatment.

Now, with the knowledge of how microwave heating differs from conventional heating, it is possible to examine industrial processes that use conventional heating sources and determine how a change to a microwave heating source would effect the establishment and monitoring of the process to assure microbiological safety.

III. INDUSTRIAL AND COMMERCIAL PRODUCTION OF MICROBIOLOGICALLY SAFE FOODS

Thermal processing for food safety seeks to destroy microorganisms. The type of microbial destruction depends upon the goal of the process and can be classified as heating for long-term storage or heating for immediate consumption. Long-term storage of food encompasses commercially sterile products or refrigerated pasteurized products. Heating for immediate consumption must address destruction of microbial pathogens.

A. Sterilization

There are two approaches to sterilization. In-container sterilization treats the container and the food together resulting in a commercially sterile product. It is the most widespread and familiar type of industrial food processing for preservation practiced. The other is aseptic processing where the food and container are sterilized separately. In either case, wherever heat is used for sterilization, microwave energy can potentially be used as the heat source. Because modern aseptic sterilization practice involves pumpable foods, the safety issues it faces are the same as pasteurization and will be discussed later in the chapter. The only difference between pasteurization and aseptic sterilization is the target temperature and hold times necessary to achieve the desired destruction of microorganisms.

In-container sterilization is accomplished most commonly through elevated pressure steam sterilization, or retorting. In order to understand the challenges in adapting steam sterilization processes to microwaves, it is therefore necessary to understand the current state of retorting operations.

1. Traditional Steam Sterilization

Steam sterilization, as it is practiced today, has benefited from many decades of scientific inquiry. This has resulted in an in-depth understanding of the process, its effects on microorganisms, and the steps required to ensure the safety of all food so processed. Traditional retorting involves foods processed for shelf stability (not refrigerated) in hermetically sealed (airtight) containers. Government regulations exist on how specific types of foods are to be processed in such containers. The regulations appear in the *U.S. Government Code of Federal Regulations* (CFR). One set is for FDA-regulated processes [2] while another is for USDA-regulated processes [3]. The application of one set over the other depends upon the meat content of the particular food being processed. If food product formulations contain more than 3% raw meat or poultry, or 2% cooked meat or poultry, the process falls under USDA jurisdiction. Otherwise FDA regulations apply. FDA regulations also apply to all seafood and pet food processing.

The specific foods targeted by these regulations are low-acid foods that are defined as any food having a water activity above 0.85 and a pH above 4.6. At a water activity of above 0.85, *Staphylococcus aureus*, a pathogenic organism, may grow. At a pH above 4.6, *Clostridium botulinum,* a spore-forming anaerobe, may grow, producing the toxic metabolite causing botulism.

The goal of processing low-acid foods for shelf stability in hermetically sealed containers is commercial sterility. Commercial sterility is defined as a state whereby ". . . microorganisms capable of reproducing in the food under normal nonrefrigerated conditions of storage and distribution; and viable microorganisms (including spores) of public health significance" are absent [2]. The regulations outline how processes are to be established so that commercial sterility is achieved, and how destruction will continue to be ensured throughout each sterilization cycle for each can in the cycle during normal production. Reestablishment is necessary for any changes in the process such as a new product formulation, a new process temperature, a change in initial temperature or a change in equipment used to deliver the thermal treatment (i.e., the pressure vessel, also known as the retort). The regulations spell out which process variables are to be monitored (e.g., steam temperature) during processing and recorded for future reference. The regulations also mandate monitoring and recording of critical factors (CFs). CFs are additional process variables that may compromise commercial sterility if they are not controlled during a process. The purpose of monitoring is to ensure control of process variables (e.g., steam pressure). The purpose of record keeping is to allow a review of past processes and to demonstrate that adequate thermal treatments have been given. In cases where an adequate thermal treatment was not given, record keeping is used to prevent the distribution of underprocessed product.

The procedures for process establishment demonstrate that a particular process produces a commercially sterile product. The most straightforward means of demonstrating this for traditional steam retorting is to record the temperature history at the coldest point in the product. If the lethality (thermal treatment) delivered at this point is sufficient, then the processor is assured that it is at least sufficient throughout the rest of the food.

Because lethality occurs at a range of temperatures and is a function of time at each temperature, lethality accumulates throughout the entire process cycle, including heating up to the operating temperature and postprocess cooling. Total lethality is therefore an integrated quantity. By using process time and cold spot temperature data, the lethality is obtained through direct graphical integration of temperature-time curves. Research in this area has shown that this can be simplified and put into equation form which requires only the process parameters and not the entire time-temperature curve. There are a large number of calculation procedures. The traditional industry method of calculation is the Ball formula method [4–6]. This formula converts the lethality obtained for the actual temperature-time data into lethality at an equivalent time corresponding to $121.1°C$ ($250°F$). Since the cold spot temperature is used, it is the processor's responsibility to experimentally find the cold spot within the product during processing. For pure conduction heating this will most likely be at the geometric center of the food. The location of the cold spot in convection-heated foods depends on the product and the process.

Alternatively, or in conjunction with mathematical evaluation discussed in the previous paragraph, inoculated test packages [6] can be used to validate a process. The test packs, containing the food product of interest, are inoculated with a calibrated microbial spore crop. The particular microorganism used for validation must have thermal death kinetics directly related to *C. botulinum*. It also should be easier and safer to use than the organism it is replacing. For the case of validating a process for *C. botulinum* destruction, one such organism is *Clostridium sporogenes*. This particular organism has the added benefit that its spores are several times as resistant as *C. botulinum* to thermal treatment. This provides an additional safety margin.

After the test packs have been inoculated with a known level of inoculum, they are processed at various temperatures or times. The packs are then incubated to allow for any microbial growth. Although a safe process is indicated by no growth, it is valuable to know the boundaries on time and temperature that separate no-growth from growth processing. This is known as bracketing. The boundaries for time can be obtained by varying process duration at a fixed temperature. Likewise, the boundaries on temperature can be obtained by varying process temperature at a fixed duration.

2. Microwave-Based Sterilization

The advent of retortable plastic containers allows the possibility of microwave-based sterilization. Such a process must present the same level of safety that is required of conventional sterilization processes. As described above, lethality for conventional steam sterilization can be calculated by the Ball formula method to aid in process establishment. At present, however, there is no such calculation method for microwave processes.

The Ball formula also assumes that the location of the coldest temperature, the cold spot, within a food does not change. Because of shifting energy patterns encountered in a container undergoing microwave heating, this assumption does not generally hold for microwave heating. If containers that are processed simultaneously do not receive the same thermal treatment, then the cold spot among all of the containers must be known for the Ball formula method to be applicable. Shifting energy patterns within a microwave system, however, can make the cold spot move among containers. For process calculations to be used in the case of microwave heating, therefore, a shifting cold spot must be taken into account. At this time there are no existing calculations that take into account shifting cold spots. Although the Ball formula method is empirical, it has been shown to be connected to a simplication of the solution to the theoretical homogeneous heat equation [4]. Microwave heating requires an additional term in the theoretical heat equation to account for the internal heat generation that microwave energy causes, making the heat equation more difficult to solve. Closed-form solutions are available for the nonhomogeneous heat equation [7, 8]. They have not been simplified, however, into a Ball-type of formula for process calculation.

When calculations cannot be used to establish a process, inoculated test packages, described above as a validation step, may be used to directly demonstrate process adequacy [9]. This approach is completely compatible with microwave-based thermal processing. However, care must be taken when using microbial validation procedures to ensure that an appropriate experimental design and data analysis is completed. The primary assumption for inoculated pack testing is that each test pack has been given the same thermal treatment. Thus to use, for example, the traditional most probable number (MPN) method of analysis [10] for an inoculated pack test in a microwave sterilization system, the processor would need to demonstrate that there does not exist a variation in lethal treatment between test samples. If a system does impart to each test sample a different quantity of energy, a characteristic minimally treated test sample would have to be statistically determined. A procedure such as a distribution-free tolerance limit test [11] could be used.

3. Monitoring and Records

Once a process is established, provisions must be made for process monitoring during normal production. A calibrated mercury-in-glass (MIG) thermometer and a temperature-time recording device [2] are required instrumentation for a pure steam sterilization process. A pressure gauge for measuring conditions within the retort is recommended but not required. The MIG thermometer acts as the reference instrument for the recording device, which produces a graph of temperature as a function of time at a single point within the retort throughout each process cycle. The single-point temperature is representative of the temperature throughout the retort due to, in part, the dispersive nature of the steam field within the retort. An approach toward uniform temperatures is the result of this dispersive nature. The approach toward thermal uniformity is further aided by devices such as vents, bleeders, and spreaders that are built into retorts. Thus, a single-point temperature measurement is sufficient for the purpose of process control and recording. This extends to retorts using steam/air mixtures, but with added mixing devices and pressure records to maintain the viability of single-point temperature measurement. From the value of this temperature and the processing time the degree of processing can be inferred.

Identifying control variables, the monitoring of which would ensure that a process has successfully and consistently provided the correct degree of microbial lethality, poses a challenge for microwaves not encountered with steam. A complication inherent in the use of microwave energy is that the microwave field, unlike the steam field, is not dispersive with no natural tendency toward uniformity. It is thus inconceivable that a measure of field strength at some point, or even multiple points, in the field itself would demonstrate consistently adequate processing as temperature measurement does in steam retorting. Also, in contrast to conventional sterilization that has benefitted from decades of research, only a small number of papers concern themselves with microwave sterilization. Smaller still are the number of papers dealing with the food safety issues of microwave sterilization. Microwave retorting has been shown to be a viable process for shelf-stable foods [12–14]. Additionally, sufficient experimental tests exist for process establishment to demonstrate a safe process for a given microwave oven, process duration, and power level. However, there have been no reports on process monitoring.

Commercialization of microwave-based sterilization would benefit greatly from an increased understanding of how it can be monitored. In steam retorting, process establishment and validation are based on the cold spot temperature and monitoring is linked to this value. However, shifting energy patterns make the location of the cold spot in microwave-heated food difficult, perhaps impossible, to know. This necessitates a different approach that would lead to adequate monitoring. One possible approach is the use of temperature distribution in a microwave-heated food instead of a single temperature measurement. The temperature distribution could be represented by a simple minimum-to-maximum temperature range. Equations to predict temperature range have been developed [15] and potentially provide a tool for the prediction of process performance. Their experimental validation, however, is necessary and may be possible through the use of chemical marker technology [16]. Chemical markers may also be used to directly establish a temperature range. Whether it is predicted and validated or experimentally established, the temperature range can then be used by adding it to the target lethality temperature of the process to obtain an adjusted target temperature. Process effectiveness could then be gauged by measuring a few temperatures in the heating food. When only one such temperature exceeds the adjusted target temperature, then all temperatures should be at least at the level of the actual target temperature. From the safety point of view, this method is conservative. Whereas the cold spot approach involved finding its exact location to indicate adequate processing, any spot that exceeds the adjusted target temperature indicates adequate processing.

Lastly, and most importantly, a link must be made between the temperature range and a measurable process parameter, such as power delivered to the microwave oven. Processing time and delivered power could then be the equivalent to the schedule time and schedule temperature for a steam retort.

Microwave heating is one of several technologies that pose challenges to safety assessment when they are applied to sterilization. Other technologies include pulse light, pulsed electric field, electron beam technology, and high hydrostatic pressure. Regardless of any particular new technology, the Food and Drug Administration encourages developers of sterilization processes employing these technologies to interact with the FDA throughout process development [17].

B. Pasteurization/Aseptic

The first part of this discussion deals with milk pasteurization. Juice pasteurization is also discussed with a final part on the challenges particulates pose in any flow pasteurization or aseptic process. Aseptic processes differ from pasteurization processes only in the target temperature and hold time.

1. Milk

The public is most familiar with pasteurization through the consumption of milk. Raw milk is both a source of extensive nourishment and a vehicle of disease. Because of this the Pasteurized Milk Ordinance (PMO) was developed. The most recent revision [18] describes pasteurization as:

> ... the process of heating every particle of milk or milk product, in properly designed and operated equipment, to one of the temperatures given in the following chart and held continuously at or above that temperature for at least the corresponding specified time:

Temperature	Time
83°C (145°F)	30 min
72°C (161°F)	15 s
89°C (191°F)	1.0 s
90°C (194°F)	0.5 s
94°C (201°F)	0.1 s
96°C (204°F)	0.05 s
100°C (212°F)	0.01 s

Pasteurization equipment is designed and operated such that milk is pasteurized on a continuous basis. This is accomplished by heating the milk, then passing the milk through a long tube. The combination of the time it takes for milk to traverse this tube, called the residence time, and the temperature of the milk at the tube outlet corresponds to one of the temperature-time combinations shown in the above table. The tube outlet temperature is measured by a calibrated thermometer that is part of a flow control scheme to prevent underprocessed milk from exiting the system. When the temperature is below the target value, a flow diversion valve at the holding tube outlet diverts the underprocessed milk back for reprocessing. Only when the temperature reaches the target value is the flow released.

Unlike the formidable problems encountered in ensuring the safety of in-container microwave-based sterilization, microwave milk pasteurization can take advantage of the fluid nature of milk. Induced convection resulting from turbulence can be designed into a microwave pasteurization system easily. Natural convection also aids in distributing heat energy. A proper design would give a temperature invariant across the cross section of the hold tube entrance. Also unlike in-container sterilization, monitoring poses no barrier to the use of microwaves because all monitoring is done external to the heating portion of the process. Thus measurement devices are not exposed to the microwave field, allowing for the use of conventional devices that could not function within a microwave field. For example, temperature measurement, critical to ensuring completeness of pasteurization, is monitored away from the heating section, thus allowing for the use of an electronic temperature sensor. Thus, given an adequately designed system, there are no impediments to the use of microwaves in milk pasteurization.

A number of literature studies have looked at microwave pasteurization of milk. In one study, a continuous flow microwave heating system using a consumer microwave oven was constructed and used to demonstrate that exit temperature could be accurately controlled by flow rate [19]. The purpose of the study was to quantify the influence of milk constituents on microwave heating of milk and to determine the suitability of the apparatus as a heat source for pasteurization. The study found that the system could rapidly heat milk and creams to any desired temperature below 100°C. However, there was no test of the system for actual pasteurization.

In another study [20], thermal denaturation of milk protein was measured in both microwave heat treatment and conventional heat treatment. Higher quality, taken as lower denaturation, resulted from microwave heating when the exit milk temperature was 85°C. A difference in denaturation between the two heating treatments was not detected at 80°C. Thus, the authors reasoned that temperature gradients within the milk in contact with the heat transfer surface in conventional heating were large enough to cause a degradation in quality (greater denaturation of milk protein) at the higher temperature but not at the lower temperature. Although not explicitly stated, it appeared that higher temperatures could be employed in pasteurization with microwaves while retaining the quality that was available only at lower temperatures with conventional heating. The issue of safety was addressed by Aktas and Ozilgen [21], who found sufficient lethality of microorgansisms in milk having undergone microwave heating. Jaynes [22] compared a microwave pasteurizer operating at 72°C with a 15 s hold time with conventional pasteurization at 62.8°C (30 min hold time) and found comparable reductions in standard plate and coliform counts. Villamiel et al. [23] also showed, with both goat and cow milk, that pasteurization via microwaves is an efficient and mild approach.

Heddleson and Doores [24] conducted a review of studies appearing in the literature on microwave heating and its elimination of food-borne pathogens. The studies span a breadth of microwave heating applications, including pasteurization of milk. Another review by Sieber et al. [25] concentrated solely on the heat treatment of milk in consumer microwave ovens. From a microbiological point of view, good mixing appeared important in achieving uniform pasteurization. The review by Sieber et al. also touched on other aspects of safety with respect to the microwave heating of milk, including infant formula, again with the emphasis on good mixing for accurate measurement of temperature. One unusual study included in their review reported on the formation of toxic compounds in milk due to microwave heating. The study attempted to show through a comparison of conventional heating of milk in a water bath at 80°C and microwave heating of milk that microwave energy caused the formation of toxic compounds. However the experimental conditions with respect to microwave heating involved milk under pressure to achieve temperatures of 174–176°C. Aside from the inapplicability of using water bath-heated milk as the "control," the review authors commented that the chemical changes in the microwave-heated milk of the proportion reported are not surprising given the high temperatures achieved. A similar study [26] did indeed convincingly show that microwaves promote increased yields of lactulose, furosine, and epilactose over conventional heating when temperatures for both reached the 70°C mark and over and for durations of 10–30 min. These temperature-time combinations exceed those required for pasteurization. It is therefore unlikely that these results would impact standard pasteurization of milk using microwave energy.

Microwaves have also been applied to the very-high-temperature, very-short-time pasteurization of milk. In a review by Young and Jolly [27], work was reported on pasteurization involving microwave heating of a jet of milk to 200°C for 170 ms (including a 40 ms come-up time) and then being cooled by jets of cold sterile milk. According to the authors, the final product could not be distinguished from the raw product in terms of quality.

From these reviews and studies, the emerging picture demonstrates that given adequate mixing and adherence to the hold temperature-time relationship defined in the PMO, microwave energy is a viable heat source for pasteurization.

2. Juice Pasteurization

Juice pasteurization is another area where microwaves could make a significant impact. The outbreaks during 1996 of *E. coli* 0157:H7 in unpasteurized fruit juices brought the issue of juice pasteurization to national attention. Regardless of whether it is perception or reality that unpasteurized juices taste better, juice producers desire a high-quality product, but safe product. Nikdel [28] cites a number of reasons why microwave energy produces a superior juice product over conventional heating. Of these, the most important is the absence of steep temperature gradients in the juice that occur at the heat transfer surface when conduction is used to deliver heat to the food. As with milk, microwave-based pasteurization of juice offers no impediments to safety assurance.

3. Aseptic Particulates

Numerous safe processes for homogeneous liquid food have been developed. Pumpable foods containing particulates, such as corn, peas, chunks of potato, and other finite food pieces, however, present challenges for safety regardless of heat source. Particulates can be a barrier to heating, protecting microorganisms they may contain. Due to hydrodynamic reasons, particulates will traverse an aseptic or pasteurization process at different rates than the transporting medium. Also, not all of the particulates themselves will have the same residence time in the process. This brings up a problem similar to that discussed previously where containers in an in-container microwave sterilization process do not experience the same energy field, and therefore receive different thermal treatments. A solution is a statistical determination of a characteristic minimally treated product, which for an aseptic particulate process is a minimally treated particulate. The formidable nature of this determination was of such magnitude that only recently was an aseptic particulate process, utilizing conventional heating, established and submitted to the FDA for review. This first filing for a particulate process was accepted by the FDA on May 31, 1997 [29]. Although no commercial aseptic particulate products have previously existed in the United States [30], interest appears to exist in their production.

The safety challenges being presented by these products are being addressed by research in three areas. The first, mentioned previously, involves the statistical design and analysis of a particulate aseptic process to be able to determine, with high certainty, that all parts of the product are being adequately processed [11]. The second area involves measuring residence time and modeling the aseptic process [31]. The third area consists of the familiar biological validation [32]. All of this work is independent of the heat source, making microwaves a viable alternative without further complicating the safety issues. By virtue of the penetrability of microwaves and of the liquid nature of food products in aseptic heating, microwaves have the potential to be a superior heat source, in terms of both safety and quality, for aseptic products.

4. Solid Foods

The last consideration of pasteurization involves solid foods. Here, however, the limitations of microwave heating are not easily accommodated. Unlike liquid foods that can be agitated, solid foods must rely on heat conduction to smooth out uneven heating patterns. The issues of concern for solid food pasteurization are the same as those for microwave reheating of food.

C. Heating and Reheating for Immediate Consumption

The potential for illness as a result of inadequate microwave heating was reported in two literature studies. Gasner and Beller [33] reported on an outbreak of salmonellosis in Alaska in 1992 directly associated with microwave-reheated roast pork. The pork was leftover from a picnic and was given to a number of households. Of the 43 people of those households, 30 ate reheated pork while 13 did not reheat the pork prior to consumption. Of the 30 who ate reheated pork, 10 ate the pork reheated in a microwave oven, while the remaining 20 ate the pork conventionally reheated (via standard oven or skillet). Of the 20 individuals who used conventional reheating, none became ill. Of the 10 individuals who ate microwave-reheated pork, all became ill. Of the 13 individuals who did not reheat the pork prior to consumption, 11 became ill. The 2 individuals that did not become ill were under 5 years of age and ate only one or two bites. It was also reported that two food preparers in the microwave reheating cases were questioned as to the temperature of the pork and both said that it was "hot."

In another incident [34], a 500 W microwave oven was used to heat a rice dish which was implicated in an outbreak of salmonellosis. Although both of these incidents occurred in consumer microwave ovens, they demonstrate what can go wrong when the unique heating characteristics of microwaves are not taken into account.

The use of commercial microwave ovens is regulated by individual states. Many states take their guidelines, however, from the 1999 Food Code [35]. Parts of the Food Code pertaining to microwave heating are 3-401.12, 3-403.11, 3-501.13, 4-501.13, and 4-602.12. Of these, the parts that deal directly with food safety as a result of cooking with microwaves are 3-401.12 and 3-403.11(B):

> 3-401.12 Microwave Cooking.*
> Raw animal foods cooked in a microwave oven shall be:
> (A) Rotated or stirred throughout or midway during cooking to compensate for uneven distribution of heat;
> (B) Covered to retain surface moisture;
> (C) Heated to a temperature of at least 74°C (165°F) in all parts of the food; and
> (D) Allowed to stand covered for 2 minutes after cooking to obtain temperature equilibrium.
>
> 3-403.11 Reheating for Hot Holding.*
> (A) Except as specified under ¶¶ (B) and (C) and in ¶ (E) of this section, potentially hazardous food that is cooked, cooled, and reheated for hot holding shall be reheated so that all parts of the food reach a temperature of at least 74°C (165°F) for 15 seconds.
> (B) Except as specified under ¶ (C) of this section, potentially hazardous food reheated in a microwave oven for hot holding shall be reheated so that all parts of the food reach a temperature of at least 74°C (165°F) and the food is rotated or stirred, covered, and allowed to stand covered for 2 minutes after reheating.

The asterisk after the section headings denotes that all of the provisions in the section are of critical importance. The portions of the 1997 Food Code pertaining to microwave heating were unchanged in the 1999 Food Code. However, with respect to microwave heating, the 1997 Food Code did have one important change with respect to the 1995 Food Code. The corresponding sections in the 1995 Food Code pertaining to the microwave heating of foods (3-401.15 and 3-401.11) included the added provision of a 14°C (25°F) overheat. Thus, instead of a target temperature of 74°C (165°F), food had to reach a minimum of 88°C (190°F). Before the release of the 1997 Food Code, a debate ensued among experts and industrial representatives on whether the overheat was required. Critics of the overheat stated that the literature studies used to establish this recommendation [36–41] are outdated and newer designs in microwave ovens obviate the need for overheat. Moreover, these references provide only qualitative evidence of the need for special safety considerations for microwave heating. No quantitative evidence existed to support the actual minimum temperature and standing time designations. This point was examined by the Retail Food Protection Branch of the FDA for possible modification in the 1997 Food Code. Ultimately, before the release of the 1997 Food Code, the overheat was eliminated on the basis that a final temperature of 74°C (165°F), deemed sufficient per the Food Code for conventional cooking, was sufficient regardless of the method of heating. The remaining concept distinguishing microwave from conventional cooking is the postmicrowaving 2 min standing time of food at ambient conditions, while covered, to allow temperature equilibration.

The concern that has been shown about the overheat, however, indicates the possible problems that may be encountered in using the microwave to heat solid foods. Unlike liquid foods that experience natural convection and may be mechanically agitated as well to help distribute heat energy, solid foods rely only on conduction for distribution of heat energy. The rate at which microwaves penetrate the food is orders of magnitude greater than the rate at which conduction can distribute the heat. If solid food is exposed to a microwave field of uniform strength, the dependence on conduction would have no consequence. Uniformity in microwave heating is not the case, however. The parts of the food where the local microwave field strength is high heat quickly. The parts where the field strength is low do not experience the same quick temperature rise, relying instead on conduction to transport heat from surrounding high temperature parts. As a result of the fast heating, lethality, defined as the integrated temperature-time exposure, could be less in the microwave oven than in the conventional oven. Heating to the same temperature takes longer in a conventional oven, thus increasing the temperature-time exposure. This led to the perceived need for a higher minimum temperature requirement for microwave heating. The cooler areas requiring conduction to receive heat led to the need for a two minute standing time post microwave heating. Data does not currently exist that can be used to directly establish reasonable and efficacious minimum temperatures and standing time recommendations.

IV. CHEMICAL MIGRATION

The selective nature of microwave heating allows the use of plastics and paper as food containers for heating. Food contact with containers, however, can lead to migration of container components, especially in heating situations. The advent of high-temperature-resistant nonglass or nonmetal containers makes this a problem for conventional heating as well. Of special concern for microwave heating is the use of susceptors. The most prominent use of susceptors is found in microwave popcorn bags. A susceptor is incorporated in the container wall and is capable of absorbing large amounts of microwave energy, converting it into thermal energy. To effect this conversion, susceptors attain sufficiently high temperatures to impart certain quality-associated benefits to the food. However, the elevated temperature can lead to problems as well. Temperature-enhanced migration into the food of existing components of the container is one such problem. The possibility of the creation of decomposition products from the packaging and their migration into the food is another. Elevated temperatures can also cause the physical breakdown of layers of the container wall that act as barriers to migration. Because migrating substances are considered indirect food additives, they are subject to regulation. An understanding of the migration of these components should lead to a better formulation of packaging material or more careful choices of the type of package.

Chemical migration results from food contact with packaging materials. Plastics and paper are ubiquitous in their use as food contact packaging, especially where microwave heating is concerned. Both plastic and paper contain many different substances as a result of the use of additives to enhance properties or as by-products of manufacture. For example, to enhance the properties of plastics, fillers, stabilizers, antioxidants, and plasticizers are added. Also, complete polymerization of plastic monomer, the single chain unit of the eventual polymer chain, is not attained in manufacturing. Therefore, some monomer and oligomers (short polymer chains) will be present. These substances are not bonded to the polymer or fiber matrix of the final product and may migrate to the food in contact with a container made with that polymer or paper. Due to the potential migration of the packaging components to the food, they are considered food additives.

Studies at room temperature have demonstrated the migration of packaging substances into food. Figge [42], for example, demonstrated the migration of additives from various plastic films into contacting edible oils and fat simulants at room temperature. Bishop and Dye [43] discussed various plastic additives and their potentially deleterious effects on health. They also demonstrated that higher temperatures increased the migration of a plasticizer from commercially available plastic film wrap into food. Ashby [44] studied migration from polyethylene terephthalate (PET), noting increases in migration with temperature.

Regulations were established for the use of food-contact packaging in heating situations [45, 46]. Although these regulations provide for use of such packaging up to 121°C and beyond, they were not intended for the extreme, and unanticipated, temperature food-contact packages encounter as a result of microwave susceptor technology.

Lentz and Crosset [47] showed that susceptor temperatures could reach as high as 280°C with a food load at the food/susceptor interface. The susceptor alone, with no contacting food to draw away heat, reached 316°C. Susceptor construction of PET coated with aluminum and laminated to paper causes the problem that, with such high temperatures, the breakdown and cracking of the PET can occur, bringing adhesives into direct contact with food making them a food additive. A number of protocols were published between 1988 and 1993 for the measurement of nonvolatile [48–53], and volatile constituents [54] of susceptor packaging. Work was done both in the United States and in England [55]. Data obtained clearly demonstrated the migration of both volatile and nonvolatile suseptor constituents into food and food simulants at measurable levels. However, studies by both FDA and the regulated industry show that volatile chemicals migrate to food at such low levels as to pose no safety concerns. FDA laboratories have also shown that the extremely small fraction of food in the daily diet that comes in contact with susceptor packaging (current market volume data show less than 0.1% of foods in the daily diet come in contact with susceptor packages) results in an insignificant dietary concentration of nonvolatile chemicals. To date, FDA's review of data on heat susceptor packaging has not uncovered any further safety problems.

Susceptor manufacturers responded to reports of migration by choosing different formulations of susceptor substrates to eliminate compounds that could adversely affect public health, or by adopting better susceptor package designs to decrease levels of migrants through greater high-temperature stability. Newer technology has resulted in a nonmetalized susceptor which can be printed directly on packages, eliminating some of the components of conventional susceptor packaging [56]. It is not known if this technology poses any migration problems.

However, regardless of the technology, any substance reasonably expected to become a component of food are food additives and as such are regulated under Title 21 of the CFR [57]. The use of any food-contact substance should comply with regulation, exemption, prior sanction, no-objection letter, or be generally recognized as safe (GRAS) as delineated in the CFR. Substances such as metals are reviewed on a case-by-case basis. Thus, metallized susceptors, if not separated from a food by a functional barrier, would be subject to review by the FDA.

V. OPERATIONAL SAFETY CONSIDERATIONS

Another aspect of microwave safety is operational in nature. A number of physical events occur from the time the oven is turned on to the time the food is removed that could compromise the safety of the operator. The most obvious is the potential for exposure of the operator to microwave leakage from the oven. Another is the interaction of microwaves with objects in the oven. Depending on the object, be it a foodstuff, container, or utensil, microwaves may generate locally concentrated thermal energy or electromagnetic fields. These can cause burns in a number of ways as well as possible ignition of the oven contents. This section examines these operational considerations.

A. Leakage

The most conspicuous safety concern in microwave ovens is microwave leakage. Exposure to a high enough level of microwave energy will cause burns. To limit the exposure of anyone operating a microwave oven, CFR [58] mandates that "The equivalent plane-wave power density existing in the proximity of the external oven surface shall not exceed 1 milliwatt per square centimeter at any point 5 centimeters or more from the external surface of the oven, measured prior to acquisition by a purchaser, and, thereafter, 5 milliwatts per square centimeter at any such point." With this regulation in place, electronic devices are designed to deal with this known limit [59]. This regulation also makes other provisions to avoid exposure to excessive microwave energy. These involve user instructions, service instructions, warning labels and details on safety interlocks.

Subchapter C of the Federal Food Drug and Cosmetic Act [60], entitled Electronic Product Radiation Control, serves to further protect the public from unnecessary exposure to radiation from electronic products. One of the results of this act was the establishment of FDA's Winchester Engineering and Analytical Center (WEAC) in Winchester, Massachusetts. There, radiation-emitting electronic products are tested to determine their compliance with the performance standards mentioned previously [58]. These products are not limited to new items. They also include products subject to consumer/user complaints, recalls, or corrective actions.

Even with limits and testing, there was concern in the 1970s that the 5 mW/cm^2 limit was not sufficient owing to an apparently "tighter" limit, 0.01 mW/cm^2, promulgated by the then Soviet Union throughout its sphere of influence [61]. A comparison of only the values of the limits implied that one was stricter than the other. The fact that both numbers have the same units added further to the confusion. The difference between the two limits lies in how they are measured. In the United States the limit is set for *leakage* of microwave radiation from the oven. In the Soviet Union the limit is set for *exposure* of the operator to microwave radiation. This exposure is obtained by averaging over 24 h as well as over the entire surface area of the body. Thus, the two limits are comparable only if most of the surface area of the operator's body is 5 cm from the oven surface and the oven is operating continuously for 24 h. At greater distances the power flux diminishes greatly, owing to the inverse relationship between power level and the square of the distance from a power source. Thus the actual exposure that an operator would obtain based on the United States limit for leakage, if distance and time were properly taken into account, would actually be less than the limit set by the Soviet Union. It should be pointed out, however, that an *exposure* standard for the sum of all electromagnetic radiation impingement has not been promulgated in the United States.

B. Arcing and Fires Caused by Metal

Unlike open flame, steam or other tangible sources of thermal energy, microwaves are intangible, supplying energy that is turned into heat as a result of interaction with food components. It is easily understood that combustible products burn when placed in contact with an open flame, for example, but the conditions that lead to the same result in a microwave oven can be difficult to understand.

One phenomenon associated with ignition in a microwave oven is arcing. Arcing is a significant static discharge between close objects at different electrical potentials. A material usually associated with arcing is metal. However, there is some confusion about metallic objects in the oven, which will be discussed below. Metals are conductors which allow easy passage of electrons. The electrons respond instantaneously to an impinging microwave field. The currents so created, however, create another electric field, which in effect is the reflection of the impinging field. The reflected field depends on the shape of the conductor. Sharp points tend to concentrate electromagnetic energy, while blunt edges and smooth surfaces give more uniform fields. Arcing occurs when two such metallic objects are near each other. Usually the surrounding air acts as an insulator, preventing the movement of charged particles. However, the electric field created by the proximity of two metallic objects causes the usually insulating air molecules to ionize in the small gap between the two objects. The creation of ions allows the passage of current when the electrical potential between the two points is different, and an electric arc is seen. Given the proximity of a combustible substance, the arc may be sufficient to ignite it. One common example is a paper twist-tie. Its use brings its two ends together. Its two ends, being sharp points, intensifies the electric field that quickly leads to arcing and ignition of the paper. Metal-to-metal proximity is thus to be avoided.

A single piece of metal, away from the oven cavity walls, does not constitute an arcing hazard. Because metal is electrically conducting, a single piece, such as a spoon, will not be observed to arc because points that would be close enough to cause an arc are at the same electrical potential. Arcs have been observed, however, on gold rimmed plates. This is due to the gold not being uniformly deposited, thus creating electrically separate points. When the two points are close enough and not at the same electrical potential arcing can occur.

There are still questions about whether *any* metal can go into a microwave oven. Metallic objects in the oven reflect the microwaves back toward the magnetron, increasing the electric field near it. This creates heat to a level that could damage the magnetron. Magnetrons in ovens manufactured before 1970 employed glass seals and were potentially susceptible to such heating. The magnetron in modern microwave ovens employs ceramic seals and are otherwise built to better withstand heating caused by microwaves reflected from the oven.

C. Arcing and Fires Caused by Food

Material usually not associated with arcing is food itself. However, if food has pointed ends and contains ionic solutes, which are influenced by electromagnetic fields, then there is the potential for arcing as well. Hot dogs are one such food that has the potential for arcing. Vegetables that are cut with sharp edges may also arc when placed near each other or a conducting surface.

Although arcing itself may only cause quality problems, its adjunct, fire, may also be caused in food. One of the most effective demonstrations of this was given, albeit tongue-in-cheek, on an Internet site [62]. A grape was prepared by slicing it not quite in half. A small part of the skin was left intact, creating a small bridge between the two halves. If the halves are placed such that their only contact is through this bridge, when microwaved, a fire erupts. What is interesting is that it is not a chance event. The reason for fire is that the bridge acts as an electrical wire, conducting ionic solutes between the two halves due to the inevitable electrical potential difference between them. Like any metal wire, if the current is too strong, it will overheat. In this case, however, the substance that is conducting will also readily burn. The result of the demonstration is a short-lived pyrotechnics display, ending when the two halves, under the force of the sparks, are pushed far enough apart to prevent further arcing. One of the important points of this demonstration is that once fire broke out, microwave energy enhanced the flame by accelerating the ions in the gas.

It is easily conceivable that a prepared food that is processed in quantities on the commercial or industrial scale, in the latter case possibly out of sight of an operator, could cause a fire if attention is not paid to considerations of how the food is prepared. Given the knowledge of how metals and foods cause arcing, microwave-processed foods should be assessed as to their possibility for this type of occurrence.

D. Burns and Boil Over

Any device used for heating food is capable of producing burns. However, because there are no open flames or hot contact surfaces in microwave heating, the false impression can be given that burns are not an important consideration.

Although food is the primary absorber of energy in a microwave oven, the container or plate may also absorb energy, becoming hot as well. Therefore, oven manufacturers advise the use microwave-safe containers. Instructions for home microwave ovens [e.g., 63] suggest that containers be tested if there is doubt that they are microwave safe. A standard test involves microwaving on high for 1 min the container in question along with 1 cup of water in a glass receptacle. The water is a safety measure that prevents excess energy from being reflected back to the magentron due to low or no absorption of the microwaves by the container. If the container is comfortable to touch after microwaving, it is microwave safe. However, this test protects the operator only from containers that will heat due to microwave exposure. All containers, microwave safe or not, can become hot as a result of heat transfer from the contacting food. It is too easy to forget this possibility because in many cases it is not a factor.

Burns may occur due to accidental contact with the heated food. A spill is one possibility. However, a phenomenon called bumping, usually only a nuisance, can cause contact between the operator and very hot food. Microwave energy has unique properties that contribute to this phenomenon. Bumping, also known as splattering or popping, is the result of localized heating, or hot spots, that cause the creation and sudden release of water vapor. The sudden release is explosive in nature and food may be ejected from the container. As previously mentioned, it is primarily a nuisance in that food becomes deposited on the oven cavity walls. However, it can occur at the moment a container is taken from microwave oven. There may exist potentially explosive pockets of vapor held in check by the weight of the food itself. Movement shifts the weight which can lead to explosive vapor release. A study was performed in an effort to quantify bumping and determine what factors influence it [64]. This paper showed how common and what wide variety of foods can lead to bumping. It offered the possible explanation of vapor build up in nonporous foods where the containing food suddenly ruptures.

Boil over and bumping is exacerbated by the design of the containing vessel and the shape of the food itself [65]. Food has the ability to bend microwaves in the same manner that glass can bend light waves, which is called diffraction. Regardless of their initial direction, impinging microwaves on an egg, for example, will be diffracted toward the center of the egg. When a food fills a container of circular cross-section, microwaves impinging on the side of the container will also be directed toward the center. The degree of diffraction is a function of the dielectric properties of the food (see Chapter 3 for more on dielectric properties). Resonances of the electromagnetic field within the food also contribute to center heating. This occurs when electromagnetic waves are not entirely absorbed during passage through food and are reflected back upon themselves when they reach an interface opposite to the point where they entered the food. The result of these phenomena is an increase in microwave energy density toward the center of the spherical or circular food. This can create significant rates of heating that far outpace those away from the center, leading to a sudden increase in water vapor pressure that cannot be gradually released. The eventual result is an explosion, exemplified by the well-known phenomenon of the exploding microwaved egg. To lessen the chance of this type of accident, any type of food with an intact shell (e.g., clams), skin (e.g., potato) or membrane (e.g., egg yolk) should have its covering pierced to prevent the increase of internal pressure to explosive levels.

Another possibility is scalding when infant formula in bottles or baby food in bottles is heated in a microwave oven. A test of the baby bottles contents alone would indicate that the bottle contents were warm, but ultimately masking the intense heating at the bottle center. This can occur in jars of baby food as well. Thus it is now widely made known to shake or stir these containers thoroughly, then test for temperature.

VI. SUMMARY

Microwave energy as a thermal source provides food processors with advantages over traditional means of heating. Simply replacing traditional sources of thermal energy with microwave energy, however, must be carefully approached because of microbiological and operational safety considerations. In some cases, pasteurization of liquid foods for example, a simple replacement of the heat source is possible because safety is assessed directly in the food and after the process is completed. In other cases, however, processes must be monitored as heating occurs. Safety assessment of each run must be made indirectly, being inferred by some measurable process parameter. Even with simple reheating, an understanding of the similarities and differences between conventional heating and microwave heating leads to safer handling of the food, in this case not only microbiologically, but with respect to the operator as well.

Research is necessary to continue the understanding of the interaction between microwave energy and the materials, both food and its packaging, that are used in the area of thermally processed foods. Not only will this enhance the safety of microwave use, but will increase its adoption throughout the various facets of thermally processed foods.

REFERENCES

1. R. F. Schifmann. Microwave technology—a half-century of progress. Food Product Design 7:32–56 (1997).
2. Code of Federal Regulations. Title 21 CFR Parts 108, 113 and 114, U.S. Government Printing Office, Washington, D.C. (1994).
3. Code of Federal Regulations. Title 9 CFR Parts 318.3 and 381.3, U.S. Government Printing Office, Washington, D.C. (1994).
4. C. O. Ball and F.C.W. Olson. Sterilization in Food Technology: Theory, Practice and Calculations. McGraw-Hill, New York (1957).
5. N. G. Stoforos. On Ball's formula method for thermal process calculations. J. Food Proc. Eng. 13:255–268 (1991).
6. National Canners Association Research Laboratories (now National Food Processors Association. Laboratory Manual for Food Canners and Processors: Vol. One—Microbiology and Processing. AVI Publishing Company, Inc., Westport, Conn. (1968).
7. J. Dolande and A. Datta. Temperature profiles in microwave heating of solids: A systematic study. J. Microwave Power Electromagn. Energy 28:58–67 (1993).
8. G. J. Fleischman. Predicting temperature range in food slabs undergoing long term/low power microwave heating. J. Food Engineering 27:337–351 (1996).
9. T. R. Mulvaney, R. M. Schaffner, R. A. Miller, and M. R. Johnston. Regulatory review of scheduled thermal processes. Food Technology, 32:73–75 (1978).
10. J. T. Peeler and F. D. McClure. Most probable number determination, in Bacteriological Analytical Manual, 7th ed., Appendix 2. FDA, Washington, D.C. (1992).

11. M. Digeronimo, W. Garthright, and J. W. Larkin. Statistical design and analysis. Food Technology 51:52–56 (1997).
12. J. A. Ayoub, D. Berkowitz, E. M. Kenyon, and C. K. Wadsworth. Continuous microwave sterilization of meat in flexible pouches. J. Food Sci 39:309–313 (1974).
13. E. M. Kenyon, D. E. Westcott, P. LaCasse, and J. W. Gould. A system for continuous thermal processing of food pouches using microwave energy. J. Food Sci 36:289–293 (1971).
14. W. F. Hermans. Continuous microwave sterilization of ready meals in 3-compartment trays, Proceedings, Microwave and High Frequency. Göteborg, Sweden, Sept. 28–30 (1993).
15. G. J. Fleischman. Predicting temperature range in food slabs undergoing short term/high power microwave heating. J. Food Engineering 40:81–88 (1999).
16. A. Prakash, H.-J. Kim and I. Taub. Assessment of microwave sterilization of foods using intrinsic chemical markers. J. Microwave Power and Electromagnetic Energy 32:50–57 (1997).
17. FDA developing model policy for aseptic process filings, Spinak tells NFPA seminar. Food Chemical News 38(37):17–19 (1996).
18. Grade "A" Pasteurized Milk Ordinance. U.S. Government Printing Office, Washington, D.C. (1995).
19. T. Kudra, F. R. Van de Voort, G. S. V. Raghavan, and H. S. Ramaswamy. Heating characteristics of milk constituents in a microwave pasteurization system. J. Food Sci 56:931–934, 937 (1991).
20. R. Lopez-Fandiño, M. Villamiel, N. Corzo and A. Olano. Assessment of the thermal treatment of milk during continuous microwave and conventional heating. J. Food Protection 59:889–892 (1996).
21. S. N. Aktas and M. Ozilgen. Injury of *E. Coli* and degradation of riboflavin during pasteurization with microwaves in a tubular flow reactor. Lebensm. Wiss. Technol. 25:422–425 (1992).
22. H. O. Jaynes. Microwave pasteurization of milk. J. Milk Food Technol. 38:386–387 (1975).
23. M. Villamiel, R. Lopez-Fandiño, N. Corzo, I. Martinez-Castro, and A. Olano. Effects of continuous flow microwave treatment on chemical and microbiological characteristics of milk. Z. Lebensm. Unters. Forsch. 202:15–18 (1996).
24. R. A. Heddleson and S. Doores. Factors affecting microwave heating of foods and microwave induced destruction of foodborne pathogens—a review. J. Food Protection 57:1025–1037 (1994).
25. R. Sieber, P. Eberhard and P. U. Gallman. Heat treatment of milk in domestic microwave ovens. Int. Dairy Journal 6:231–246 (1996).
26. M. Villamiel, N. Corzo, I. Martinez-Castro, and A. Olano. Chemical changes during microwave treatment of milk. Food Chemistry 56:385–388 (1996).
27. G. S. Young and P. G. Jolly. Microwave: The potential for use in dairy processing, Austral. J. Dairy Tech 45:34–37 (1990).
28. S. Nikdel, C. S. Chen, M. E. Parish, D. G. MacKellar, and L. M. Friedrich. Pasteurization of citrus juice with microwave energy in a continuous-flow unit. J. Agric. Food Chem. 41:2116–2119 (1993).

29. S. Palaniappan and C. E. Sizer. Aseptic process validated for foods containing particulates. Food Technology 51(8):60–68.
30. J. W. Larkin. A workshop to discuss continuous multiphase aseptic processing of foods. Food Technology 51(10):43–44.
31. S. K. Sastry, Measuring the residence time and modeling a multiphase aseptic system. Food Technology 51(10):44–48.
32. J. E. Marcy. Biological validation of aseptic multi-phase foods. Food Technology 51(10):48–52.
33. B. D. Gessner and M. Beller. Protective effect of conventional cooking versus use of microwave ovens in an outbreak of Salmonellosis. Am. J. Epidemiol 139:903–909 (1994).
34. M. R. Evans, S. M. Parry, and C. D. Ribeiro. Salmonella outbreak from microwave cooked food. Epidemiol. Infect. 115:227–230 (1995).
35. Food Code, U.S. Department of Health and Human Services, Public Health Service, Food and Drug Administration, Washington, D.C. (1999).
36. J.A.G. Aleixa, B. Swaminathan, K. S. Jamesen, and D. E. Pratt. Destruction of pathogenic bacteria in turkeys roasted in microwave ovens. J. Food Sci. 50:873–875, 880 (1985).
37. S. E. Craven and H. S. Lillard. Effect of microwave heating of precooked chicken on *Clostridium perfringens*. J. Food Sci. 39:211–212 (1974).
38. C. A. Dahl, M. E. Matthews, and E. H. Marth. Fate of *Staphylococcus aureus* in beef loaf, potatoes and frozen and canned green beans after microwave heating in a simulated cook/chill hospital food service system. J. Food Prot. 43:916–923 (1980).
39. Food and Drug Administration. Food preparation—product temperature criteria for microwave cooking of pork, pork products and beef roasts (5/3/84). R.F.P. Prog. Inform. Manual. 2-403(b) (1984).
40. C. A. Sawyer, S. A. Biglari, and S. S. Thompson. Internal and temperature and survival of bacteria on meats with and without a polyvinylidene chloride wrap during microwave cooking. J. Food Sci. 49:972–973 (1984).
41. C. A. Sawyer, Post-processing temperature rise in foods: Hot air and microwave ovens. J. Food Prot. 48:429–434 (1985).
42. K. Figge. Migration of additives from plastics films into edible oils and fat simulants. Food Cosmet. Toxicol 10:815–828 (1972).
43. C. S. Bishop and A. Dye. Microwave heating enhances the migration of plasticizers out of plastics. J. Environ. Health 44:231–235 (1982).
44. R. Ashby. Migration from polyethylene terephthalate under all conditions of use. Food Additives and Contaminants 5:485–492 (1988).
45. H. C. Hollified. Food and Drug Administration studies of high-temperature food packaging. Food and Packaging Interactions II (S. J. Risch and J. H. Hotchkiss, eds.), ACS Symposium Series No. 473. American Chemical Society, 1991, pp. 22–36.
46. Code of Federal Regulations. Title 21 CFR Part 177. 1390. U. S. Government Printing Office, Washington, D.C. (1994).
47. R. R. Lentz and T. M. Crossett. Food/Susceptor interface temperature during microwave heating. Microwave World, 9:11–16 (1988).

48. L. Castle, A. Mayo, C. Crews, and J. Gilbert. Migration of poly(ethylene terephthalate) (PET) oligomers from PET plastics into foods during microwave and conventional cooking and into bottled beverages. J. Food Prot. 52:337–342 (1989).
49. L. Castle, S. M. Jickells, J. Gilbert, and N. Harrison, Food Addit. Contam. 7:779–796 (1990).
50. T. H. Begley and H. C. Hollifield. Liquid chromatographic determination of reactants and reaction by-products in polythylene terphthalate. J. Assoc. Off. Analyt. Chem. 72:468–470 (1989).
51. T. H. Begley and H. C. Hollifield. High-performance liquid chromatographic determination of polyethylene terphthalate oligomers in corn oil. J. Agric. Food Chem. 38:145–148 (1990).
52. T. H. Begley, J. L. Dennison, and H. C. Hollifield. Migration into food of polyethylene terephthalate (PET) cyclic oligomers from PET microwave susceptor packaging Food Addit. Contain. 7:797–803 (1990).
53. T. H. Begley and H. C. Hollifield. Migration of dibenzoate plasticizers and polythylene terephthalate cyclic oligomers from microwave susceptor packaging into food-simulating liquids and food. J. Food Prot. 53:1062–1066 (1990).
54. T. P. McNeal and H. C. Hollifield. Determination of volatile chemicals released from microwave-heat-susceptor food packaging. J. AOAC Intl. 76:1268–1275 (1993).
55. D. H. Watson and M. N. Meah (eds.). Food Science Reviews, Vol. 2: Chemical Migration from Food Packaging. Technomic Publishing Co., Inc. Lancaster, PA (1996).
56. M. E. Kuhn. Unzapped potential: Can new packaging technology reheat the microwaveable foods market?. Food Processing 57:49–55 (1996).
57. Code of Federal Regulations, Title 21 CFR Parts 170 to 189, U.S. Government Printing Office, Washington, D.C. (1994).
58. Code of Federal Regulations, Title 21 CFR Part 1030. 10, U.S. Government Printing Office, Washington, D.C. (1994).
59. C. R. Buffler. Microwave Cooking and Processing. Van Nostrand Reinhold, New York (1993).
60. Federal Food, Drug and Cosmetic Act. U.S. Government Printing Office, Washington, D.C. (1993).
61. M. Osepchuk. Review of microwave oven safety. Journal of Microwave Power 13:13–25 (1978).
62. P. R. Michaud. Fun with grapes—a case study, http://www.sci.tamucc.edu/~pmjchaud/grape/ (1994).
63. Microwave Cooking for Today's Living. Goldstar Part # 4B75126A (1992).
64. Y. C. Fu, C. H. Tong, and D. B. Lund. Microwave Bumping: Quantifying Explosions in Foods During Microwave Heating. J. Food Sci. 59:899–904 (1994).
65. T. Ohlsson and P. O. Risman. Temperature distribution of microwave heating spheres and cylinders. J. Microwave Power 13:303–310 (1978).

Index

Activation energy, 192
Active packaging, 160
Agency approval of oven design, 250
    CISPR (EU), 251
    FCC, 251
    FDA, 250
    IEC, 250
    NSF (National Sanitary Foundation), 252
    OSHA, 250
    UL (Underwriters Laboratories), 252
Amadori products, 176
    in breads and biscuits, 180
    lack of decomposition, 181
Applicators (*see* Cavity, Components of microwave oven)
Arcing:
    causes of 490
    in food, 491
    in metals, 491
Aroma producing compounds, 185
Aroma release, 181
    factors that influence, 181
        internal vaporization, 181
        transport of volatiles by steam, 181
        moisture content, 181

[Aroma release]
    food matrix interactions, 183
    and Henry's law, 184
    in mixtures of aromas, 183
    methods to measure, 181
    and partition coefficients, 183
    and product development, 184
    ways to minimize loss, 186
Aromas, 173 (*see also* Maillard reaction, in food systems)
    defined, 181
    generation under microwave heating, 175
Arrhenius equation, 192
Aseptic processing of particulates, 485
Attenuation coefficient, 5
Attenuation constant, 71
    units of, 72
Attenuation in decibel, 72

Bacon cooking, system of, 260
Bacon precooking, 305
    economic importance, 306
    reasons for improved quality, 305
    systems for, 306

Bacterial destruction and injury, 191
  of *Aspergilus niger,* 199
  of *B. subtilis,* 197, 202
  of *Bacillus cereus,* 197, 198, 202
  of *Campylobacter jejuni,* 197
  of *Clostridium perfringens,* 197
  of *C. sporogegenes,* 198
  enhanced effect defined, 196
  of *Enterobacter cloacae,* 201
  of *Enterococcus,* 197
  of *Escherichia coli,* 197, 198, 199, 200, 202, 206
  of *L. plantarum,* 202
  of *Lactobacillus plantarum,* 203
  of *Listeria monocytogenes,* 197, 198, 200, 201, 205
  nonthermal effects defined, 195, 196
  of *P. fluorescens,* 198
  of *Pediococcus sp.* NRRLB-2, 354
  of *Penicilium sp.,* 199
  of *Pseudomonas aeruginosa,* 198
  of *Rhizopus nigricans,* 199
  of *Saccharomyces cerevisiae,* 203
  of *Salmonella typhimurium,* 199
  of *Salmonella,* 197, 201, 206
  of *Staphylococcus aureus,* 197, 200, 202
  of *Streptococcus faecalis,* 201
  thermal effects defined, 195, 196
Baking, 312–319
  advantages, 313
  chemically leavened products, 317–319
    cake baking, 318
    doughnut processing, 318
  yeast-leavened products, 313–317
    brown-and-serve, 317
    proofing, 313–315
Ball formula method:
  for conventional sterilization, 478
  inapplicability for microwave sterilization, 479
Biscuits, flavor development, 180 (*see also* Drying)
Boil-in-bag (*see* Packaging, passive)

Boil over, safety problem, 492
Boundary conditions (*see also* Maxwell's equations):
  interface between two materials, 10
  at the walls of a microwave cavity, 11
Bread (*see* Pasteurization of, Product development, proprietary technologies)
Breads, flavor development, 180
Brown-and-serve products, pasteurization of, 329 (*see also* Baking)
Brownies and cookies (*see* Product development, proprietary technologies)
Browning:
  ingredients for, 375
  prevented by wet food surface, 178
Browning compounds (composition of), 185
Browning dishes (*see* Packaging, active)
Bumping, 493
Burns, 492
Butter tempering, 312

Cake baking (*see* Baking)
Cake mixes (*see* Product development, proprietary technologies)
Cavity (*see also* Components of a microwave oven):
  cylindrical, 14
  modeling, 18, 33–63
    boundary conditions, 62
    excitation, 63
    governing equations, 62
    modes in an empty rectangular, 16
  multimode, 14
    assumption of exponential decay in, 15
  single-mode, 14
Cereal cooking and drying, 325
CFR (U.S. Government Regulations), 477
Characteristics of ovens (*see* Oven systems)
Chemical bonds (breaking of), 195

Index

Chemical markers, 293
 in measurement of sterilization, 141, 293
 as time-temperature indicators, 293
Chemical susceptors, 375
Chicken precooking, 303
 savings in, 305
 systems for, 304
Choke (*see* Components of microwave oven)
Cobalt chloride (*see* Measurement of E field)
Cold spot, measurement of (*see* Measurement of surface temperature)
Cold spot, shifting during microwave sterilization, 479
Cole-cole equation, 74
Color development, 178
 by adding Maillard-active ingredients, 185
Combination cooking, system of, 266
Combination heating, 61 (*see also* Snack drying)
Combination oven, to enhance aroma, 186
Commercial sterility, 477
Components of microwave oven (industrial and domestic):
 cavities and applicators, 234, 241–242
 controls, 240, 248
 design principles, 233
 feeds, 237, 243
 leakage suppression in seals, 238
 matching feed, 244
 oven cavities, 234, 242
 power supplies, 239, 246
 safety, 240, 248
 seals, 238, 244
 sensors, various, 248
 power sources, 226
 gyrotron, 226
 klystron, 226
 magnetron, 226
 magnetrons (high power) at 2.45 GHz, 230
 magnetrons (reliability of), 231
 magnetrons for industrial use (915 MHz), 229, 232

[Components of microwave oven]
 magnetrons for ovens (2.45 GHz), 227
 magnetrons for ovens, trends in, 232
 microwave tubes, 226
 trends and availability, 273
Computer-aided engineering:
 electromagnetics, 33–63
 heat and mass transfer, 163
 inclusion of focusing, 164
 limitations, 165
 pressure driven transport, 165
 variable penetration depth, 164
 various levels of detail, 163
Computer simulation (*see* Modeling)
Conductivity, 70
Containers, microwave safe, 492
Control of temperature (*see* Temperature control in a microwave oven)
Cooking (*see* Reheating)
Cook value, measurement of, 292
Corner overheating, 42
Cost (operating):
 bacon cooking system, 260
 liquid heating system, 264
 potato chip processing system, 262
 tempering system, 259
Coupling of electromagnetics and heat transfer, 153
 computational procedure, 154
 revealing migration of cold points, 157
 when significant, 155
CPET (*see* Packaging, passive)
Crackers (*see* Drying)
Crisping, prevented by wet food surface, 178
Critical factors, 477

Debye equation, 73
Delta T theory, 181
Design of ovens:
 effect of position and design of feed, 25
 manufacturing tolerance and, 25
 using computer simulation (*see* Modeling)

Dielectric constant, 3, 70
Dielectric loss factor, 3, 70
Dielectric properties (*see also* Heat transfer):
  composition dependence, 83
  density dependence, 76
  effect of storage, 99
  of food with different composition, 359
  frequency dependence, 73
  measurement, 78
  of mixtures, 77
  temperature dependence, 76
Dielectric properties of various materials:
  at freezing temperatures, 85
  at sterilizing temperatures, 85
  baked foods, 103
  cereal grains, 87
  cheddar cheese, 98, 99
  corn (yellow-dent field), 87
  dairy products, 97
  data compilations, 106
  fish and seafood, 99
  fruits and vegetables, 93
  meats, 103
  milk and its constituents, 97
  nuts, 93
  processed cheese, 99
  rice (rough), 91
  salt solutions, 81
  tylose, 85
  water, 75
  water (adsorbed), 76
  water (free and bound), 84
  wheat (hard red winter), 87
Dielectric relaxation, 3
Dough formulations (*see* Product development, proprietary technologies)
Doughnut processing (*see* Baking)
Drying, 148, 320–325
  advantages, 322
  biscuits and crackers, finish drying of, 325
  cereal, 325
  effect of heating non-uniformities, 149

[Drying]
  pasta, 324
  potato chips, 322–324
    finish drying, 322
    nonfried, 323
    puffed, 323
    snack, 324
  at various power levels, 149
Dual ovenability (*see* Packaging, passive)
D-value, 193

Edge overheating, 42
Edible susceptors, 376
Effective loss factor (*see* Dielectric loss factor)
Efficiency:
  attainable, 274
  of consumer ovens, 266
E field measurement (*see* Measurement of electric field)
Electric field, 3
Electromagnetic fields:
  fundamentals, 2–18
  inside an oven, 34
  magnitude and uniformity (*see* energy absorption)
Electromagnetics, coupling with heat transfer, 153
Electromagnetic spectrum, 1
Emissions:
  in-band and communications use, 251
  out-of-band, 251
  regulation (*see* Agency approval)
Energy absorption:
  aspect ratio (food) effect, 41
  dielectric properties (food) effect, 37
  magnitude in an oven, 36, 54
  shape (food) effect, 40
  surface area (food) effect, 36
  uniformity in an oven, 36, 54
  volume (food) effect, 36
Energy deposition in food in an oven:
  food and oven factors affecting, 33
  magnitude, 33
  uniformity, 33

Index

Energy distribution, oven factors, 54
  effect of feed location, 59
  effect of mode stirrer, 57
  effect of oven size and geometry, 55
  effect of placement, 55
  effect of turntable, 56
  in an empty cavity, 54
Enhanced effect, 196
  in enzyme inactivation, 208
Enzyme inactivation, 191, 206
Enzyme inactivation (specific):
  pectin methylesterase, 206
  soybean lipoxygenase, 206
  wheat germ lipase, 206
Establishing a process, using innoculation, 479
Explosions, 145
Exponential decay, finite or thin slab, 9 (*see also* Plane waves)

F_0 (equivalent time), 193
F_0 values, measurement of, 293
Fabry-Perot interferometer, in fiberoptic thermometry, 285
FDA-regulated processes, 477
FDTD, 19
FEM, 20
Fiberoptic probes (*see* Measurement of electric field, Measurement of temperature at a point)
Finish drying (*see* Drying)
Fire, causes of, 490
First-order reaction, 137
Fish tempering, 309–312 (*see also* Meat tempering)
Flavor, 173 (*see also* Aroma release, Maillard reaction)
Flavor development, 178
  in breads and biscuits, 180
  by adding Maillard-active ingredients, 185
  in cakes, 179
  using susceptor packaging, 185
Flavor release (*see* Aroma release)

Focusing, 44
  effect of penetration depth, 46
  effect of wavelength, 46
  experimentally observed, 46
  in frozen foods, 46
  in meats and vegetables, 46
  numerically calculated, 46
  in sphere of large radius, 45
  in sphere of small radius, 44
Food code:
  for microwave cooking, 486
  for microwave reheating, 486
Food components, interactions with microwaves, 357–377
Freeze drying, 150
  effect of power level, 151
  moisture leveling, 151
  temperature and moisture profiles, 151
Frequencies:
  available, 272
  domestic microwave oven, 2
  and efficiency, 273
  industrial processing, 2
  industrial, scientific and medical (ISM), 1
  ISM applications, 272
  ISM bands and communication, 274
  microwave, 1
Fringing field applicators, 61
Frozen foods (heating of) (*see* Heat transfer in solids)
Fungi, 197
Future ovens, 59
  combination heating, 61
  fringing field applicators, 61
  phase control heating, 60
  variable frequency oven, 60

Glass for packaging (*see* Packaging, passive)
Gyrotron (*see* Components of microwave oven)

Heating (*see* Reheating)
Heating rate, factors affecting (*see* Energy absorption)

Heating rates:
    low- and high-loss foods compared, 51
    water and corn oil compared, 51
Heat and mass transfer coefficients, 124
Heat and moisture transport, 115
    boundary conditions, 122
        convective heat gain/loss, 122
        evaporative heat loss, 123
        radiative heat gain/loss, 123
    coupling with electromagnetics, 119
    exponential decay of energy, 120
    governing equation, 121
    Lambert's law, 119
    Lambert's law (applicability of), 121
    nonuniformities, 117
        descriptions, 118
        due to property variation, 117
        rates of heat generation, 117
        surface condition (*see* Boundary Conditions)
    surface transfer coefficients, 124
    various transport mechanisms, 122
Heat transfer:
    coupling with electromagnetics, 153
    effect of changes in food temperature, 132, 157
    effect of microwave frequency, 132
    in liquids, 129
        in a container, without agitation, 129
        in a tube (continuous flow), 129
        superheating (inhibition of boiling), 132
    nonuniformity over time, 157
    in solids, 125
        frozen foods (thawing and tempering), 133
        temperature profile for no diffusion, 127
        temperature profile in a thick slab, 127
        temperatures and time for thawing, 135
        three characteristic temperature profiles, 128
    thermal runaway, 132

Henry's law, 184
History:
    industrial processes and technology, 216, 220–221, 224, 226
    microwave oven, 215, 221
Home use (*see* Principles for home use)
Hot air addition, 160
Hot spot, measurement of (*see* Measurement of surface temperature)
Hydrophobicity and aroma release, 183

Inactivation (*see* Bacterial destruction)
Industrial applications, early, 217
Industrial processes, 299–337
    barriers to successful adoption of, 331
    future of, 330
    history of, 301
    improving the likelihood of success, 332
    list of installed systems, 331
    overview, 302
    potential future applications, 334
Infant formula, safety issue, 493
Infrared heat addition, 161
    infrared penetration effect, 161
Infrared heating and food safety, 473
Infrared thermography (*see* Measurement of temperature over a surface)
Ingredient interactions, 355–395
Ingredients of food:
    alcohol, effect of, 361
    browning agent, 375–376
        necessary ingredients, 375
    crisping agent (*see* Browning agent)
    edible susceptors, 376
    emulsions and suspensions, effect of, 361
    food polymers and plasticizers:
        gums, ionic and nonionic, 371
        microparticulated components, 373–374
        proteins and protein hydrates, 370
        starches, native and modified 368–370
    and inability to predict properties from composition, 361

Index

[Ingredients of food]
  and interactions with microwaves, 357–377
    dispersions, 359
    emulsion, 359
    hydration dynamics, 359
    macromolecular complexes, 359
    layered foods, 374
      pizza, 374
      selection of ingredients for, 374
    oligosaccharaides, effect of, 361
    polyols, effect of 361
    relative effects of various components, 363
    salt, effect of, 361
    sugar, effect of, 361
    water, 362–363
      alcohol-water mixtures, 368
      bound, 361
      bound to food polymers, 363
      electrolytes, 367–368
      nonelectrolytes in, 364
      pH and ionic strength, effect of, 364
      sugars and alcohol, 365
      sugars in, 364
Injury of bacteria, 204
Innoculation for validation, 478
Introducing microwaves to new processes, 334
Ionic conduction, 3

Juice, pasteurization of, 484

Kinetic parameters (*see* Kinetics)
Kinetics:
  bacterial destruction, 192
  determination of parameters, 193
    continuous-flow heating, 194
    non-isothermal heating, 194
  factors affecting, 196
    heterogeneity of food, 197
    reduced come-up time, 196
  first-order, 192
Klystron (*see* Components of microwave oven)

LDPE (*see* Packaging, passive)

Leakage, suppression in seals, 238
Liquid crystal (*see* Measurement of E field)
Liquid heat exchangers, system of, 264
Lossy materials, 2
Low-acid foods, 477

Magnesium flurogermanate, in fiberoptic thermometry, 285
Magnetic component in food, 7
Magnetic Resonance Imaging (*see* Measurement of internal temperature)
Magnetron (*see* Components of microwave oven)
Magnetron, modeling, 21
Magnitude and uniformity of heating:
  food factors, 36
  oven factors, 54
Maillard reaction under conventional heating, 178
Maillard reaction under microwave radiation, 173
  acceleration in a closed system, 175
  consecutive reactions, 175
  in food systems, 178
    baked and roasted products, 179
    chemical composition of volatiles, 179
    difference with model systems, 178
    effect of short time scale, 181
    effect of wet food surface, 178
    factors affecting aroma generation, 179
  in model systems:
    altered distribution of products, 176
    chemical composition and yields, 177
    compared with conventional heated, 177
    effect of electrolytes and pH, 176
    effect of salts, 176
    no fundamental difference with conventional, 177
    solvent mediated, 176
  parallel (competitive) reactions, 175

506 **Index**

Maintaining a desired temperature (*see* Temperature control in a microwave oven)
Market:
   for consumer oven, 299
   for industrial oven, 299
Maxwell's equations, 4
   Ampere's law and, 4
   boundary conditions, 10
   electric field strength, 4
   Faraday's law and, 4
   Gauss's law and, 4
   magnetic field strength and, 4
   solutions in a finite slab, 9
   total current density and, 4
Measurement of cook or sterilization values, 292
Measurement of dielectric properties, 78
Measurement of E field (*see* Measurement of electric field)
Measurement of electric field, 280–282
   using Cobalt chloride, 280
   by equating energy absorption, 280
   using liquid crystal sheets, 280
   using non-enzymatic browning, 280
   using probe based on energy absorption, 281
   using a thermal fax paper, 280
   using a thermoset polymer, 280
Measurement of internal temperature using MRI, 291
   comparison with infrared images, 291
   system description, 291
Measurement of moisture content, 272
Measurement of moisture loss, 295
Measurement of moisture profiles, 295
Measurement of temperature, 282–288
   array of thermocouples, 283
   fiberoptic probes:
      based on fluorescence, 285
      based on interferometry, 285
   thermocouples, 282
   thermocouples modified for microwave, 282
Measurement of temperature over a surface, 288–291
   infrared thermography, 288

[Measurement of temperature over a surface]
   accuracy issues, 290
   comparing thermal images, 290
   development of, 288
   innovations in, 289, 290
   measuring transient temperature, 290
   showing non-uniformity of heating, 290
Meat-filled products (*see* Product development, proprietary technologies)
Meat processing, 303–309
Meat tempering, 309–312
   effective ways of, 310
   problems in, 309
Metals in microwave (*see* Arcing)
Microbe, injury, 204 (*see also* Bacterial destruction)
Microwave assisted synthesis, 176
Microwave oven, picture showing components, 242
Microwave safe containers, 492
Migration of chemicals, 487–489
   from packaging, 488
   regulations, 488, 489
   temperature reached for susceptor, 488
Milk, pasteurization of 481–484
Modeling (analytical), 18
Modeling (numerical), 18–27
   adaptive meshing, 22
   boundary conditions for a cavity, 62
   commercial software, 23
   difficulties with industrial oven, 25
   domain decomposition, 22
   domestic oven design, with, 23
   excitation for a cavity, 62
   finite difference time domain method (FDTD), 19
   finite element method (FEM), 20
   future directions, 22
   governing equations for a cavity, 62
   industrial oven design, with, 25
   mode stirrer, 25
   multigrid techniques, 23
   packaging design, with, 26

Index

[Modeling (numerical)]
  parallel processing, 23
  turntable, 25
Modes in cavity:
  analytical modeling, 18
  numerical modeling, 18
Modes of propagation, 14
  in an empty rectangular cavity, 16
  TE or transverse electric, 14
  TEM or transverse electromagnetic, 14
  TM or transverse magnetic, 14
  in waveguides, 14
Mode stirrers and heating uniformity, 158
Moisture equilibration (*see* Moisture leveling)
Moisture leveling, 148
Moisture loss, measurement of (*see* Measurement of moisture loss)
Moisture profiles, measurement of (*see* Measurement of moisture profiles)
Moisture transport, 141
  comparison with conventional heating, 143
  description using effective diffusivity, 141
  explosions and puffing from pressure, 145
  focusing effect and, 148
  in low and high moisture, 145
  increased loss due to heating non-uniformity, 149
  moisture leveling, 148
  moisture loss, 145
  moisture profiles, 145
  pressure profiles, 145
  pumping effect in high moisture foods, 145
  temperature profiles, 145
Molds, 197
Monitoring, 480–481
  of conventional sterilization, 480
  challenges in microwave sterilization, 480
MRI (*see* Measurement of internal temperature)

Multicompartment dinner (*see* multicomponent foods)
Multicomponent foods, 50
  heating rate and uniformity in, 51

Nonthermal effects (*see also* Bacterial destruction):
  in enzyme inactivation, 206
  and nonuniformity of temperature, 204
  some explanation, 209
Non-uniformity of heating, as demonstrated by infrared thermography, 290

Operational safety (*see* Safety)
Order of reaction, 192
Oven:
  industrial use, 215 (*see also* Oven systems)
  power sources available, 273
  selection for performance testing, 380
Oven systems (detailed descriptions of), 252
  consumer ovens, 252
  commercial ovens, 255
  industrial ovens, 259
    bacon-cooking system, 260
    combination cooking, 265
    liquid heating system, 264
    potato chip processing, 262
    tempering system, 259

Packaging, 397–470
  design using models (*see* Modeling)
  key functional requirements, 398
Packaging, active:
  advantages over passive, 407
  browning dishes, 411
    drawbacks of, 411
    safety when overheated, 411
  defined, 407
  disposable surface-heating packages, 414
  options in, 409
  slot line heaters, 419
  surface heating, creation of, 409

[Packaging, active]
　susceptors, 412–418
　　evaluating procedures for, 420
　　metallized film, 412
　　patterned metallized film, 416
　　printable coatings, 418
　　regulatory compliance, 419
　　self-limiting metallized film, 414
　　successful applications, 420
　　　enrobed sandwiches, 422
　　　frozen pizza, 420
　　　frozen pot pies, 424
　　　popcorn, 421
　　　waffles, 422
Packaging, passive, 399–407
　defined, 399
　outlook, 406
　performance requirements, 399–401
　various materials for, 401–406
　　crystallized polyester trays (CPET), 406
　　glass, 401
　　LDPE coated paperboard, 404
　　LDPE trays, 405
　　paper and polymer composite, 404
　　paper, paperboard and film structures, 403
　　PET coated paper, 404
　　plastic coated paper, 404
　　polymer, 405
　　polypropylene extrusion coated paper, 404
　　polypropylene trays, 406
　　polystyrene trays, 406
Pancake (*see* Product development, proprietary technologies)
Paper for packaging (*see* Packaging, passive)
Partition coefficients and aroma release, 183, 184
Pasta (*see* Drying, Pasteurization, Product development, proprietary technologies)
Pasteurization, 136, 325–330, 481–485
　of bread, 329
　of brown-and-serve products, 329

[Pasteurization]
　compared with conventional, 326
　distribution of F_0 values, 138
　distribution of temperatures, 138
　of fresh pasta, 328
　industrial systems, 140
　of juice, 484
　migration of cold points, 157
　of milk, 481
　　conventional heating, 481
　　microwave heating, possibilities of, 482–484
　of ready meals, 327
　validation of lethality, 141
　why microwaves can be advantageous, 137
Patents in packaging, selected from years 1949–1999, 444
Penetration depth, 7, 72
　expression for, 7
　explanation for, 7
　values for food materials, 120
Performance testing of products, 380–382
　abuse testing, 381
　heating instructions, development of 381
　selection of ovens and testing protocol, 380
Permittivity:
　complex, 3, 70
　of free space, 3
Phase constant, 5
Phase control heating, 60
Pizza (*see also* Product development, proprietary technologies):
　microwaveable dough, 388
　packaging (*see* Packaging, active)
　selection of ingredients, 374
Plane waves, 5
　absorption, 12
　exponential decay in, 6
　reflection, 12
　refraction, 13
　semi-infinite slab, 6
　solution, 5
　transmission, 12

Index

Polymer for packaging (*see* Packaging, passive)
Polypropylene (*see* Packaging, passive)
Polystyrene (*see* Packaging, passive)
Popcorn, packaging (*see* Packaging active)
Potato chip processing, system of, 262
Potato chips (*see* Drying)
Poultry processing, 303–309
Power cycling and heating uniformity, 158
Power density, 3
Power dissipated, 3, 70
Power distribution (*see* Energy distribution)
Power factor, 70
Power output:
  of commercial ovens, 268
  of consumer ovens, 266
    effect of tube warmup, 266
    effect of variation in line voltage, 266
    standard method of measurement, 267
    standards for increased efficiency, 268
  of industrial ovens, 269
Power penetration depth (*see* Penetration depth)
Power sources (*see* Components of microwave oven)
Precooking (*see* Individual items)
Pressure development from heating, 145
Principles for home use, 339–354
  adjusting cooking time:
    for change in oven wattage, 343
    for change in weight, 342, 344
  containers, effect of, 344
  covers, effect of, 344
  converting directions from conventional heating, 350–351
  adjustments for ingredients, 351
  general principles, 350
  factors affecting cooking time, 341, 342

[Principles for home use]
  relationships to conventional cooking temperature, 340
  shielding with foil, 346
  simple test to determine hot and cold areas, 340
  simple test to determine wattage, 340
  slow cooking, 343
  standing time, 342
  various foods:
    meats, 349–350
    vegetables, 349
  various processes:
    arranging and stirring, 346
    browning and crisping, 346
    defrosting, 347
    reheating, 348
    softening and refreshing, 348
Product development:
  and aroma release, 184
  factors to consider, 377–380
    food geometry, 379
    impedance matching, 377
    penetration depth, 378
    state of water and dissolved constituents, 378
    use of active packaging, 380
    volume and surface area of food, 379
  and ingredient interactions, 355–395
  and interaction of food components, 357–377
  and oven factors, 356
  proprietary technologies, 382–389
    aroma coating, 383
    batters and breadings, 383
    bread, 384
    brownies and cookies, 385
    cake mixes, 385
    ready-to-eat cereals and snacks, 386
    dough formulations, 387
    meat-filled products, 388
    pancake, 388
    pasta, 388
    pizza, 388

[Product development]
   sauce for cooking, 389
   stuffing mix, 389
   sweetener coating for popcorn, 389
Propagation constant, 5
Puffing, 145 (*see also* Drying, Product development, ready-to-eat cereals)

Quality improvement, 158
  by controlling heating uniformity and rate, 158
    active packaging, 160
    combination ovens, 159
    mode stirrers, 158
    movable wall cavities, 160
    oven design changes, 159
    power cycling, 158
    turntables, 158
  by controlling moisture loss and distribution, 160
    use of hot air, 160
    use of infrared heat, 161

Rate constant, 192
Ready meals (*see* Pasteurization)
Ready-to-eat cereals and snacks (*see* Product development, proprietary technologies)
Record keeping (*see* Safety)
Records (*see* Monitoring)
Regulations (*see* Agency approval)
Reheating, 485–487
  food code related to, 486
  potential for illness, 485
Resonance (inside a food), 50
  in a cylinder under plane wave, 50
  in a slab under plane wave, 50

Safety, of active packaging, 411 (*see also* Migration of chemicals)
Safety, in equipment design:
  in domestic oven, 248
  in industrial processing, 238, 240
Safety, microbiological, 471–497
  as related to uniformity of heating, 472–476
  in direct contact heating, 472

[Safety, microbiological]
  in radiant (infrared) heating, 473
  in microwave heating, 474–476
  difficulties in ensuring, 475–476
  effect of evaporative cooling, 123
  in industrial foods, 476
  in microwave sterilization, 479
  monitoring of processes for, 478
  record keeping for, 478
  in steam sterilization, 477
    regulations in the United States, 477
Safety, operational, 489–493
  arcing and fires, 490–492
  boil over, 492
  bumping, 493
  burns, 492
  explosion in food, 493
  leakage, 489
  scalding (in case of infant formula), 493
Sauce for cooking (*see* Product development, proprietary technologies)
Sausage patties:
  advantages of, 308
  systems for, 308
Sensors:
  in a domestic microwave oven:
    moisture, 248
    temperature (surface), 248
    various, 271
    weight and cook by weight, 248
  in an industrial oven, arcing sensor, 271
Skin depth, 6
  typical values, 8
Snack (*see* Drying)
Snell's law, 12
Spencer, Percy, 216
Speed cooker, 62
Spherical food:
  heated in plane waves, 39
  heated in a cavity, 40
  volume effect, 40
Sterilization, 329–330 (*see also* Pasteurization)

Index

[Sterilization]
    measurement of, 292
    microwave:
        challenges in monitoring, 480
        validation using innoculation, 479
Stuffing mix (*see* Product development, proprietary technologies)
Successful applications, criteria for, 300
    (*see also* Industrial processes)
Superheating, and acceleration of chemical reactions, 174
Susceptors addition, 163
    temperature profiles in presence of, 163
Susceptors (*see* Packaging, active)
Sweetener coating for popcorn (*see* Product development, proprietary technologies)

Temperature control (set-point) in a microwave oven, 293
    domestic oven, 293
    research applications, 294
Temperature rise, factors affecting (*see* Energy absorption)
Tempering, 132, 309–312 (*see also* Meat tempering, Fish tempering, Butter tempering)
    system of, 259
Thawing, 132 (*see also* Tempering)
Thawing time, 134
    as affected by aspect ratio, 134
    as affected by power levels, 134
    as affected by volume, 134
Thermal effects (*see* Bacterial destruction)

Thermal image (*see* Measurement of surface temperature)
Thermal properties, 124
    density, 124
    specific heat, 124
    thermal conductivity, 124
Thermistor for temperature control, 293
Time-temperature indicators, 293
Turntables and heating uniformity, 158
Types of ovens (*see* Oven systems)

Uniformity of heating, improvement through oven design, 270
USDA-regulated processes, 477

Validation:
    of conventional sterilization, 478
    of microwave sterilization, 479
Vapor transport, effect on aroma release, 182
Variable frequency oven, 60
Volatiles, transport by steam, 181

Waffles, packaging (*see* Packaging, active)
Water:
    consideration in product development, 368
    dielectric properties of, 75, 76, 84
    heating rates compared with oil, 51
    interaction with other ingredients, 362–368
Weight loss, measurement of (*see* Measurement of moisture loss)

Yeasts, 197

Z-value, 193